Food Industry Quality Control Systems

Food Industry Quality Control Systems

Mark Clute

CRC Press
Taylor & Francis Group
Boca Raton   London   New York

CRC Press is an imprint of the
Taylor & Francis Group, an **informa** business

CRC Press
Taylor & Francis Group
6000 Broken Sound Parkway NW, Suite 300
Boca Raton, FL 33487-2742

© 2009 by Taylor & Francis Group, LLC
CRC Press is an imprint of Taylor & Francis Group, an Informa business

No claim to original U.S. Government works
Printed in the United States of America on acid-free paper
10 9 8 7 6 5 4 3 2 1

International Standard Book Number-13: 978-0-8493-8028-0 (Hardcover)

This book contains information obtained from authentic and highly regarded sources. Reasonable efforts have been made to publish reliable data and information, but the author and publisher cannot assume responsibility for the validity of all materials or the consequences of their use. The authors and publishers have attempted to trace the copyright holders of all material reproduced in this publication and apologize to copyright holders if permission to publish in this form has not been obtained. If any copyright material has not been acknowledged please write and let us know so we may rectify in any future reprint.

Except as permitted under U.S. Copyright Law, no part of this book may be reprinted, reproduced, transmitted, or utilized in any form by any electronic, mechanical, or other means, now known or hereafter invented, including photocopying, microfilming, and recording, or in any information storage or retrieval system, without written permission from the publishers.

For permission to photocopy or use material electronically from this work, please access www.copyright.com (http://www.copyright.com/) or contact the Copyright Clearance Center, Inc. (CCC), 222 Rosewood Drive, Danvers, MA 01923, 978-750-8400. CCC is a not-for-profit organization that provides licenses and registration for a variety of users. For organizations that have been granted a photocopy license by the CCC, a separate system of payment has been arranged.

Trademark Notice: Product or corporate names may be trademarks or registered trademarks, and are used only for identification and explanation without intent to infringe.

Library of Congress Cataloging-in-Publication Data

Clute, Mark.
 Food industry quality control systems / Mark Clute.
    p. cm.
 Includes bibliographical references and index.
 ISBN 978-0-8493-8028-0 (hardback : alk. paper)
 1. Food industry and trade--Quality control. I. Title.

TP372.5.C58 2009
664.0068'5--dc22   2008026560

Visit the Taylor & Francis Web site at
http://www.taylorandfrancis.com

and the CRC Press Web site at
http://www.crcpress.com

*For Dana,
my strength, love, and life.*

Contents

The Author xix
Preface xxi

Chapter 1   Introduction 1

Top-Down Quality Management 3
The Cost of Quality 5
The Role of Management in Quality Control 6

Chapter 2   Quality: Role and Function 11

The Role of Quality 11
    Establishing the Standards or Baseline 11
    Measuring Compliance against the Baseline 12
    Reporting Noncompliance 13
    Sharing the Experience 14
The Quality Control–Production Relationship 14
    The Secretive and Compartmentalized Relationship 15
    The Divisive or Adversarial Relationship 15
    The Conciliatory Team-Building Relationship 15
The Quality Control–Vendor Relationship 16
The Quality Control–Customer Relationship 17
The Quality Control–Regulatory Agency Relationship 18

Chapter 3   Quality Control Systems Development Overview 19

HACCP Manual Cover 23
HACCP Manual Spine Label 24
Quality Control Manual Cover 25
Quality Control Manual Spine Label 26
Pest Control Book 1 Cover 27
Pest Control Book 1 Spine Label 28
Pest Control Book 2 Cover 29
Pest Control Book 2 Spine Label 30
Other Programs Manual Cover 31
Other Programs Manual Spine Label 32
Receiving Manual Cover 33
Receiving Manual Spine Label 34
Shipping Manual Cover 35
Shipping Manual Spine Label 36
Manual Tab Section Labels 37

Chapter 4 Book 1: Hazard Analysis Critical Control Point (HACCP)
              Program 39
Program Type: Required 39
HACCP Program Planning—Theory 41
HACCP Program Planning—Implementation 44
Book 1: Section 1 45
    HACCP Overview 45
Book 1: Section 2 46
    Process Flows 46
Book 1: Section 3 46
    Product Description 46
Book 1: Section 4 47
    Hazard Analysis 47
    Chemical Risk Analysis 48
        Example 49
    Microbiological Risk Analysis 50
        Example 53
    Physical Risk Analysis 54
        Examples 55
    Allergenic Risk Analysis 55
        Example 57
Book 1: Section 5 57
    Critical Control Point Determination 57
        Example 1 60
        Example 2 60
        Example 3 61
Book 1: Section 5a 61
    Control Limit Establishment 61
    Corrective Actions 62
Book 1: Section 6 62
    HAACP Plan Summary Sheet 62
        Example 63
Book 1: Section 7 63
    Verification 63
Book 1: Section 8 64
    Auditing 64
Book 1: Section 9 65
    Training 65
Supplemental Materials 66
    General Overview 67
    Development and Implementation Team Roster 70
    HACCP Team Meeting Attendance 71
    Supplier Data Sheet 72
    Program Description Worksheet 73
    Process Flow Master List 74
    Process Flow Template 75

Contents   ix

        Risk Assessment Worksheet for Chemical Risk Analysis............................76
        Risk Assessment Worksheet for Microbiological Risk Analysis 77
        Risk Assessment Worksheet for Physical Risk Analysis.............................78
        Risk Assessment Worksheet for Allergens Risk Analysis 79
        Product/Process Control Point Evaluation Worksheet 80
        HACCP Plan Summary Chart..81
        HACCP Team Yearly Auditing Meeting Attendance..................................82
        Verification Documentation...83
Examples...84
        Supplier Data Sheet 85
        HACCP Principle 1..86
        Process Flow for Chopped Product A 86
        Process Flow for Baked Product A...87
        Product Description Worksheet Complete 1 88
        Product Description Worksheet Complete 2 89
        Chemical Risk Analysis Complete 90
        Microbiological Risk Analysis Complete...91
        Physical Risk Analysis Complete...92
        Allergens Risk Analysis Complete 93
        Product/Process Control Point Evaluation Worksheet Complete 1..............94
        Product/Process Control Point Evaluation Worksheet Complete 2..............95
        Product/Process Control Point Evaluation Worksheet Complete 3..............96
        HACCP Plan Summary Chart..97

Chapter 5   Quality Control Program: Overview...99

Theory..99
Application...99
Examples..100
Supplemental Material...101
        Quality Control Program: General Overview 102
        Quality Monitoring Scheme
Examples..107
        Quality Monitoring Scheme Examples 108

Chapter 6   Organizational Chart..113

Program Type: Required..113
Theory..113
Application...113
Supplemental Materials 115
        Organizational Chart 116

Chapter 7   Good Manufacturing Practices Program 117

Program Type: Required..117
Theory..117

Application 128
  GMP Standards 128
  Daily Line Check Sheet 131
Supplemental Materials 133
  Sanitation Standard 1 134
  Sanitation Standard 2 135
  Sanitation Standard 3 136
  Sanitation Standard 4 137
  Sanitation Standard 5 138
  Sanitation Standard 6 139
  Sanitation Standard 7 140
  Sanitation Standard 8 141
  Sanitation Standard 9 142
  Sanitation Standard 10 143
  Sanitation Standard 11 144
  Sanitation Standard 12 146
  Sanitation Standard 13 147
  Sanitation Standard 14 148
  Verification Documentation of GMP Standards Training 149
  Good Manufacturing Rules 150
  Verification Documentation of Employee GMP Annual Review 152
  Good Manufacturing Rules: Visitors 153
  GMP Audit Procedure/Schedule 155
  Good Manufacturing Program 156

Chapter 8   Pest Control Program 157

Program Type: Required 157
Theory 157
  Sanitation 158
  Mechanical Control 158
  Cultural Control 159
  Biological Control 159
  Chemical Control 159
Application 164
Examples 164
Supplemental Materials 165
  Rodent and Pest Control Program 166
  Compensation Worksheet 168
  Pest Control Contract Example 169
  Commercial Applicator License – Company 171
  Certificate of Insurance 172
  Label Example 1-3 173
  MSDS Example 1-3 177
  Plant Diagram Example 184

Contents xi

       Rodent/Pest Inspection Form 185
       Commercial Applicator License – Technician 186
       Pest Control Service Report Example 187

Chapter 9 Allergen Program 189

Program Type: Required 189
Theory 189
Application 192
Supplemental Materials 194
       Risk Analysis Worksheet for Allergen Food Hazards 195
       Allergen Information Request Form 196
       Allergen Inclusion Chart 198
       Daily Plant Sanitation Inspection Form 199
       Allergy Testing Form 200
       Allergen Testing Program: Initial Training Verification 201
       Allergen Testing Program: General Overview 202

Chapter 10 Weight Control Program 205

Program Type: Required 205
Theory 205
Application 206
Examples 207
Supplemental Materials 208
       Weight Control Program Overview 209
       Weight Control Chart 210
Examples 211
       Weight Control Chart Example 212

Chapter 11 Inspection Program 213

Program Type: Required 213
Theory 213
Application 215
Supplemental Materials 217
       Inspection Team Roster 218
       Monthly Inspection Form 219
       Internal Inspection Program: General Overview 220
       Daily Plant Sanitation Inspection Form 222

Chapter 12 Sanitation Program 223

Program Type: Required 223
Theory 223
Application 229

Supplemental Materials 231
    Master Sanitation Schedule 232
    Master Sanitation Schedule: Daily Production 235
    Master Sanitation Schedule: Daily Non-Production 236
    Master Sanitation Schedule: Weekly Master 237
    Master Sanitation Schedule: Monthly Master 238
    Master Sanitation Schedule: Yearly, Quarterly, Semi-Yearly 239
    Sanitation Procedure Template 240
    Sanitation Procedure Training Verification 241
    Sanitation Program General Overview 242
    Master Sanitation Schedule: Daily Production 243
    Master Sanitation Schedule: Daily Non-Production 244
    Master Sanitation Schedule: Weekly Master 245
    Master Sanitation Schedule: Monthly Master 246

Chapter 13 Metal Detection Program 247

Program Type: Required 247
Theory 247
    Placement 247
    Size 248
    Testing 248
    Rejection 249
Application 250
Supplemental Materials 252
    Metal Detection Program: General Overview 253
    Metal Detector Operation 255
    Metal Detection Sheet 257
    Foreign Material Investigation Log 258
    Metal Detection Training Verification 259

Chapter 14 Regulatory Inspection Program 261

Program Type: Required 261
Theory 261
Application 264
Supplemental Materials 265
    Regulatory Inspection Program: General Overview 266

Chapter 15 Lot Coding Program 269

Program Type: Required 269
Theory 269
Application 271
Supplemental Materials 273
    Lot Coding Program: General Overview 274
    Lot Code Explanation 275

Contents   xiii

Chapter 16 Customer Complaint Program 277

Program Type: Required... 277
Theory... 277
Application.. 280
Supplemental Materials 281
    Customer Complaint Form 282
    Customer Complaint Flow Diagram... 283
    Customer Complaint Log 284
    Customer Complaint Program: General Overview 285

Chapter 17 Receiving Program.. 287

Program Type: Required... 287
Theory... 287
Application.. 289
Supplemental Materials 290
    Receiving Log.. 291
    Receiving Log Directions.. 292
    Verification Documentation: Receiving Ingredient/Packaging Training 294
    Receiving Program: General Overview... 295

Chapter 18 Shipping Program.. 297

Program Type: Required... 297
Theory... 297
Application.. 298
Supplemental Materials 299
    Shipping Container Inspection Log.. 300
    Shipping Log Directions.. 301
    Shipping Program: General Overview 302

Chapter 19 Specification Program 303

Program Type: Optional 303
Theory... 303
    Raw Material Specifications... 303
    Processing Specifications 304
    Finished Goods Specifications 304
Application.. 305
    Raw Material Specification 305
    Finished Goods.. 305
Supplemental Materials 306
    Ingredient Specification.. 307
    Finished Product Specification 309
    Certificate of Compliance... 310

Chapter 20 Recall Program 313

Program Type: Required 313
Theory 313
Application 317
Supplemental Materials 322
    Recall Program Overview 323
    Recall Communication 326
    Emergency Recall Notification List 327
    Raw Material Recall Sheet 328
    Finished Product Recall Sheet 329
    Effectiveness of Product Recall Sheet 330

Chapter 21 Supplier Certification Program 331

Program Type: Required 331
Theory 331
Application 333
Supplemental Materials 336
    Continuing Food Guarantee 337
    Request for Evidence of Insurance—Vendors 338
    Supplier Risk Categorization Sheet 339
    Ingredient Supplier and Co-Packer Quality and Safety Survey 340
    Ingredient Supplier and Co-Packer Quality and Safety Survey 341
    GMP Checklist 342
    Food Safety Audit Report 348
    Summary of Audit Findings 350
    Food Safety/GMP Rating Analysis 351
    Food Safety Systems 352
    Food Safety and GMP Assessment Rating System 368
    Supplier Certification Program: General Overview 369
    Supplier Data Sheet 370

Chapter 22 Hold/Defective Material Program 371

Type of Program: Required 371
Theory 371
Application 373
Supplemental Materials 374
    Hold Notification 375
    Release Notification 376
    Destruction Notification 377
    Hold Log 378
    Release Log 379
    Destruction Log 380
    Hold/Defective Material Program Cover 381
    Hold/Defective Material Program Spine Label 382

Contents xv

    Manufacturing Deviation Report 383
    Hold/Defective-Material Program: General Overview 384
    Hold Tag 386

Chapter 23 Glass, Hard Plastic, and Wood Program 387

Type of Program: Required 387
Theory 387
Application 388
Supplemental Materials 390
    Glass/Plastic Inspection Chart 391
    Verification Documentation: Glass, Plastic, and Wood Training 392
    Glass, Hard Plastic, and Wood Control Program: General Overview 393

Chapter 24 Loose-Material Program 395

Program Type: Required 395
Theory 395
Application 397
Supplemental Materials 398
    Color Scheme Chart 399
    Loose-Material Control Training Verification 400
    Loose-Material Program: General Overview 401

Chapter 25 Microbiology Program 403

Program Type: Optional 403
Theory 403
Application 405
Supplemental Materials 406
    Microbiology Program: General Overview 407
    Microbiological Testing Log 408

Chapter 26 Security/Biosecurity Program 409

Program Type: Required 409
Theory 409
    Management 409
    Human Element—Staff 411
    Human Element—the Public 412
    Facility 413
    Operations 414
Application 414
Supplemental Materials 416
    Security Program: General Overview 417
    Security/Biosecurity Training Verification 420
    Key Check Out Check In Log 421

Visitor Sign In Sheet 422
Visitor Sign In Log Cover 423
Visitor Sign In Log Spine Label 424

Chapter 27 Kosher Program 425

Program Type: Optional 425
Theory 425
Application 426
Supplemental Materials 427
    Receiving Kosher List 428

Chapter 28 Organic Program 431

Program Type: Optional 431
Theory 431
Application 434
Supplemental Materials 436
    Organic Formulation Submission Form 437

Chapter 29 Environmental Responsibility Program 439

Program Type: Optional 439
Theory 439
    Management 439
    Hazardous Materials 440
    Wastewater 440
    Hazardous Waste Handling 440
    Air Emissions 440
    Waste Management 440
Application 441
Supplemental Materials 442
    Environmental Policy 443
    Environmental Checklist 444

Chapter 30 Environmental Testing Program 447

Program Type: Optional 447
Theory 447
Application 450
Supplemental Materials 451
    Environmental Testing Program: General Overview 452

Chapter 31 Outside Audits 459

Program Type: Optional 459
Theory 459

Contents   xvii

Chapter 32 Social Responsibility Program 463

Program Type: Optional 463
Theory 463
Application 464
Supplemental Materials 465
      Social Responsibility Statement 466

Chapter 33 Continuing Food Guarantee Program 467

Program Type: Required 467
Theory 467
Application 467
Supplemental Materials 469
      Continuing Food Guarantee 471
      Food and Drug Guarantee 472

Chapter 34 Contract Laboratory Testing Program 473

Program Type: Optional 473
Theory 473
Application 474
Supplemental Materials 475
      Chain of Custody 476

Chapter 35 Record Keeping 477

Program Type: Optional 477
Theory 477
Application 477

Chapter 36 Other Forms 479

Program Type: Optional 479
Theory 479
Application 479
Supplemental Materials 480
      Confidentiality Agreement 481
      Nonconforming Material Report 483
      Log of Work-Related Injuries and Illnesses 483
      Summary of Work-Related Injuries and Illnesses 485
      Injuries and Illnesses Incident Report 486
      Credit Application 487
      Carrier Seal Policy 489
      GMO Inquiry Letter 490
      Extended Nutritional Information Inquiry Form 491
      Halal Inquiry Letter 492

Affidavit for Ingredients 493
Broker Agreement 494

Index 501

The Author

Mark Clute is a food science and technology graduate from Oregon State University. During his 20 years of experience in the food industry, he has developed numerous HACCP programs and quality control systems. His work has included various aspects of the food industry, including fresh, frozen, and retorted fish; vegetable processing; jams and jellies; sauces; candy; frozen deserts; yogurt; juices and pop; spices and spice blends; nuts; extracts; ice cream inclusions; and baked goods.

From his early days working in the food industry he recognized that there was no compendium of program information for small to midsized companies or for food science professionals. After years of looking for these reference books he wanted to create a book that provided a manual of how to assemble all the required programs, with supporting documents and forms needed to meet strict regulatory compliance. It is his hope that by implementing the programs contained herein, *Food Industry Quality Control Systems* will contribute to producing safe and wholesome products.

Preface

Recall, recall, recall...these are the words that today's food manufacturers and processors hear from the federal government on a seemingly daily or weekly basis. From tainted packaged spinach, food-borne illness at restaurants, *Escherichia coli* in ground beef from several locations and in pepperoni on frozen pizza, poisoned pet foods, salmonella in white-chocolate baking squares and pot-pies, and hard plastic in soup, the food industry has recently provided its consumers a sordid litany of contaminated and adulterated products. Although these recalls involved microbiological, physical, chemical, and allergen risks, companies from large to small have failed to institute the proper controls necessary to monitor, isolate, or prevent these types of defects. This failure is generally attributable to the lack of a systematic approach to implementing a quality control program, either by a lack of labor resources or, more commonly, by a lack of time and information resources. To help fill this time and information void, a new and exciting resource is available for food company owners, managers, and quality control professionals: *Food Industry Quality Control Systems*.

This book captures the big picture of a company's quality control systems and outlines the basic methods of establishing a thorough quality control program. It provides the necessary customizable forms and documents needed for a thorough and efficient implementation. *Food Industry Quality Control Systems* leads the reader through a complete information collection step followed by a thorough microbiological, allergen, physical, and chemical risk analysis of the ingredients, products, and processes. This provides the basis for a hazard analysis critical control point (HACCP) program plan development to be used as the foundation of the entire quality control system.

Next, the reader is taught how to construct the five reference manuals—HACCP, Quality Control, Pest Control Book 1, Pest Control Book 2, and Other Programs and supplementary manuals, receiving and shipping, which house the program documents and supporting materials. For each of the subprograms and functions, this edition contains a separate chapter devoted to the theory and application of the topic. To assist the reader, all of the necessary forms for implementation at the end of each chapter are housed conveniently on a compact disc located in the back of the book. This CD is formatted to mimic the construction of each of the manuals and their specific documents are clearly referred to throughout the text. In those cases where a completed example is helpful, these are also provided in separate folders on the CD.

There is no absolute or single method for the creation of a quality control program and no one program can fit all of the various situations encountered by today's food professionals; likewise there is no single resource that summarizes the underlying required programs that make up a basic quality control program. However, today's budding food science students and quality control personnel working in the food industry or within regulatory or inspection services need a single resource to consult or to recommend to clients that provides this foundational information. *Food Industry Quality Control Systems* satisfies all of these requirements.

1 Introduction

Comprising farmers, manufacturers, processors, distributors, wholesalers, retailers, restaurants, food service establishments, and customers, today's food industry is a complex multitiered system of producers and users. The food industry is regulated by the U.S. Department of Agriculture (USDA), the Food and Drug Administration, state departments of agriculture, health departments, and numerous other governmental bodies that establish an ever mounting mass of regulations; however, it is up to each individual company to interpret the laws and regulations and establish systems and programs that ensure a safe food supply. In the midst of this overwhelming regulatory pressure, today's food companies are also faced with low margins, globalization, increased market competition, a shrinking trained employee pool, and an ever changing customer distribution pattern. All of these external pressures have forced most companies to cut corners where they can, usually in personnel, training, or quality systems development.

These personnel and program cuts at some point bring each company to experience a costly mandatory or voluntary recall or a product withdrawal from the marketplace. In 2005, the Food and Drug Administration's enforcement division performed 261 recalls: 131 class I recalls, 96 class II recalls, and 34 class III recalls. In 2006, the FDA performed 194 total recalls or field corrections of food and food products: 96 class I recalls, 67 class II recalls, and 31 class III recalls. These recalls implicated all facets of food products, including soda pop and juices, baked goods, spices and condiments, fresh and processed fruits and vegetables, sauces, dry mixes, dressings, ice cream and other dairy products, oils, and various fresh and processed fish or seafood. Companies that recalled products varied in size from large multinational firms, such as Coca-Cola, Pepsi, Kraft, Frito Lay, Interstate Brands, Dole, and Nabisco, down to small local companies such as Carlson Orchards, WestFarm Foods, Starway Inc., Golden Eagle Smoked Foods, and Aquafarms Catfish, Inc.; they were from all parts of the country.

The Food and Drug Administration instituted recalls or field corrections of products due to various food safety issues, including microbiological, foreign material, labeling, shelf life, chemical contamination, improper processing, and packaging issues. These recalls were separated as class I, a situation in which there is a reasonable probability that the use of or exposure to a violative product will cause serious adverse health consequences or death; class II, a situation in which use of or exposure to a violative product may cause temporary or medically reversible adverse health consequences or where probability of serious adverse health consequences is remote; or class III, a situation in which use of or exposure to a violative product is not likely to cause adverse health consequences. In most cases, these recalls were voluntary, although in some extreme cases legal action was required.

Recalls or field corrections for microbiological reasons involved pathogens such as *Listeria* found in tuna salad, queseo seco cheese, salmon, various sandwiches, ice cream, blue cheese, and sheep cheese; *Escherichia coli* O157:H7 in romaine prepackaged salads and alfalfa sprouts; *Staphylococcus aureus* in frozen peas; and *Salmonella* in frozen clam meat, butter, onions, nonfat dry milk, basil, pasteurized whole eggs, cake-batter ice cream, tomatoes, potato strips, frozen fish, sesame seed paste, almonds, alfalfa sprouts, cherry nut mix, black pepper, toffee, and fudge. There were also recalls or field corrections due to the presence of nonpathogens such as yeast in flavored drinks, mold in sucrose solutions, yogurt, and purified water.

Foreign material recalls and field corrections were based on the presence of nonfood items found as part of the final product. These included metal fragments in iron tablets; glass from thermometers in bread; glass in sauce, hashbrowns, and cashew nougat cookies; plastic from buckets in ice cream; and filth in eggplant garlic spread,.

Many of the products returned were due to issues surrounding the label, including:

undeclared colors such as red #40 and red #1 in picante sauce; red #49 in spiced apple almonds; yellow #5 in ice cream and in potato chips; yellow #5, red #40, and blue #1 in candy; red #10 and red #1 in picante sauce; and yellow #5 in pineapple pie;

undeclared ingredients such as eggs and whey in muffin mix; eggs in crab cakes; eggs and milk in oatmeal cookies; peanut butter in ice cream; wheat in walnut cheesecake; dairy ingredients, eggs, and hazelnuts in cookies; walnuts in brownies; eggs and buttermilk in honey Dijon dressing; pecans in streusel cake; almonds in caramel candy; sugar in sugar-free angel food cake; wheat in ice cream; milk in soymilk; peanuts in candy; sodium caseinate in calcium tablets; peanuts in ice cream; pistachios, almonds, pecans, and soy nuts in pecan/caramel clusters; eggs in sausage gravy with biscuits; milk in whipped topping; milk in biscotti; and wheat in trout;

unapproved ingredients such as ponceau 4R in jelly and in dried ginger and Sudan 1 in Worcestershire sauce;

lack of ingredient statements on candy; and

the mislabeling of a case where the product inside the package did not match the statement of identity on the label, such as olive oil that was actually soybean oil, water labeled as soda pop, raw clams labeled as cooked, and noneviscerated fish.

Shelf-life recall issues mainly focused on products exhibiting degradation due to microbial spoilage during or after their intended shelf life. Chemical contamination caused recalls due to sulfites in apricots, dried mushrooms, dried dates, raisins, soup mix, pears, dried tomatoes, sweet potato slices, dried bellflower root, dried vegetables, and dried apples; patulin in apple cider; ciprofloxacin, enrofloxacin, and fluoroquinolone antibiotic drug residues in fish; histamine in ahi/yellowfin tuna and escolar; pesticide residues in celery root; and lead in plates. Other products were manufactured via a process that did not conform to USDA-certified hazard analysis critical control point procedures such as orange, apple, grapefruit, carrot, lemonade, and watermelon juices, and Atlantic salmon, "good manufacturing practices"

Introduction 3

(GMPs) that were poor, such as anchovies that were improperly stored or improperly acidified to control pathogens or products that were decomposed prior to the "best by" date.

Historical analysis shows that there are no companies exempt from the possibility of having to conduct a recall or voluntary product recall. The common factor underlying each of these recalls is that they most likely could have been prevented with a systematic, documented, and verifiable quality control program. This type of program is one that every food manufacturer and handler should install in the developmental stages of the company, continually enhance, and use as a tool to lower risk, mitigate regulatory involvement, and ensure that the food products are safe from the farm to the consumer. Although intermediate managers and online employees can do their part to build small pieces of the overall system, it takes a systematic approach to ensure that all possible risks that affect the safety of the food are mitigated. To maximize effectiveness, the systematic quality control program must be supported through the use of a top-down quality management commitment.

TOP-DOWN QUALITY MANAGEMENT

Every company, whether it is large or small, old or new, or produces few or numerous products, has a defined management structure. Someone, the top decision maker, is ultimately responsible for everything that occurs at the company. In addition to determining the direction of the company, what products will be sold, where they will be sold, at what price to sell them, when to expand, what equipment to buy, whom to hire, when to hire, from what vendors to purchase, how to get the products to market, product packaging, and numerous other daily, monthly, and yearly decisions, the top decision maker must decide the company's standard of quality. What will be acceptable? What defects are allowed and what are not? Are there set quality specifications or are the finished product specifications moving targets? Will there be a quality program? Who will be responsible for the quality program? Who is responsible for quality? All of these questions are underlying components of one of the most important decisions made at management's highest level: What is the commitment of management or ownership to quality? Will top management provide the personnel and resources needed to create and maintain a thorough quality control program or will it grant only the minimum support needed to get through a governmental or customer food safety audit and accept the consequences of this action, up to and including potentially killing a consumer? As part of this decision, management or ownership must decide who will run the quality program, where the ideas regarding quality will come from, and just exactly what management's or ownership's responsibility to the quality control program is.

In most companies, the quality program is overseen by a single individual usually titled quality control manager, quality assurance manager, technical services director, vice president of quality, or something similar denoting the area responsibility and level of accountability. This person traditionally is schooled in a science discipline such as food science and technology, foods and nutrition, microbiology, or holds an interdisciplinary degree that encompasses classes in microbiology, food safety, nutrition, chemistry, biochemistry, and engineering. He or she is tasked with the responsibility of creating a quality control program that is multifaceted; customer, supplier, government, and competitor interactive; interdisciplinary; nontechnology based but technology supported; and internal interdepartmentally supportive. It is his or her responsibility always to be on the cusp of regulatory requirements promulgated by federal, state, and local governments and to have enough knowledge to be able to understand the basics of what makes a food safe; enough curiosity to seek new, improved, or alternate methods of measurement; and enough strength and courage to stand up for what is right even in the face of daunting opposition. The quality control manager does not, however, need to know everything about quality control or quality control systems; he or she just needs to have a strong, effective set of resources to bring to bear on any problems. In light of the magnitude of this responsibility and the scope of knowledge needed, management must ask themselves where do all of the "quality" or food safety ideas come from?

Each food manufacturing company has its own culture and identity based on many factors, including age, location, products, ownership structure, and knowledge and experience of employees. Although each factor contributes to the structure and function of the quality department, the role and responsibility of bringing ideas that improve or support the quality function of the company to the table lie almost exclusively with the quality control manager, who is to seek out, analyze, and implement new concepts and technologies that replace or strengthen existing quality control systems. With the overwhelming amount of regulation and ever increasing liability that a company faces, even before producing the very first product, the concepts and technologies that the modern quality control manager must discern and understand are highly complex and require a basic knowledge of food chemistry, allergens, food law, sanitation, microbiology, engineering, packaging technology, environmental science, safety practices, product development, and computer applications. He or she also must have a working knowledge of how to implement and manage kosher, organic, vegan, and halal certification, or any combination thereof.

Due to the emphasis on moving undergraduates into graduate programs designed to develop researchers or academians, most colleges that offer food science and technology degrees focus on the underlying theories of food engineering, food chemistry, microbiology, food law, and general food production. However, with the number of food production companies in the world, the regulatory focus on food safety and the rapid dissemination of recall information to the public, there is a surprising lack of practical application materials and teaching devoted to understanding and building a thorough and complete quality control system that encompasses but is not limited to programs in the following areas:

hazard analysis critical control point (HACCP);
GMP (Good Manufacturing Program);
pest control;
allergen;
weight control;
inspection;
sanitation;
foreign object detection;

Introduction

regulatory inspection;
lot coding;
product traceability;
customer feedback;
shipping and receiving;
product specifications;
product withdrawal;
supplier certification;
defective material; and
other non-food-safety issues such as biosecurity policy and registration, environmental policy, social responsibility policy, and kosher, organic, vegan, and halal certifications.

Where does the quality control manager obtain information to build a successful quality control program?

In today's technology-driven society, the modern quality control manager has numerous outlets to gather the resources and information required to build and refine a quality control program. The Internet, trade magazines, books, seminars, classes, inspectors, other quality control managers and personnel, trade shows, and networking are methods and opportunities to gain tips and ideas for how to build, design, and implement quality control systems within the food plant. They each provide a piece of information on a concept, system, topic, or application that the quality control manager needs to implement as a component or piece of a component of the overall quality control system. The limitations of these types of sources are that their focus usually is on a single topic, subject, or concept, such as glass control, rodent or insect identification, postsanitation testing, equipment surfaces, metal detectors, lot coding, or employee hygiene; is relegated to the underlying theory or system, single product, package, or ingredient such as pest exclusion, the cost of quality, case specification development, or raw material testing; or is just a general overview of the big picture such as developing and implementing an HACCP program or defining managements commitment to quality.

This piecemeal availability of information forces the quality control manager and the company to expend a lot of time, money, and energy to gather the knowledge and resources to design and implement the needed quality systems. Amid the high complexity of the system required, the time and money needed to gather the information and set up the program, and the skills demanded to custom design the programs and implement them, is it any wonder that many top decision makers fail to make one of the single most important decisions related to the long-term health and financial success of their company—that is, to hire a quality control manager and provide him or her the tools needed to be successful in the job?

THE COST OF QUALITY

Every food company strives to manufacture and distribute products that are safe and wholesome for its consumers. As part of this process, top management determines the company's commitment to the process and program of quality. It must determine how much money the company will spend and how many people hours it will require on a constant basis. In short, top management must determine the cost of quality.

Many food industry and non-food-industry people, governmental analysts, and lawmakers have tried to place a number on the cost of quality. These estimates are usually risk based and utilize an uneasy balancing act between how much money is being spent on the quality program versus how much will be lost if a product that has a quality defect is sold. The rationalization goes something like this: If the company spends X amount on people and resources to support the quality program, is that more money than it would spend if it had a defect that resulted in a customer complaint? This logic takes on a life of its own when subtle changes due to variations in types of defects, the amount of product in the marketplace, shipping costs, litigious customers, governmental involvement, customer goodwill, and loss of life are factored in. It is not uncommon to see monetary awards to plaintiffs for amounts from in excess of tens of thousands of dollars up to millions, depending on how serious the defect was and whether the jury felt that the company's management had taken the appropriate steps to prevent the defect from occurring in the first place.

One of the most overlooked aspects when analyzing the cost of quality is how to get the most out of the dollars spent while still minimizing the greatest risk. In considering this, two questions arise: When should money be spent on quality programs and where should the money go that, when applied, will yield the greatest reduction of opportunity for a defect to get to a consumer? There is an old saying in the quality control business that goes something like this: "It's not *if* a problem will happen, but *when* the problem will happen." Therefore, the only fundamentally sound response is to formulate a quality control program from the moment the very first product is conceived or as soon as possible after that. This program needs to be comprehensive and proactive. Failure to institute a quality control program will lead to potential food safety issues up to and including legal liability. Companies that are proactive recognize that, in the realm of food safety, "he who hesitates is lost."

Whether a company is brand new, relatively new, or long in the tooth, there is a recognition that quality control programs are not built in a day and that the most important pieces must be built first. This is the science of prioritization. All companies are not alike. They produce various products that contain various risks using various technologies. The company's location, product matrix, packaging, production environment, personnel, and market all play a significant role in helping to decide how and when to build the components of a successful quality control program. It is up to the quality control manager to determine the order of development and scope of each building block.

THE ROLE OF MANAGEMENT IN QUALITY CONTROL

Hire a quality control manager, prioritize the work load, and then what? This is a common question asked by food companies' managers; the answer encompasses many varied responses. These usually include, singly or in combination, walking away and letting the quality control manager do his or her job with little or no direction or communication, micromanaging the quality control manager, continually reprioritizing the work, eliminating some of the systems needed, establishing an ineffective and inappropriate reporting function, failing to provide the proper balance of authority and responsibility, or failing to provide the necessary monetary or material resources.

Unfortunately, these responses are usually counterproductive to the successful development of the quality control system by stymieing the progression of implementation; reducing, eliminating or undermining the authority of the quality control manager; or limiting the amount of quality time spent on system development. When management walks away and lets the quality control manager do the job with little or no direction or communication, the quality control systems are completed piecemeal due to the lack of understanding of the process intricacies or product parameters. Because production line and product differ vastly from company to company, the learning curve for the quality control manager is steep as he or she grapples with the fundamental question of how to set up a comprehensive system that encompasses all of the little nuances without having seen all of the products run.

Quality control managers that are hired and then micromanaged end up with quality systems that are incomplete. They spend their time attempting to complete the demands of the moment that management has placed upon them and never have an opportunity to flesh out the systems. Management's quality demands tend to be focused on current projects, new products, or immediate quality-related issues, many of which can be handled in the short term by existing personnel. This type of approach to building a quality control program fails to address the inherent risks from the process and each product.

When management continually reprioritizes the quality control manager's work load it leads to systems that are incomplete. The manager's time is spent building a piece of a system, getting interrupted, refocusing, building a piece of another system, getting interrupted, refocusing, and so on until there is a paralysis. Systems fail to get completed, implemented, or verified; because they are incomplete they lend themselves to failing to prevent major quality problems, leading to financial and/or legal repercussions.

Sometimes the response of management is to limit the types and scope of systems needed for a complete quality control program. This is potentially dangerous because each system interacts with and supports the other systems in the program. A simplistic example of this is what would occur if management did not implement good manufacturing policies, but did implement a thorough microbiological testing program. Employees would bring in bacteria from outside the facility and possibly contaminate the products. The microbiological testing program would determine that there was a contaminant, but without the supporting good manufacturing program to prevent the bacteria from entering in the first place, the problem would continue.

Company management that provides an ineffective or inappropriate reporting system sets the company up for the significant potential of employee unrest and product liability. Ineffective or inappropriate reporting systems are characterized by the quality control manager reporting to production or operation management. This reporting system leads to a definite conflict of interest because production's mission is to get as much product produced in as little time as possible, but the quality function's duty is to implement a comprehensive quality control program that ensures that a perfect product is manufactured every time. The competing roles of each department mean that if the quality function reports to the production or operation function, quality will always suffer. This is a common pitfall in small to medium-sized companies that do not have a set, defined reporting structure.

Another major pitfall management makes when hiring a quality control manager to develop a quality system is failing to provide a balance between the responsibility of creating the programs and the authority needed to implement and enforce the systems. When this balance is skewed toward having more responsibility and less or no authority, the quality control manager can create the systems, but implementation and enforcement become a problem. Resistance is usually from production employees that have not been exposed to quality systems or production managers that do not want their "territory" to be invaded. This resistance leads to the quality control manager fighting to implement even the smallest advancements, all the while receiving complete resistance and no support. When this balance is skewed toward having no responsibility and total authority, the quality control manager does not feel the need to implement the systems expeditiously and those that are implemented end up taking a "my way or the highway" approach, thus alienating production employees and management alike. In either scenario, responsibility and authority that are out of balance lead to incomplete or poorly implemented systems, hard feelings, and employees who do not care about quality.

The last major pitfall facing management when hiring a quality control manager is failure to provide the monetary or material resources needed to build the programs. As with any road to progress, a significant investment must me made in time and resources to ensure that the most thorough program is implemented, thus reducing the greatest amount of risk. Often, management takes shortcuts as a means to limit the short-term, out-of-pocket expense. This view limits the thoroughness of each program by not providing needed materials, such as rodent traps, testing equipment, computer technology, outside testing services, office products, or online personnel, and leads to food safety areas that are potential contamination points, as well as increased liability.

What is the role of management within the quality control environment after the manager is hired? It is a role that many food manufacturing and distribution companies either do not consider or, if they do, fail to commit to in its entirety. This role has two equally important but distinct parts: setting the organizational chart and communication and support.

When management decides to hire a quality control manager, the first consideration given must be to the company reporting structure. As with the military, there must be clear lines of reporting so that communication moves from one level of the company, such as the processing area, to other areas, such as the sales department, purchasing, or upper management. The quality control manager needs defined lines of accountability above and below him or her so that when quality issues arise, systems need to be implemented, materials need to be purchased, employees need to be trained, or other situations occur, there is a designated employee with whom to discuss each situation and know that conflicts of interest will not crop up and that action will take place. It is traditional within a food company to establish the reporting structure for the quality department to be directly to the president or top technical person. By reporting in this direction the company removes the possibility of a conflict of interest.

Conflicts occur when the quality control manager reports to operations, because of the need to get the most products out the door versus the responsibility of producing the highest quality product, to the finance area, because of the need to skimp on program expenses versus the responsibility to build a complete food safety program, or to product development because of the need to focus on new products in lieu of focusing on existing products. Each of these conflicts limits the speed of implementation and thoroughness of any newly built systems, so it is imperative that management correctly establishes an organizational chart that recognizes these conflicts *prior* to hiring a quality control manager.

The second and equally important role of management in quality control is that of communication and support. This role is split between verbal and written communication and other miscellaneous forms of support. During the course of a quality control manager's daily and weekly activity, it is essential that he or she has unfettered access to upper management. This access is needed because, during the system development, upper management must be fully aware of what systems are being implemented, their effects on personnel, and any miscellaneous expenses that have been or might be incurred. If this verbal communication is strong, the quality control manager will successfully install the needed systems. If it is weak or nonexistent, he or she will be able to install only a minimal, nonfunctioning program framework.

The companion to the verbal communication is the written communication from upper management. This takes the form of e-mails, faxes, or memos of inquiry as to the progress of the development of the program; offers to help; and notification of conditions that may hamper or help the implementation process. As with the verbal communication, good written communication will enhance the successful implementation, while weak written communication leaves everyone guessing about what has happened, will happen, or needs to happen.

The final types of support that upper management lends to the quality control manager are grouped together as miscellaneous and include monetary, material, intervention, and human support. Often, during implementation, the quality control manager needs special or odd types of support such as equipment, training materials, meals for the employees for a "job well done," intervention between the quality department and production employees to facilitate a smooth transition, and extra employee help when needed. Each request to management for these additional types of support should be carefully considered, and the implications of providing or withholding the requested support should measured.

In summary, the role of management in quality control is to be an agent for change through its constant support of the quality control manager. Failure to do so will yield devastating results that extend long term and have far reaching financial consequences to the health of the company. Constant and aggressive support for the implementation of the quality control program leads to the production of high-quality, inherently safe products and the respect and admiration of customers, employees, suppliers, and other companies.

2 Quality
Role and Function

THE ROLE OF QUALITY

As a food company hires a quality control manager and begins the process of implementing the quality control program, the role of quality control within the company should be examined. What is the quality control department's job within the company? What is the relationship between quality and production? Between quality and vendors or suppliers? Between quality and customers? Between quality and regulatory agencies? Each of these relationships has distinguishable characteristics and unique dynamics that lead to the overall success or failure of the quality control manager and program.

Before evaluating the relationships between quality control and others, it is important to state and evaluate the responsibilities of the quality control department. In most food companies the quality control department becomes involved with all areas of the company, including operations, sales, marketing, and research and development. These diversions take the quality control manager away from the four basic functions of quality control:

establishing the standards or baseline;
measuring compliance against the baseline;
reporting noncompliance; and
sharing the experience.

ESTABLISHING THE STANDARDS OR BASELINE

The first and primary responsibility of the quality control manager is to establish the standards or baseline by looking at the current state of the product or manufacturing process and determining what the product or process should look like. This process involves asking many questions such as:

What should the starting, intermediate, and final weight of the product be?
What color attributes should the product have?
What shape, size, or other dimensions should the product have?
What should each level of packaging look like?
Are there temperature attributes related to the product?
How should the product be palletized?
What lot-coding system should be used?
What size should the lot code be and where should it be placed?
Are there any physical or microbiological requirements for the product?

What is the chemical or physical makeup of the product?
Are these microbiological attributes?

From each of these questions and the numerous others specific to the product being manufactured, the quality control manager determines what the individual attributes and characteristics of the final product should be. He or she then determines what the baseline requirement for the attribute is. Because no physical measurement is absolute, each of these will have a tolerance specification—a plus or minus, if you will. This baseline becomes the "attribute specification" that, when combined with all the attribute specifications, makes up the final "internal product specification." These internal product specifications are in contrast to the "customer product specifications" given to each customer or potential customer and are used exclusively by operations, product development, and quality control as a means of verifying that the process is in control and that the products produced are consistently up to the standards set for them. Each product manufactured must have an internal finished product specification. In some companies this takes the form of a "processing" specification.

MEASURING COMPLIANCE AGAINST THE BASELINE

Once the finished product specification is determined, the next basic function of the quality control manager is to determine how to measure compliance with each of the attributes against the baseline. This will involve a myriad of measurement techniques, such as weighing or measuring each product either on or off the line, pulling samples for in-house or outside laboratory analysis, checking the packaging properties, evaluating case counts, or conducting an organoleptic evaluation. In some cases the measurement responsibility will be placed on the supplier in the form of a certificate of analysis. The measurement techniques chosen need to be specific to the individual product attribute as set forth on the internal final product specification and should be tailored specifically to the product process. For instance, if the attribute to be measured is soluble solids (Brix), it may not be possible to evaluate this attribute for each finished product passing down the line; therefore, a process-specific measurement needs to be established (i.e., one that evaluates a statistically relevant quantity of samples). In similar fashion, each measurement that is taken should be documented on a product evaluation and measurement recording form that allows the operator or technician to record each measurement.

These measurement recording forms should contain the attribute to be measured, the specification for the attribute, how it should be measured, a place for recording the measurement, and the action to be taken if the measurement falls outside the specification. Each of these components is critical to its successful measurement, documentation for future reference, and any follow-up action that is taken. The listing of the attribute on the form tells the technician or operator specifically what he or she should be measuring at that specific point in the process. Listing the required specification lets the technician or operator know what the outside control limits (upper and lower) for the attribute are and gives a basic idea of what the target reading should be. This helps to prevent operator or technician errors due to "experience guessing"—guesses based on the idea that the operator or technician has done or seen the measurement before—or "supposition guesses"—guesses based on what the operator or technician supposes the measurement should be. The specification may be listed separately or as a component of charting the measurement results.

The third component of the recording forms (how the attribute should be measured) gives the operator or technician a designated method for conducting the measurement. Sometimes this is a clear, concise method that may be detailed on the recording form (such as "using a calibrated caliper"), but sometimes the measurement method may be too complex to be listed in its entirety. In this case a separate measurement procedure must be written, numbered, and maintained as part of the quality control manual. For each measurement taken by the operator or technician, properly documented training must be given to ensure the consistency between the employees conducting the measurement. If the operator or technician is not properly trained to exercise the measurement, the opportunity for product variance increases up to and including producing a product that is out of specification.

For each specification that the operator or technician needs to measure, the recording form must contain spaces to record the sample time, sample number, measurements, other parameters specific to the sample, and comments made regarding the sample. These spaces need to be of adequate size and space so that the data can be clearly read or interpreted during the recording time and at any time in the future. Too often, the operator or technician is given too little space to record all of the data and/or comments needed to be made at the time, so when the data are reviewed, it becomes very cumbersome to understand what actually occurred.

The final component of the measurement form is one of providing a place for the operator or technician to write notes or comments. These might include production information such as down-time notes, how many operators are on the line and who they are, line speeds, etc. This is the location for documenting actions taken on the line in response to a quality defect. Because this is so important, care must be taken to note any action taken, who was notified, what was done, any timing information, and any other relevant information for future reference.

Reporting Noncompliance

When a quality defect is found and documented, the technician assumes the third role of quality control, which is to report the defect. This function usually contains four parts: notification to others of the defect, follow-up to make sure the defect does not occur again, documenting how the problem was fixed, and changing the processing specification as needed.

Notification to others of any defect can be simple or complicated. For instance, if the defect is just a matter of the label applicator not applying a label to each package, then a simple notification to the line operator will suffice (although if this is a constant occurrence, a different and elevated reporting procedure should be used). On the other hand, if the defect is a microbiological problem that has reached the customer, then multiple parts of the company must be notified, including ownership and top management. Who and how to notify are defect specific and care should be taken by the technician to notify those people with a need to know but to limit the open discussion of product defects as a means to protect the brand. Often, product defect information gets openly discussed with employees and then they discuss it with or around people from outside the company. This opens the company to regulators and competitors knowing inside information and being able to use it as a means to apply regulatory pressure or to gain a competitive advantage against the company.

After the proper defect notification is complete, the technician needs to follow up with the person or persons designated to fix the problem. This may be someone in any department other than quality control and may include more than one person or department. The person responsible for conducting the follow-up may be the technician; however, in any case, the follow-up and any supplementary action taken need to be documented for future reference.

Subsequent to notification, follow-up, and documentation, the current processing specification must be evaluated with an eye toward making adjustments based on preventing the defect from happening in the future. Although this change does not have to be complicated, it needs to address a specific area of the process that can be measured and documented and will give the operator guidance. The new processing specification needs to be disseminated to the appropriate departments and the old specification collected and destroyed. It is imperative that there are no old specifications lying around when a change is made because they will inevitably be found and used and the defect will occur again.

Sharing the Experience

The final component of reporting the defect is sharing the experience. This involves discussing the problem and solution with other members of the operations and quality control staffs through one-on-one discussions, group discussions, written communications, or formal training sessions so that each team member gains the experience of defect recognition, solution determination, and implementation for as many situations as possible. Sharing these experiences helps to build the knowledge base within the company so that when defects occur, employees leave or are on vacation, or new employees are trained, there are any number of people who can solve or prevent defects from happening.

Another method of sharing the experience is to create a situation logbook. The logbook is a central binder, or log, that quality control or operations employees can write in that outlines the problem, the solution, and other pertinent facts. It is available to all people in the company and serves as a ready reference should problems arise and as a training guide for new employees. The logbook allows for the experience to be shared forward into perpetuity.

THE QUALITY CONTROL–PRODUCTION RELATIONSHIP

One of the most important and fundamental relationships within a company is that between quality control and production. To understand this relationship one must first understand the roles each plays within the company. Production's responsibility is to produce as much product as fast as possible with a minimum amount of waste and a minimum amount of labor. On the other hand, quality control's responsibility is to verify that every product is manufactured within the agreed upon specifications at all times. As two diametrically opposite and fundamentally exclusive tasks, the format this relationship takes within a company can determine if the company will be successful or fail. Three basic types of the production–quality relationships are found within companies: secretive and compartmentalized, divisive or adversarial, or conciliatory and team building.

THE SECRETIVE AND COMPARTMENTALIZED RELATIONSHIP

A secretive and compartmentalized quality control–production relationship is characterized by each department working separately. No or limited information is shared with the other department and neither department goes out of its way to help or support the other. Because each department sees itself as an isolated island within the company, it has no need to share or receive information. Problems are solved in a compartmentalized setting with an eye toward quality or toward production, but generally not both. Not much information is shared with upper management because neither department wants to look uninformed. This type of relationship does not promote quality improvement, improved production efficiency, or employee growth. Employees working under this system tend to feel stifled by the company and disrespected by employees outside their own departments. Upper management generally is unaware of this type of relationship until a crisis occurs, at which time the natural tendency is to blame the other department for not sharing information or working as a team.

THE DIVISIVE OR ADVERSARIAL RELATIONSHIP

The divisive or adversarial quality–production relationship is characterized by an underlying tension between departments, sometimes leading to outright anger and subversion. Vital information is routinely withheld in the hopes of making the other department look bad or incompetent, and there is a blatant disregard for the position and responsibility of the other department role and responsibility. Quality control seeks to demand strict and complete adherence to standards and measurements without taking into consideration that the company makes money only when production is producing. Production seeks to dispense with all quality standards or measurements and demand that production go as fast as possible to produce as much as possible. This type of relationship generally leads to a few high-quality or many low-quality products being produced. It damages the company finances and employee moral and leads to great internal strife.

THE CONCILIATORY TEAM-BUILDING RELATIONSHIP

The final type of quality–production relationship is that of a conciliatory team-building one. It is characterized by an open sharing of quality and production issues, specifications, measurement techniques and requirements, production requirements, maintenance schedules and requirements, production and quality staffing requirements, and any of numerous other pieces of information needed by one or both of the departments to produce the maximum amount of products successfully with the highest level of quality. Problems are solved in a team-oriented environment, with both departments able to express their needs and ideas. Department managers and employees feel comfortable interacting and sharing with the other department and grow personally and professionally. This type of relationship is the only one of the three that allows the company to grow at a rapid pace and still maintain quality products, great efficiency, and high employee morale.

Each of these quality–production relationships exists in numerous food companies throughout the world. It is critical for upper management to observe and monitor this fundamental relationship within the company and take whatever action is needed, up to and including staffing changes, to facilitate the conciliatory team-building relationship. If management allows either of the other two types of relationships to exist, the company is destined to flounder through poor quality or poor production. It will be tough to retain employees because they will feel disrespected and underappreciated; customer complaints will increase because quality will suffer, and the possibility of legal and regulatory intervention will increase.

THE QUALITY CONTROL–VENDOR RELATIONSHIP

Because of the quality control department's system-building activities, the department has numerous opportunities to interact with vendors and suppliers of ingredients, packaging, equipment, and other supplies. These opportunities for interaction occur usually either in the hazard analysis phase of the HACCP development or during various crisis situations. This relationship is characterized by the quality control employee having the upper hand, or control, within the relationship because the vendor wants desperately to make a sale. Care should be taken during this interaction because many times suppliers are willing to do whatever is asked or is needed to make that happen. This includes providing or offering to provide property, meals, entertainment, or other gratuities in return for a favorable recommendation of their products or services to the purchasing department or other buyers within the company or an acceptance of a substandard ingredient. This is *graft;* not only is it unethical but it also undermines the entire company's reputation within the industry, not to mention the reputation of the employee accepting a gratuity. Often, employees who take vendor gratuities find themselves in the precarious position of having to do something for the vendor to keep him or her from informing upper management of their misconduct. Therefore, during this interaction, the quality control employee must be careful always to maintain a position of the highest integrity. Care should be taken by the quality control employee to establish a firm no-graft line with each vendor so that it is clear to all observers that each transaction is untainted.

Other components of the quality–vendor relationship involve communication and respect. Communication in the form of e-mail, telephone, and fax needs to be complete and timely. If the vendor needs information to initiate or complete a request, the quality control employee needs to provide it in a responsible, timely, and effective manner and then follow up as needed. Often, the quality control employee is very busy and as such lets things go or "pushes them to the back burner" until they are urgent. Finally, when the employee realizes that the information is needed, he or she calls the vendor and expects the vendor to drop all other customers and "jump through hoops" to provide the needed information. Although there are times when expediency is needed, this should not be the normal practice of any quality control employee. It is not only unfair to the vendor but also very disrespectful and, in the long term, may damage the vendor's relationship with the company and the employee's reputation in the industry.

Along a similar vein, the quality control employee should expect the vendor to provide requested information in a complete and timely manner. When a request is made of a vendor, it should include the specifics of what is needed, when it is needed, to whom it should be provided, and in what form the information should be provided. A written record of the request should be made and a follow-up timeline noted. If the vendor fails to provide the needed information on the agreed upon schedule, the quality control employee needs to discuss the issue respectfully with the vendor. If it continues to occur, further measures need to be taken that might include discussing the issue with others in the vendor's company or discussing the issue with the internal purchasing department. When the vendor does not respond as requested and the quality control employee strives to solve the problem, he or she must remember that his or her actions reflect on others' perceptions of him or her and the company.

THE QUALITY CONTROL–CUSTOMER RELATIONSHIP

During the course of daily quality control activities, the quality control employee interacts with customers who have found defects, have technical questions, or have complaints associated with the products they purchased. This communication generally takes place via telephone but may take the form of letters, e-mails, or face-to-face interaction. In each situation quality control employees must remember that how they present themselves to the customer, no matter how irate the customer is, will set the tone for whether the customer's expectations are met and whether the customer and his or her friends ever buy the product again.

For every complaint, the quality control employee should document the issue for further follow-up and give the customer a specific outline of what will be done and when it will be done. This timetable is very critical to the customer. If nothing is done or if something is not done in a timely manner, the customer will feel frustrated and may increase his or her demands up to and including legal action. This is not good. Extreme care should be taken to treat the customer with the utmost care and respect. To this end, many companies have created special departments to handle customer inquiries exclusively and train their employees extensively in how to provide and document high-quality service.

It is the responsibility of the quality control employee to communicate to upper management any customer-related complaints, no matter how small. This communication may be a quick e-mail or a formal monthly report. It is imperative that upper management knows what is happening with customers so that it may plan marketing strategies, improve production and quality systems, or make alterations to the product or package.

THE QUALITY CONTROL–REGULATORY AGENCY RELATIONSHIP

Every food plant or distributor has one or more regulatory agencies that oversee its compliance to laws and regulations. These regulatory agencies may include the U.S. Department of Agriculture (USDA), the Food and Drug Administration (FDA), State Departments of Agriculture, state, county, or local departments of health, or any combination of these entities. Each of these may at any time contact or visit the company for the purpose of a scheduled inspection, a surprise audit or to follow up on a formal customer complaint. It is the responsibility of the quality control manager to act as the company representative to these agencies. This interaction involves three components: planning, communicating, and personal interaction.

Planning involves having a written plan outlining how regulatory agencies and inspections will be handled when an inspector or other agent enters the building. This plan involves all aspects of the interaction up to and including adequately training company employees on what to say and do when they are in the presence of the inspector. As a general rule, a systematic, well-thought-out plan will save time, energy, and money in the long run. This plan is part of the quality control manual.

Communications between the Quality Manager and regulatory agencies takes the form of emails, faxes, or phone calls. Normally, this is to respond to a regulatory agency or solicit information from one. These communications should be conducted in a friendly and professional manner. The agency must feel that the company is conducting itself with the highest ethical and food safety standards. If the quality control employee is disrespectful or secretive with the agent, the agent may conclude, judgmentally, that the company is hiding something. This might possibly lead to targeting the company for increased and more thorough inspections. It is the underlying but unspoken task of any inspector or agency to find something wrong; when he or she does find something wrong, how he or she has been treated prior to finding the offense can dictate the types of response given when things arise. With the availability of the Internet today it is easy to access regulatory agencies' Web sites and gather most needed information without personal contact.

Personal interaction usually occurs when the regulator visits or inspects the company facility. When this occurs, care must be taken to treat the regulator with the utmost respect and courtesy. This includes greeting them in a timely manner, giving them your complete attention, and being responsive to their requests. When contact occurs, always maintain a professional and conciliatory demeanor.

3 Quality Control Systems Development Overview

Food Industry Quality Control Systems is a comprehensive guide for the development of the programs and systems needed to build and manage a complete basic food safety system. It includes separate sections for each of the regulatory required foundational programs, such as good manufacturing practices (GMPs); pest control; inspections; fill control; sanitation; metal detection; allergen control; receiving; shipping; glass, wood, and hard plastic control; bioterrorism; and loose material, as well as the regulatory suggested hazard analysis critical control point (HACCP) program. Also included is information on additional programs that the quality control department usually administers or in which it is involved, such as kosher, environmental, social responsibility, organic, vegan, and security/biosecurity. To support the reader with systems implementation, a compact disc is provided that provides all of the basic forms for documenting the general program, auditing the process, and verifying the program's continuity and integrity. Each of these forms is designed so that it can be customized for the individual situation and company. They are saved as Microsoft Excel, Microsoft Word, Microsoft PowerPoint, or Adobe Acrobat files.

As an aid to personnel and companies implementing quality control programs, the supporting quality control program sections are divided into two parts: theory and application. The former outlines the basic theory behind the program, including how it fits into the scheme of the overall quality control program, the reasoning behind the need for the program, and the resources required. The latter walks the reader through how to implement the program successfully, and where the forms are located, how to fill them out, and where to file them. For simplification, "quality control manager" should be interpreted by the reader as the designated person tasked with implementing the quality control programs.

The development of a complete quality control program utilizes a systematic plan that involves developing a HACCP program as a means to determine the hazards and risks to which the product and process are exposed, as well as a quality control program containing all the underlying subprograms that support the HACCP program and other departments within the company. The types and styles of supporting programs will vary depending on the raw materials required; the product being manufactured or processed; how it is packaged; and the product's distribution channel. For example, if the finished product is only packaged on a scale by hand into bulk industrial boxes, there is no need for a weight control program because each is presumably exactly on weight. Likewise, if the products are not produced under kosher supervision or organic certification, these programs are not needed. When determining what programs to develop, the quality control manager should err on the side of developing all the needed programs because not having one could expose the company and its products to risk.

As the programs are developed and standard operating procedures (SOPs) are written, the quality control manager needs to build a focused centralized document-keeping system that promotes open access for all who need to evaluate, inspect, or review any program. One easy method for storing and maintaining the program documents is by the use of three-ring binders. Five separate binders should be set up and labeled as: HACCP Manual, Quality Control Manual, Pest Control—Book 1, Pest Control—Book 2, and Other Programs Manual. In addition, two supplemental three-ring binders should be created for the receiving and shipping logs. It is recommended that both the spine and the cover be labeled for viewing ease. Spine and label samples may be found on the CD under Supplemental Forms and at the end of this chapter. To each binder add tabs for the appropriate sections as designated in the following list. Sample file folder tabs for each of these sections are located on the CD under Supplemental Forms and at the end of this chapter. They are color coded (CD) for easy recognition, but may be changed to any desired color or just printed in black.

Book 1: HACCP Manual
    Section 1: Program Overview
    Section 2: Process Flows
    Section 3: Product Descriptions
    Section 4: Hazard Analysis Sheets
    Section 5: Critical Control Worksheets
    Section 6: HACCP Plan Summary Sheet
    Section 7: Verification Statement
    Section 8: Auditing Statement
    Section 9: Training
Book 2: Quality Control Manual
    Section 1: Overview
    Section 2: Organization Chart
    Section 3: Good Manufacturing Practices
    Section 4: Pest Control Program
    Section 5: Allergen Program
    Section 6: Weight Control Program
    Section 7: Inspection Program
    Section 8: Sanitation Program
    Section 9: Metal Detection Program
    Section 10: Regulatory Inspection Program
    Section 11: Lot Coding Program
    Section 12: Customer Complaint Program
    Section 13: Receiving Program
    Section 14: Shipping Program
    Section 15: Specification Program
    Section 16: Recall Program
    Section 17: Supplier Certification Program
    Section 18: Defective Material Program
    Section 19: Glass/Hard Plastic/Wood Program

Section 20: Loose-Material Program
Section 21: Microbiological Program
Book 3: Pest Control—Book 1
  Section 1: Outside Pest Control Contract
  Section 2: Company License
  Section 3: Company Insurance
  Section 4: Labels
  Section 5: MSDSs
  Section 6: Facility Map
  Section 7: Applicator's License
Book 4: Pest Control—Book 2
  Section 1: Facility Map
  Section 2: Pest Control Sheets
  Section 3: Contract Applicator's Sheets
Book 5: Other Programs Manual
  Section 1: Security/Biosecurity Program
  Section 2: Kosher Program
  Section 3: Organic Program
  Section 4: Environmental Responsibility Program
  Section 5: Environmental Testing Program
  Section 6: Outside Audit Program
  Section 7: Social Responsibility Program
  Section 8: Continuing Food Guarantee Program
  Section 9: Contract Laboratory Testing Program
  Section 10: Record Keeping

After these books have been developed, they should be located conspicuously in the quality control manager's office for open access to all employees or inspectors.

All program documents that are developed should follow a clear, concise format that includes a description of the program, a statement of responsibility, details of the program, and a method for program verification. The description of the program outlines the reason for the program, indicates whether it is mandated by a regulatory agency, and offers a brief outline of the program. Questions concerning why the company needs the program, what hazard or risk the program will address, and what the implementation of the program will involve should be asked. If the program is mandated, a regulatory reference should be made, such as "code of federal regulations 21CFR110.80(a)." The statement of responsibility explains who is directly responsible for administering the program and who the backup is in case something befalls the person of primary responsibility or he or she is just not available. This is especially important in companies where managers travel or have many responsibilities. In some cases the responsibility for administering and enforcing the program may be split between various departments or individuals. If this is the case, this should be clearly delineated so that any confusion may be alleviated when issues arise.

The details of the program are outlined using an explanation of how the program is set up and how it works. When this section of the program document is written, questions concerning what is being measured, how it is being measured, where the measurement takes place, who will conduct the measurement, and how measurements will be documented should be answered. Care should be taken to include the basics of how the program is implemented, administered, and audited. Normally, an outline consisting of each basic step is listed. Do not include specific, detailed steps because this might necessitate numerous changes to the program document as the products and processes change over time. When a form is created for measuring purposes, a separate set of specific directions, or a standard operating procedure, should be developed and placed in the book next to its corresponding form.

Each program document or form needs to include an identifying number that reflects its place in the overall quality control program record system. Each should also have an effective date and a version number. If the document is revised, a note should be added that explains how and/or why the document was revised. When using Microsoft products, the effective date, revision date, reason for revision, and version number can be effectively placed in the footer section using an 8-point font.

Program and document development can be a very time-consuming and costly exercise. Many times, companies spend hundreds of thousands of dollars and many years in development as they struggle to contain the risks associated with ingredients, finished products, suppliers, employees, and customers. This time and effort can be minimized by utilizing a comprehensive resource such as *Food Industry Quality Control Systems*.

HACCP Manual

HACCP MANUAL

Quality Control Manual

QC MANUAL

Pest Control
Book 1

PEST CONTROL BOOK 1

Pest Control
Book 2

PEST CONTROL BOOK 2

Other Programs Manual

OTHER PROGRAMS

Receiving Manual

RECEIVING MANUAL

Shipping Manual

SHIPPING MANUAL

Quality Control Systems Development Overview

HACCP Book Labels Program Overview Process Flows

Product Descriptions Hazard Analysis Sheets Critical Control Worksheets

HACCP Plan Summary Sheet Verification Statement Auditing Statement

Training **Quality Control Manual Labels** Overview

Organizational Chart Pest Control Program Allergen Program

Good Manufacturing Practices Inspection Program Sanitation Program

Weight Control Program Microbiological Program Temperature Control Program

Metal Detection Program Lot Coding Program Customer Complaint Program

Regulatory Inspection Program Shipping Program Specification Program

Receiving Program Recall Program Supplier Certification Program

Defective Material Program	Glass/Plastic/Wood Program	Loose Material Program
Pest Control Book-1 Labels	Outside Pest Control Contract	Company License
Company Insurance	Labels	MSDS's
Facility Map	Applicators License	
Pest Control Book-2 Labels	Facility Map	Pest Control Sheets
Contract Applicators Sheets		
Other Programs Manual	Security/Biosecurity Program	Environmental Program
Social Responsibility Program	Kosher Program	Environmental Testing Program
Non-GMO Program	Organic Program	Vegan Program
Contract Laboratory Program	Continuing Food Guarantee Program	Third-party Audit Program
Record Keeping		
Receiving Manual	Log Sheets	Kosher List
Ingredient Statements	**Shipping Manual**	Log Sheets

4 Book 1
Hazard Analysis Critical Control Point (HACCP) Program

PROGRAM TYPE: REQUIRED

The basic underlying and fundamental program for the development of a functional and successful food safety program is the hazard analysis critical control point program, commonly referred to by its acronym, HACCP. A HACCP program provides the management of a food company or distributor a systematic approach to evaluate products and processes for physical, microbiological, chemical, and allergenic hazards that might occur and pose a food safety hazard as well as direction to install corrective measures to prevent them from occurring. This approach is required by the federal government for animals and animal products (9CFR417), fish and fishery products (21CFR123), and juice (21CFR147); however, it is recommended for all products by the U.S. Food and Drug Administration (FDA), the U.S. Department of Agriculture (USDA), the Food Safety and Inspection Service (FSIS), other regulatory agencies, and industry groups.

The creation of the HACCP plan utilizes all of the departments and expertise of the company and the inherent knowledge of the products and processes. As the foundational program, the HACCP plan helps to determine which other supporting programs need to be developed. The following section provides a basic step-by-step instruction for building a HACCP program. Each step explains what is to be done and how it is to be accomplished. Prior to beginning the development of this program and in preparation for collecting and storing all of the forms generated during the HACCP program development, a three-ring binder should be prepared. It should be labeled "HACCP Manual" and contain 9 labeled tabs—Section 1: Program Overview; Section 2: Process Flows; Section 3: Product Descriptions; Section 4: Hazard Analysis sheets; Section 5: Critical Control Worksheets; Section 6: HACCP Plan Summary Sheet; Section 7: Verification Statement; Section 8: Auditing Statement; and Section 9: Training.

The responsibility for directing the development of the HACCP plan within the company usually is assigned to the quality control manager because, to implement a complete HACCP plan, this person must be able to recognize and evaluate the big picture. He or she must be able to view and understand the interactions between all of the components of the production and distribution flow, including major aspects such as the building structure, warehousing methods, processing line construction and sanitation, packaging, ingredients, and distribution methods, as well as minor aspects, such as supplier food safety practices, pest prevalence, employee practices, maintenance practices, and many other nuances. Furthermore, he or she must have a basic understanding of microbiology, chemistry, equipment, allergens, statistics, engineering, and other disciplines. Each must be recognized as a small piece of the HACCP evaluation and of the overall quality program. Typically, based on the educational background of the quality control manager, he or she is the person best suited for this duty, although this is not always the case. When choosing the HACCP plan implementation leader, always choose the person who understands the business and products being manufactured.

In preparation for the development of the HACCP plan, a HACCP team is formed. This should include employees from some or all of the various departments within the company, including production, maintenance, sales, warehousing, quality control, research and development, sanitation, microbiology, operations, purchasing, and accounting. Each represented department brings a unique perspective to the hazard evaluation and contributes to the development of a basic HACCP knowledge base within the company. In some small companies, this team may consist of the quality control manager, an operations manager, and perhaps one or two other people. This is satisfactory as long as the people on the team have a complete understanding of the products and processes. In larger companies, the HACCP team might consist of employees from operations, engineering, maintenance, sanitation, warehousing, product development, quality control, safety, accounting, and sales. This team should include local personnel involved in the operation because they are more familiar with its variability and limitations. In addition it fosters a sense of ownership of the final product. Although it is important to bring as much knowledge and food experience exposure as possible to the team, there is no set or required team format or structure. The key to having a successful HACCP team lies in the commitment of the personnel to strict adherence to completion of each of the HACCP principles and a total support of the team from management. A form to use as documentation of the members on the team can be found on the CD (Book 1_HACCP Program\Section 1_HACCP Overview:HACCP Development and Implementation Team Roster) and at the end of this chapter. It should be filled out and placed after the program document in the HACCP section of the Quality Control Manual.

Once the HACCP team is formed, its members need to meet on a regularly scheduled basis to ensure that the program is being developed in a consistent manner to completion. It is also important for the team to have a designated leader (normally, the quality control manager) and to have designated meeting plans and task assignments. So that there is a permanent record of all HACCP team meetings, attendance should be taken and summary minutes kept that include tasks and responsibilities agreed upon during the meeting. Team meeting attendance can be documented using the form on the CD (Book 1_HACCP Program\Section 1_HACCP Overview:HACCP Team Meeting Attendance) and at the end of this chapter. These meeting records should be kept in a separate file or notebook for later reference by team members or regulatory and third-party inspectors.

The responsibility of the team is to develop a HACCP plan and all the supporting materials, and then to set up the verification and auditing functions. The team follows certain steps in order to build the program:

planning:
    collect ingredient information
    collect process flow information
hazard analysis:
    fill out product description forms
    fill out hazard analysis sheets: chemical, microbiological, physical, allergen
    determine risks
critical control points:
    determine critical control points
    determine control steps
measurement:
    develop measurement methods
    implement measurement systems
verification:
    establish verification function
auditing:
    establish auditing function
training:
    share information

During the first team meeting, the team leader teaches the other members what a HACCP program is, the steps that will be taking during the development process, and how it acts as the supporting function to the overall quality control program. It is important that each team member have a basic understanding of what HACCP is, the role it plays in the overall scheme of the quality program, and the function of the supporting programs. After the committee members are trained, they are split into teams to begin the task of planning the HACCP program.

HACCP PROGRAM PLANNING—THEORY

No HACCP plan can be successful without a thorough planning phase because it provides for the creation of a document filing system and collection of the technical documents needed to conduct the analysis. The filing system provides a formal and systematic location where all the technical documents and supporting materials obtained from the suppliers are stored. It customarily is organized by product code or item number for easy reference and is separated into three parts: ingredients or raw materials, packaging materials, and finished goods. For ease of use, color-coded tabs should be used for each section (e.g., orange for raw materials and finished goods, blue for packaging, and white for finished goods). A supplier data sheet is placed in the front of each ingredient or raw material folder.

This sheet is used for documenting the complete collection of all the technical documents and contains a place to enter the supplier name, contact information, product identification, and internal company ingredient information. A place for entering emergency contacts is provided as an easy reference in the case of an emergency involving the supplier's ingredients. The balance of the supplier data sheet contains a check list of the required documents requested for each ingredient and an allergen summary chart.

The list of documents and information requested is divided into two categories: required and optional. Required documents include:

1. *Ingredient specification sheet.* This document lists the name of the ingredient and gives a general description, the supplier's ingredient number, the ingredient statement, the packaging and palletizing method, storage conditions, and shelf life at the recommended storage. It also lists any specific physical, chemical, and microbiological attributes along with the target specification, tolerance or range from the target, and the testing method used by the supplier to evaluate the attributes. This document is used as a basis for evaluating possible physical, chemical, microbiological, or allergen hazards. Understanding the ingredients contained therein aids in developing the finished product ingredient statement and in determining handling and storage methods after purchase. Care should be taken to verify that each of the required components is on the ingredient specification sheet. If all components are not, the supplier should be contacted and a new specification sheet obtained.
2. *Nutritional statement.* This document lists the nutritional breakdown of the ingredient. With the availability of detailed nutritional databases such as the USDA National Nutrient Database for Standard Reference (http://www.nal.usda.gov/fnic/foodcomp/search/) and/or nutritional software like Nutribase Clinical by Cybersoft Inc., Food Processor SQL by ESHA Research, or Nutritionist Pro by Axxya Systems, suppliers can easily create a complete nutritional breakdown. This breakdown includes total calories, calories from fat, calories from saturated fat, total fat, polyunsaturated fat, monounsaturated fat, cholesterol, sodium, potassium, total carbohydrate, dietary fiber, soluble fiber, insoluble fiber, sugars, sugar alcohol, other carbohydrates, protein, vitamin A, percent of vitamin A present as beta-carotene, vitamin C, calcium, iron, and other essential vitamins and minerals. Because of their size, some companies will not provide the complete nutritional breakdown but will provide either a nutritional panel or an approximate analysis. In cases like this, the supplier should be asked to provide the actual formula or an approximate calculation so that the nutritional breakdown can be calculated from a database.
3. *Continuing food guarantee.* This document guarantees that the ingredient is manufactured in accordance with federal sanitary guidelines as set forth in the Food, Drug and Cosmetic Act and its subsequent amendments. It also states that the product is not adulterated and/or misbranded within the meaning of the act. The continuing food guarantee provides a layer of confidence to the manufacturer that, from a legal standpoint, the ingredient supplier stands behind what it is supplying. Each guarantee should have an open-ended revocation period or expiration date and must be signed by one of the ingredient supplier's corporate officers.
4. *Biosecurity Act compliance.* Each ingredient supplier is required to register its company under the Public Health Security and Bioterrorism

Preparedness and Response Act of 2002. The act requires manufacturers, processors, packers, distributors, receivers, holders, and importers of food to keep specific records on their suppliers and customers. This allows inspectors to trace the origins of a questionable product one step forward and one step backward at each place in the distribution chain. Compliance can be documented by providing, on letterhead, a statement of compliance signed by a corporate officer. In lieu of a compliance letter, some companies will provide the registration number.
5. *Declaration of allergens.* This statement is a list of the top eight allergen-containing ingredients: milk, egg, wheat, soybean, fish, crustacean shellfish (e.g., shrimp, lobster, crab, and crawfish), peanuts, and tree nuts (e.g., almond, walnut, and hazelnut). Some producers will provide additional ingredients that either cause minor allergenic reactions, such as seeds (e.g., celery seeds and mustard seeds) and colors (e.g., yellow #5), or aggravate sensitivities such as sulfites and monosodium glutamate (MSG). The declaration specifies whether the supplied ingredient contains the allergen, an allergen-containing ingredient is used on the same production line as the ingredient, or an allergen-containing ingredient is stored in the same warehouse as the supplied ingredient. If any of the top eight allergens is not listed, the supplier must be required to resubmit a complete declaration.

Optional documents solicited from the supplier during the planning phase include:

1. *Material safety data sheet (MSDS).* If, under normal conditions of use, the product could result in a hazardous exposure situation for downstream employees who will be working with or otherwise handling that product, then an MSDS is required. These include products such as flavors with a hazardous material carrier, like ethanol, or ingredients that pose an explosion hazard, like flour. Most food ingredients are exempt from this requirement, but this document should be requested for safety reasons.
2. *Kosher certificates.* If the finished product in which the ingredient is to be used will ultimately bear a kosher symbol, then a kosher certificate must be requested. When received, the certificate should be reviewed to ensure that the name of the manufacturer, the vendor's ingredient name, and the item number are correct and match the label on the ingredient's container. This comparison against the label can be done before the ingredient is purchased for the first time by requesting a copy of the container label or by checking the ingredient stock in the warehouse. The expiration date should also be confirmed to ensure that the supplier has provided the most current document.
3. *Organic certificates.* For those companies that sell products that comply with the National Organic Program (NOP), a certificate of organic origin needs to be obtained. The ingredient supplier obtains this from USDA-accredited certifying agencies and it generally has an expiration date or date of plan recertification. The certifying agency's accreditation should be checked to make sure it is current with the USDA. A current list of accredited agencies is found at http://www.ams.usda.gov/nop/CertifyingAgents/CertAgenthome.html.

4. *Nongenetically modified organisms (non-GMO) statement.* As a companion document to the organic certificate for those companies that produce organic products and as a standalone program for those companies that are not producing organic but want to maintain a non-GMO status for an ethical or marketing reason, a non-GMO statement should be requested from the supplier. This states that the ingredient contains no genetically modified organisms and this status is verified either through DNA testing or source determination. The statement should be signed and dated by a corporate officer because it becomes a legal document.
5. *Vegan certificate.* Similar to kosher certification, if the end product will be labeled as vegan, then a vegan certificate or statement should be requested. This statement indicates that the ingredient does not contain substances that are not acceptable for consumption by vegans.
6. *Halal certificate.* If the end product will be marketed and labeled toward the Muslim faith, then a halal certificate should be requested.

HACCP PROGRAM PLANNING—IMPLEMENTATION

The first task the HACCP team undertakes is to create a supplier file for each ingredient. The supplier data sheet on the CD (Book 1_HACCP Program\Planning:Supplier Data Sheet) and at the end of this chapter should be accessed and then stapled to the left inside cover of a manila folder. The tab on the folder should be labeled with the company's internal item number, the ingredient name, and the supplier name for quick reference. Starting with the first ingredient on the master list of ingredients, the internal item number and the ingredient number should be filled in. Next, the supplier is contacted and the following technical information is requested:

- a specification sheet that includes an ingredient statement, storage conditions, and shelf life when stored at the recommended storage conditions;
- a nutritional data sheet;
- a continuing food guarantee;
- a statement confirming that the supplier has registered with the Public Health Security and Bioterrorism Preparedness and Response Act of 2002 (the Bioterrorism Act);
- a list of emergency contacts with names and phone numbers;
- an allergen statement; and
- an explanation of the lot code and how to read it.

Depending on the type of ingredient, company, or customer base, additional information may be requested. This includes:

- a material safety data sheet (MSDS) if the ingredient contains a hazardous component such as ethanol as a flavor carrier;
- a declaration stating that the ingredient does not contain genetically modified organisms (non-GMO); and
- any or all of the following outside certifying body certificates: kosher, halal, organic, and vegan.

When the supplier mails, faxes, or e-mails the information, it should be collected at a central repository in anticipation of the team placing it in the individual ingredient file. As the information is placed in the file, the supplier data sheet is filled out. Upon receipt, each document needs to be dated with a "received on" date. This assists in the future to validate that current information is always on hand because some documents expire. It is important that all of the requested information is obtained because it is used for HACCP development, program compliance, and legal or regulatory compliance. An example of a completed supplier data sheet can be found on the CD (Book 1_HACCP Program Examples\Planning:Supplier Data Sheet completed) and at the end of this chapter. Once all of the supplier ingredient information is collected and documented on the supplier data sheets, the files are filed according to the internal item number and the team proceeds to the HACCP program development phase. It is imperative that all of the information be collected for each ingredient before moving on to the next phase of development.

Although every HACCP plan is adapted to fit the individual idiosyncrasies of the plant, products, and process, they all contain the same basic components as set forth by the federal government:

Hazard analysis	21CFR123.6c(1)	9CFR417.2c(1)
Critical control point determination	21CFR123.6c(2)	9CFR417.2c(2)
Control limit establishment	21CFR123.6c(3)	9CFR417.2c(3)
Monitoring procedure development	21CFR123.6c(4)	9CFR417.2c(4)
Corrective action development	21CFR123.6c(5)	9CFR417.2c(5)
Verification procedure development	21CFR123.6c(6)	9CFR417.2c(7)
Record keeping procedures	21CFR123.6c(7)	9CFR417.2c(6)

The implementation of each of these tasks is discussed in detail in the following sections.

BOOK 1: SECTION 1

HACCP Overview

The opening page of the HACCP book is a document outlining the overall program. This should include a paragraph explaining the purpose of the program, its regulatory foundation, and each of the seven functional parts. Within this opening paragraph, a statement of management's commitment to the development and continual improvement of the program along with the specific legal citations on which the program is based should also be incorporated.

The next paragraphs of the HACCP overview outline the parts of the HACCP program, hazard analysis, process analysis, critical control point determination, control limit establishment, monitoring procedures, corrective actions, verification procedures, and record keeping. Each of these paragraphs explains what is to be done and who will do it and references any forms needed. They should be simple and need not be overly complicated because the fine details will be covered within each specific section.

A customizable HACCP overview statement is located on the CD (Book 1_ HACCP Program\Section 1_HACCP Overview:HACCP Overview) and at the end of this chapter. It should be signed by the owner, president, or senior technical person to show management support and compliance and then placed in the HACCP program book, section 1. If this form is used, it is necessary to make sure that the text is changed as needed and also that the effective date is changed in the footer (CD).

BOOK 1: SECTION 2

Process Flows

For each product—finished good or intermediate—even if there are only minor differences, the HACCP team must diagram the process used to manufacture it. This diagram should include all handling, storage, machines, form and format changes, packaging, palletizing, shipping, and any other part of the process. Initially, these diagrams will be used to begin the process of hazard analysis. Later, they will be completed by adding the critical control point references.

As the HACCP team develops the process flow sheets, they should begin by diagramming the flow that an ingredient takes from the time it is received until it leaves as part of a finished good. A couple of examples of process flows are found on the CD (Book 1_HACCP Program Examples\Section 2_Process Flows:Process Flow Example 1 and Book 1_HACCP Program Examples\Section 2_Process Flows: Process Flow Example 2) as well as at the end of this chapter. The first one is for a chopped product and the second is for a baked good. Process flows can be created by using a computer program such as Microsoft PowerPoint or Word, or they can be drawn by hand. The important point is to create a visual representation of the ingredient flow during manufacturing to use as a reference for future analysis. CD (Book 1_HACCP Program/Section 2_Process Flows:Process Flow) and at the end of the chapter can be used as a worksheet by the HACCP team.

In many food manufacturing companies the same process is used for many products. An example of more than one product with the same process flow would be the case where the same product is manufactured and packaged on the same machines, but in different pack sizes. These similarities lend themselves to the keeping of a process flow master list. The stock process flows can be named A, B, C, etc. or 1, 2, 3, etc. A stock process flow master list sheet is located for reference on the CD (Book 1_HACCP Program\Section 2_Process Flows:Process Flow Master List) and at the end of this chapter. Three columns are on this form: the product number, the product diagram name, and the appropriate process flow diagram number or letter. This master document should be placed in Book 1: HACCP at the beginning of section 2, followed by each of the completed process flows placed in order by finished product number.

BOOK 1: SECTION 3

Product Description

After the ingredient information has been collected and the finished good process flows completed, the next step is to complete a written description of each of the finished goods. A written description of the item helps the team focus on what the item is used for, how it is processed and packaged, and how each ingredient is

Book 1: HACCP Program   47

received or prepared. Examining the way the ingredients are received and stored might be an early indicator of any risks that they might contain or risks that might develop during storage, processing, or packaging. This form is in two parts; first is a written step-by-step description of the finished good. For each finished good, it is necessary to write the flow that the product goes through, including the type of product (such as refrigerated, frozen, shelf stable, etc.), how it is manufactured step by step, how the finished product is stored (such as refrigerated, frozen, shelf stable, etc.) and shipped (such as frozen, refrigerated, dry van, common carrier, etc.), and how the customer stores and uses it (such as stored frozen, refrigerated, shelf stable). Team members should ask themselves some questions when filling out the top section of the product description form:

Will it be used as an ingredient by another manufacturer?
Is it sold as a retail product?
How is it expected to be used?
How is it expected to be stored by the customer?
How is it processed?
How is it packaged?
What kind of packaging is used?
How is it stored before shipment?
How are the raw materials and packaging handled before use?

Second, the bottom part of the form is a listing that shows the storage conditions of the ingredients when they are received. The left-hand column is where the company's raw material item number should be placed; the center column is for a description of the raw material; and, in the right-hand column, a short description of how the raw material is received or prepared should be entered. In this column, generalities such as "refrigerated," "frozen," or "dry" may be used. Located on the CD (Book 1_HACCP Program\Section 3_Product Description:Product Description) and at the end of this chapter, this form should be filled out for each finished good and placed in the HACCP book, section 3, by finished product number. A few examples of completed product descriptions are found on the CD (Book 1_HACCP Program Examples\Section 3_Product Description:Product Description Complete [1] and Book 1_HACCP Program Examples\Section 3_Product Description:Product Description complete [2]) and at the end of this chapter. The first is a simple chocolate chip cookie formula that uses both room-temperature and refrigerated ingredients. It also utilizes an ingredient that comes in a bulk format. The second is a complex steak sauce formula that blends various ingredients that come in room-temperature, refrigerated, and frozen forms. Both examples contain a product number; a description of how the final product is processed, packaged, and stored; a list of ingredients; and a description of how each ingredient is received.

BOOK 1: SECTION 4

HAZARD ANALYSIS

The hazard analysis is the most difficult and yet the most important component of the overall HACCP program. Its importance is due to the fact that this is when the team identifies steps in the food production process where hazards could occur, assesses their severity and human health risk, and determines a preventative measure. Its difficulty is in being able to recognize the potential hazards during the analysis with those who have limited experience. For some processes, the hazard analysis can be difficult for even the seasoned professional; however, through careful and thoughtful application of the physical, chemical, microbiological, and allergenic hazard forms, even the novice should be able to lay the foundation for a strong HACCP risk assessment. If, during any part of these four analyses, questions arise based on something that might lead to a food safety issue, consult a local food science department, an industry trade group, or a local regulatory officer. Each of these resources brings his or her expertise to the issue without being judgmental or bringing scrutiny on the individual or the company. That being said, as the hazard analysis is performed, many of the possible hazards can be prevented from entering the process by not absorbing the problem from the supplier. Therefore, *careful examination of each ingredient and finished product is critical.*

The basic approach to hazard analysis is to analyze each finished product for risks. By using the chemical, physical, microbiological, and allergen risk analysis sheets, the finished good is analyzed based on how it is stored (in a frozen, refrigerated, or ambient state), whether there is a potential to introduce a hazard into the product during processing, and whether, if introduced, the hazard can be removed by the customer. Based on these criteria, the finished good as a whole is determined to contain or not contain the specific hazard.

The next step is to consider each of the raw materials that make up the finished good. They should be analyzed to determine whether they contain any hazards, whether the hazard can be removed during manufacturing, whether the product can be recontaminated between manufacturing and packaging, whether it is detectable by the customer, and whether the ingredient contains a hazard that can be removed by the customer. Based on these criteria, each of the raw materials is determined to contain or not contain the hazard.

CHEMICAL RISK ANALYSIS

A form on the CD (Book 1_HACCP Program\Section 4_Hazard Analysis:Chemical Risk Analysis) and at the end of this chapter can be used to analyze the chemical risks that pertain to the finished products and the ingredients of which they are composed. A chemical hazard is defined as any chemical in the ingredient or packaging that may contribute to a food safety issue or may be a health hazard. Traditionally, these are chemicals such as pesticides on vegetative ingredients and non-GRAS (generally recognized as safe) chemicals such as food additives contained within imported ingredients or as undeclared components within flavors. Although some issues surrounding allergens can be regarded as chemical in nature, the heightened scrutiny in recent years from a regulatory standpoint relegates these to a separate allergen risk analysis.

For each finished product, the team should fill out the chemical risk analysis sheet by doing the following. In the upper section of the risk assessment worksheet for chemical food hazards, the product name, product item number, and date of

Book 1: HACCP Program

assessment are filled in. In the finished product section, a "yes" or "no" answers how the finished product is stored when in the possession of the company, the distributor, the retailer, and the customer. It must be determined if it is possible for a chemical hazard to be introduced at any point during the distribution chain and the appropriate column filled in with a "yes" or "no." If a "yes" is placed in a column, then it has to be determined if the customer can detect the hazard and remove it. If a "no" is placed in the last column, then a hazard is present in the finished good.

In the ingredients section, all of the individual ingredients contained in the product before any processing must be filled in. The ingredient statement contained on each of the suppliers' specification sheets should be carefully evaluated, looking for non-GRAS list ingredients, imported ingredients, ingredients that are agricultural in nature, and flavor ingredients that might use a chemical carrier such as ethanol. Chemical hazards are usually not found within the ingredient supply of ingredients produced in the United States, but they can be found within foreign-produced and imported ingredients.

"Yes" or "no" should be filled in to indicate whether the chemical is removed during manufacturing. An example of a chemical that can be removed is ethanol contained in a flavor; pesticides sprayed on fruits and vegetables would not be removed. If the ingredient can be contaminated between manufacturing and packaging, then a "yes" is put in this column. If the ingredient contains a hazard, is this hazard detectable or removable by the consumer? If so, a "yes" is placed in the next column. In the next to last column, a "yes" or "no" indicates if the ingredient contains a hazard and the supplier can remove it. The last column denotes whether a certificate of analysis (COA) is available from the ingredient supplier.

Once the chemical risk assessment is complete, the next step is to analyze the results and determine a course of action for control of the hazards. As a basic rule, no chemicals should be allowed to be contained in any of the ingredients. If, during the analysis, a chemical hazard is found that cannot be removed during manufacturing, then the supplier should be contacted and requested to remove it. If the supplier agrees to this, then it should be required to provide a COA that expressly indicates that the ingredient has been tested for the chemical and found to be free of it. In the unlikely event that the supplier does not wish to remove the chemical or provide the COA, then an alternate supplier or ingredient should be sourced. Care should be taken to verify that all chemicals are removed from all ingredients.

Example

How to fill out the chemical risk analysis is demonstrated on the CD (Book 1_HACCP Program Examples\Section 4_Hazard Analysis:Chemical Risk Analysis complete) and at the end of this chapter. In this case, steak sauce was analyzed and each of the ingredients was examined for possible chemical risks. Based on an examination of the supplier documentation and considering the fact that tomatoes and pineapples are agricultural products known possibly to contain pesticides, a "yes" was placed in the appropriate box for the hazard. It was also determined that the supplier could provide a chemical-free COA with every lot to guarantee that the product was chemical free.

MICROBIOLOGICAL RISK ANALYSIS

The team can use the form on the CD (Book 1_HACCP Program\Section 4_Hazard Analysis:Microbiological Risk Analysis) or at the end of this chapter to identify microbiological hazards contained within the ingredients or finished product. Microbiological hazards take many forms and are the most difficult to determine, analyze, and control. Recent national recalls of meat and spinach due to *Escherichia coli* O157:h7 and peanut butter from *Salmonella* emphasize the need to be extremely careful when conducting the microbiological risk analysis.

Bacteria, yeasts, molds, and sometimes their end products, such as toxins, are found in many forms throughout the growing, harvesting, manufacturing, and distribution environments. Detailed treatises on the morphology and prevalence of each possible microbiological hazard are available from local bookstores, colleges, or trade associations and should be consulted. When the microbiological hazard review is performed, three categories of organisms should be investigated. These are total bacterial load, pathogenic organisms, and yeasts and molds.

Total bacterial load is exemplified by measuring the amount of heterotrophic organisms. These organisms can be tested for by an aerobic plate count (APC), total plate count (TPC), or standard plate count (SPC). APC, TPC, and SPC are acronyms used fairly interchangeably by industry and testing laboratories alike, although TPC is the most common. The food industry uses this test as an indicator to determine the overall level of sanitation within the manufacturing and distribution processes and to determine whether the processing kill step was sufficient. The higher the microbial load found in the TPC is, the greater is the possibility that the processing environment is not clean or that the process was not sufficient enough to kill an adequate number of the organisms present.

Pathogenic organisms (pathogens) are defined as organisms that, when ingested, can cause sickness up to and including death. The FDA defines five classes of pathogenic organisms or toxins. These are pathogenic bacteria, enterovirulent *E. coli* group, parasitic protozoa and worms, natural toxins, and other pathogenic agents. Some examples of these organisms and the types of foods in which they are found are:

Salmonella: found in raw meats, poultry, eggs, milk and dairy products, fish, shrimp, frog legs, yeast, coconut, sauces and salad dressings, cake mixes, cream-filled desserts and toppings, dried gelatin, peanut butter, cocoa, and chocolate;
Clostridium botulinum: found in honey and improperly thermally processed foods;
Staphylococcus aureus: found in meat and meat products; poultry and egg products; salads such as egg, tuna, chicken, potato, and macaroni; bakery products such as cream-filled pastries, cream pies, and chocolate éclairs; sandwich fillings; and milk and dairy products;
Campylobacter jejuni: found in raw chicken, raw milk, and nonchlorinated water;
Yersinia enterocolitica: found in meats (pork, beef, lamb, etc.), oysters, fish, and raw milk;

Listeria monocytogenes: found in raw milk, supposedly pasteurized fluid milk, cheeses (particularly soft-ripened varieties), ice cream, raw vegetables, fermented raw-meat sausages, raw and cooked poultry, raw meats (all types), and raw and smoked fish;
Escherichia coli O157:H7: found in undercooked or raw hamburger (ground beef), alfalfa sprouts, unpasteurized fruit juices, dry-cured salami, lettuce, game meat, cheese curds, and raw milk; and
Giardia lamblia: found in contaminated water used to wash vegetables.

Many pathogens are found naturally in the environment and get introduced into the food supply by poor manufacturer sanitation and improper process controls. Pathogens are all extremely dangerous and harmful to humans and serious care must be taken by the HACCP team to ensure that they are all excluded or killed during processing. None of these organisms should typically be found in any processed ingredient purchased for use in the manufacturing process. The onus is on the supplier during growth, manufacture, processing, and distribution of the ingredient to produce it in such a manner to preclude the presence of pathogens in the ingredient and to be able to verify this. In the case of unprocessed ingredients such as raw meat, milk, or field crops, these might contain pathogens that are naturally present and their presence should be noted in the microbiological hazard analysis.

Yeasts and molds are a class of organisms that are found throughout the environment. They come in various shapes and sizes, utilize different food sources, and may or may not like oxygen. These are easily differentiated using standard microbiological methodology. Generally, these organisms are undesirable within any raw material or finished good, although there are a few notable exceptions, including probiotics added to products like kefir and yogurt, mold in cheese, yeast added as a leavening agent in baking, and yeast used in the fermentation industry.

As a basic rule, care should be taken to investigate each supplier thoroughly to understand how the ingredient is manufactured and packaged. The core supplier investigation tool for HACCP purposes is the initial document collection as outlined in the HACCP preparation section. This includes the receipt of a specification sheet that lists ingredients, storage requirements, and packaging. Each of the ingredients and subingredients within each ingredient should be considered as the microbiological risk analysis is conducted.

The CD (Book 1_HACCP Program\Section 4_Hazard Analysis:Microbiological Risk Analysis) or the form at the end of this chapter can be used to fill out the product name, product item number, and date. In the finished product section, how it is stored at each of the manufacturing and distribution steps should be indicated. Whether the product can be contaminated at any point in the process should be considered. If it can, the appropriate box should be filled in and then whether the customer applies a kill step such as heating during preparation before eating.

The next section, ingredients, is the most important risk analysis conducted as part of the HACCP program. Careful analysis can prevent the company from adopting the problems of its suppliers and allow it to gain a better understanding of the types of processing that need to be applied. As each ingredient is evaluated to determine if it contains a microbiological hazard, the following issues need to be considered:

- Ingredient complexity: Is it composed of a single ingredient or a matrix of blended ingredients?
- Ingredient source: Where did the ingredient come from? Was it grown from the earth? Was it grown in an organic or nonorganic manner? Was it grown, processed, or manufactured in or outside the United States?
- Ingredient type: Is the ingredient a commonly used ingredient with clearly recognized microbiological specifications and testing methods?
- Ingredient processing: Does it go through a manufacturing or processing step that effectively kills or significantly reduces the ingredients' microbial load?
- Ingredient packaging: Does its package lend itself to keeping the product from being contaminated during distribution?

If the ingredient is complex, then each of the subingredients, including processing aids such as carriers in flavors, desiccants in spice blends and free-flowing agents in dry vegetables, cheese, and frozen inclusions, should be evaluated from the standpoint of whether they have gone through a kill step. When the subingredients naturally contain a high microbial load, they may contaminate the end ingredient unless the final kill step is adequate to achieve a thorough kill.

Examining the ingredient's source, physically and geographically, assists the HACCP team in roughly determining a starting microbial load. The physical origin of the ingredient, whether it is an agricultural product, is important because agricultural products that have not undergone further processing inherently contain higher microbial loads. Some notable agricultural products, such as purees, variegates, whole and/or chopped, sliced, diced, or pureed fruits and vegetables, are microbial suspects. Agricultural products also have the potential to be contaminated with naturally occurring pathogenic organisms such as *Listeria, Staphylococcus,* and *E. coli.* It is not uncommon to find high total plate counts, high yeast and mold counts, and high coliform counts on plant-based ingredients. Another source-related question to be asked is what country the ingredient is from. Different countries have different regulatory requirements—some more microbiologically strict and some less. For example, countries such as Turkey, Iran, and India are the source for many spices, but the microbiological requirements for the final exported product are not as stringent as those required for American producers. Alternatively, products produced in Japan traditionally must conform to stricter microbiological requirements than those for U.S. producers.

The ingredient type is also important in evaluating whether it may or does contain a microbiological hazard. If the ingredient is highly processed or publicized, and generally recognized microbiological standards exist, it is less likely to contain a microbiological hazard. Ingredients that fall into this class include sugar, corn syrup, dried powders, vinegar, flavors, and salt.

If and how the ingredient is processed lends an insight to whether the ingredient will potentially have a microbial hazard. Those that undergo a significant kill step such as heating, irradiation, thermal processing, or pasteurization will have a lower potential than those that do not. Most processed ingredients undergo some type of kill step during their processing, but it should not be categorically assumed that, if they did, they are clean.

The last major question to consider regarding whether the ingredient contains a microbial hazard is how it is packaged. Ingredients whose containers have a controlled atmosphere headspace through the use of inert gases like nitrogen or carbon dioxide, those that are vacuum packed, and those that are heated before packaging into sealed containers tend to have fewer microbial hazards. Packaging such as corrugated cases with poly liners may potentially be contaminated during distribution because the product may shift inside the case and be exposed; such packaging also has a propensity to get damaged by handling.

The name of each of the primary ingredients in the finished product should be written in the left-hand column of the risk analysis form. Next, for each ingredient, the preceding questions should be considered and whether the ingredient contains a microbial hazard should be indicated. If there is any question whether it does or does not, it should be considered to contain one until it can be verified that it does not. It is necessary always to be safe and err on the side of caution. For those ingredients that do contain a hazard, it should be indicated in the next column whether the hazard is removed during processing. Then, whether the ingredient can be recontaminated between manufacturing and packaging should be indicated. In the next column, whether the ingredient can be recontaminated during distribution or storage should be filled in.

The final two columns on the microbiological risk analysis sheet involve controlling the defined hazards. In the second column from the right, it is necessary to fill in whether the supplier can remove the microbiological load or reduce it to a manageable level. If the hazard can be removed or controlled, this is the preferred method of control and it should be required of the supplier. The supplier should be required to provide two types of documents that address the microbiological issue. First, a new specification sheet that documents the new, lower microbiological specification should be provided. Second, the supplier should be required to provide a certificate of analysis for each lot of material sent to the company. Typical limits for each bacterium should be set in accordance with normal ranges for the ingredient. If the supplier refuses to provide the needed or desired level of testing and/or documents, the first course of action is to attempt to negotiate compliance. If or when this fails, consideration should be made to source the ingredient from another supplier. It is innately better to upset even the oldest supplier than to allow unwanted microbiological hazards into the plant. If the supplier agrees to provide a certificate of analysis, a check mark is placed in the box in the far right column of the sheet.

The effectiveness of the microbiological hazard analysis relies on the care that is taken during the evaluation of the finished good and the individual ingredients contained therein. It is essential that all possible hazards are found, documented, and systematically removed from the incoming ingredients. To this end, suppliers should be put on notice and held accountable to provide ingredients that have no or very low microbial loads and definitely no pathogens.

Example

The form found on the CD (Book 1_HACCP Program Examples\Section 4:Microbiological Risk Analysis Complete) and at the end of this chapter demonstrates how to fill out the microbial risk analysis. In this case, steak sauce was analyzed and each of the ingredients was examined for possible chemical risks. Based on an examination of the supplier documentation and considering the fact that tomatoes are agricultural products known to contain possibly high levels of mold or salmonella, a "yes" was placed in the appropriate box for the hazard. It was also determined that the supplier could provide a chemical-free certificate of analysis with every lot to guarantee that the product carried a microbial load that was within the agreed upon specification. Furthermore, the raisin paste, spice blend, and orange peel also may have higher levels of microbes, so a COA would be required from the supplier. Even though they also are agriculturally based, the supplier puts the garlic and onion powder through a kill step and provides a COA.

PHYSICAL RISK ANALYSIS

The physical risk analysis form on the CD (Book 1_HACCP Program\Section 4_ Hazard Analysis:Physical Risk Analysis) and at the end of this chapter is used to identify physical hazards contained within the ingredients or finished product. These risks are usually the easiest to remove because they are the easiest to see with the naked eye. As with the chemical and microbiological risk analyses, the physical risk analysis examines whether the finished product and the individual ingredients contain any risks. On this form, the name of the product, date of analysis, and product item number are filled in at the top.

In the finished product section, how the finished product is stored in the company warehouse, at the distributor, at the retailer, and by the customer is indicated. Next, it is necessary to determine if it is possible, based on the packaging configuration, for the finished product to become contaminated with a physical object. With modern packaging and regulations, retail packaging usually prevents any type of postmanufacturing physical contamination; however, industrial packaging does lend itself to possible physical contaminants.

On the bottom half of the form, all of the ingredients that make up the finished product should be listed. It is necessary to evaluate whether the ingredient might or does contain a physical hazard. Physical hazards usually come in three categories: growing or harvesting hazards, manufacturing or processing hazards, and packaging hazards. Growing or harvesting hazards are those associated with growing the ingredient or harvesting it from an agricultural setting. These hazards are things such as rocks, wood, stems, rodent and pest pieces, and other foreign material inherently found on the ground and picked up with the base ingredient during harvesting. When evaluating whether the ingredient may contain these types of physical hazards, it is necessary to consider how and where the ingredient was grown, if the physical hazard can be removed in its entirety by the supplier, and if the ingredient contains the hazard naturally. For example, when ground-based ingredients such as corn, peas, and soybeans are evaluated, stones, stems, leaves, and insects are included as part of the harvesting process. Generally, they are removed during processing, but in some cases the law makes allowances for low levels of hazards to stay within the food.

Book 1: HACCP Program   55

The next column is where it is noted if the hazard is removed during manufacturing or processing. This can be as easy as using screens to remove rocks, leaves, and stems, or metal detectors, magnets, or screens, to remove large and small pieces of field metal up to a combination of several physical and mechanical steps to remove all noningredient matter. Next, it should be considered whether, during the manufacture of the ingredient, it can be recontaminated between the manufacturing and packaging steps. Industrial packages such as corrugated poly-lined cases or kraft multiwalled bags usually lend themselves to possible recontamination at this point because the opening of the container is usually exposed.

In the fourth column, whether it is possible and probable for the customer to detect or remove the determined physical hazard prior to using the product should be indicated. Fill in the next column as to whether the supplier can remove the identified hazard during processing. If the supplier can, determine if it has the capability of testing for the hazard prior to shipment and, if so, whether it can provide a certificate of analysis that identifies this. Place this answer in the final column.

Examples

A completed form for a retail steak sauce product is exhibited on the CD (Book 1_HACCP Program Examples\Section 4:Physical Risk Analysis Complete) and at the end of this chapter. The top section shows that the product is stored at room temperature during all four phases of distribution: in the company warehouse, at the distributor, at the retailer, and with the customer. It also reflects that the customer may store the product in the refrigerator, probably after opening. This section also shows that a physical hazard, such as a hair, might get introduced at the customer level because the customer opens the container and would be able to detect any possible hazards upon use.

In the lower section, the analysis indicates that the tomato puree and the raisin paste may contain hazards such as stems and rocks but that these are removed during manufacture and that they are not recontaminated between manufacture and packaging. The spice blend, on the other hand, may contain hazards such as rocks and stems but also might contain metal from the milling and screening process. Although the rocks, stems, and metal are removed with screens and magnets during manufacture, it should be noted the manufacturing has the capability to remove them, so a certificate of analysis should be required on a lot-by-lot basis.

ALLERGENIC RISK ANALYSIS

The allergen risk analysis form found on the CD (Book 1_HACCP Program\Section 4_Hazard Analysis:Allergen Risk Analysis) and at the end of this chapter is used to identify allergen hazards contained within the ingredients or finished product. In the past, allergens have not been considered as hazards or were considered during chemical risk analysis. However, in light of the current emphasis by regulatory bodies for all foodstuffs to contain the proper allergen warnings, a separate risk analysis for allergens is prudent and should be conducted. This follows the same procedure as the chemical, microbiological, and physical hazard analyses but relies very heavily on the documents collected during the HACCP preparation step. Thus, the integrity of each supplier is important.

The allergen hazard analysis evaluates the "big eight" allergens of milk, wheat, soybeans, peanuts, tree nuts, eggs, fish, and shellfish as well as other allergenic components such as seeds. These all cause allergic reactions in some fraction of the population. The analysis also evaluates for the presence of other reaction-causing agents, including monosodium glutamate, sulfites, and colors. Although this exercise may seem like overkill, it provides a method for gathering the information needed to prevent unwanted allergens from entering the process, controlling the ones that are present, and labeling them correctly, thus mitigating future legal repercussions. A complete explanation of what allergens are, where they are found, and the food safety allergen program, is found later in this book.

On the top of the allergen risk analysis form, in the product name, product number, and date of analysis should be filled in. In the next section, how the product is held during the various stages of postmanufacturing distribution—frozen, refrigerated, or at room temperature—is indicated. Then, on the left side of the form, each of the ingredients used in the product is listed. The ingredients should be filled in on both the top half of the form and the bottom. If more room for ingredients is needed, a second chart can be used. Then, for each ingredient, a "yes" or "no" indicates whether it contains the allergen. This information is available by evaluating the allergen statements collected during the HACCP preparation. Care should be taken to verify that the supplier has covered all of the allergens in question on the allergen sheet because many times suppliers do not address them or just lump all of them together. Either of these supplier ruses will cause a misinterpretation of the facts and allow allergens to slip into the system.

Some of the more common misinterpretations are as follows. First, it is easy to assume that just because the supplier listed some of the allergens, those that were not listed as being in the ingredient were not listed due to an oversight. This is a false assumption because many allergens can be hidden within the ingredient's subingredients, or the supplier may know they are present and not want the allergen statement to look bad. Second, in the case of wheat and gluten allergens, some suppliers think that just because the ingredient contains no wheat it also does not contain gluten. This is a false assumption because gluten is contained in some, but not all, cereal seeds. Third, just because a supplier states that the ingredient does not contain any of the allergens does not make it true. The ingredient statement provided on the specification sheet should be examined in all cases to verify that there are no stray or undeclared allergens.

In the course of examining supplier allergen statements, two possible miscellaneous declarations might be found: (1) "processed on the same line as…" and (2) "processed in the same plant as…." The first indicates the potential direct contamination of the ingredient via contact with the same processing surface. In this case a "yes" should be placed in the respective allergen boxes denoting that the ingredient may or may not contain the allergen. The second case indicates that the ingredient might or might not have been exposed to contamination because it was stored in the same warehouse. This may be a potential problem because some allergenic substances may be airborne, although normally they are not. In this case the word "plant" should be placed in the respective allergen boxes. This hazard will be dealt with during the section on setting up the allergen control program.

After the allergen review is completed, each of the ingredients should be evaluated for the presence or lack of seeds, monosodium glutamate (MSG), sulfites, and colors. "Yes" or "no" is written in the appropriate box. Again, it is very important to review the ingredient statement to determine that there are no undeclared substances. This is especially important with the colors because many suppliers forget or fail to declare them adequately.

The four risk analyses should be filed and the hazard analysis continued until all of the finished products are have been analyzed. This can be a time-consuming process for all of the team members, but it is a critical component of building a successful quality program because it provides an in-depth look at each of the ingredients and finished products. It is important not to move ahead to the next section until all of these are complete because it is difficult to establish critical control points without knowing all of the hazards.

Example

A typical allergen evaluation for a baking mix, item number BM42170.40, is outlined on the CD (Book 1_HACCP Program Examples\Section 4:Allergen Risk Analysis Complete) and at the end of this chapter. It contains the normal baking ingredients, such as flour, oil, salt, sugar, and leavening. Each ingredient is examined for the presence of each of the allergens and other components. The supplier allergen statements were examined for whether the ingredient contained the allergen or whether the ingredient may have been contaminated during processing by being run on the same line as another allergen or being stored in the same plant as other allergens. The enriched flour was found to contain wheat allergens and to have been processed in a plant that also contained eggs. Therefore, "yes" is in the "wheat" box and "plant" placed in the "eggs" box. None of the other ingredients contained, were processed on, or were stored with other allergens.

Administrative note: Once all four of the risk analyses are complete for any given product, they can be placed behind tabs in the HACCP book if there are not many or a separate manila file can be created for each. The files can be labeled:

HACCP   Product number: _____   Product Name: _____

BOOK 1: SECTION 5

CRITICAL CONTROL POINT DETERMINATION

After the process flow diagramming and the hazard analysis are complete, the next step toward the development of the HACCP plan is to determine the critical control points. These are defined as points within the process at which a physical, chemical, microbiological, or allergen hazard can be controlled, removed, or prevented from entering the process as a means of ensuring the safety of the food. Typical critical control points would be packaging review (allergen risk), metal detection (physical risk), pasteurization (microbiological risk), baking (microbiological risk), or Elisa testing (allergen risk). Each of these involves a kill step, foreign material removal, or hazard prevention. This is contrary to control points, which are points in the process at which a quality parameter is monitored so that the overall product conforms to a set consistency limit. Examples of control points are weight control, color verification, size determination, packaging checks, and particle size. Although each of these is important, none of them rises to the level of having a bearing on the overall safety of the food; thus, they are not "critical."

A form on the CD (Book 1_HACCP Program\Section 5_Critical Control Point Determination:Critical Control Point Worksheet) or at the end of this chapter is used to evaluate the connection of the processing steps, product components, control points, and critical control points. To complete this evaluation, one can start with one of the process flows created during the HACCP planning exercise. The process flow name or number, along with the evaluation date, is written on the top of the form. The process flow name or number should match one of the process flows diagramed in the "Process Flow" section of the HACCP program development. Next, each of the steps detailed on the flow is written in the left-hand processing step column. The receipt of ingredients, packaging, and any rework streams that occur must be included. The names and/or product item numbers of all the products that use this processing line configuration are written at the bottom. All tests that are or should be completed during the processing step need to be written in the left column. These generally are line and process specific. Although it would be impossible to list all of the processing possibilities, the following are some typical processing steps from the manufacture of various food products with their correlated parameters to evaluate:

Processing Step	Parameter to Evaluate	Hazard
Ingredient receipt	Container cleanliness	None
Ingredient receipt	COA verification	Physical, chemical, microbiological, allergen
Packaging receipt	Ingredient verification	Allergen
Ingredient mixing	Consistency	None
Pasteurization	Bacterial load	Microbiological
Baking	Temperature	Microbiological
Drying	Available water	Microbiological
Packaging	Seals	Microbiological
Packaging	Weight	None
Lot coding	Accuracy	None
Grinding/chopping	Particle size	None
Fermentation	pH	Microbiological
Final product	Microbiological testing	Microbiological
Filling	Volume	None
Postpackaging	Metal detection	Physical
Milling	Magnets	Physical
Irradiation	Bacterial load	Microbiological
Warehousing	Temperature	None
Thermal processing	Time and temperature	Microbiological

Book 1: HACCP Program

As a guide to decide what parameters to evaluate for the incoming raw materials, the team can review the hazards that were determined during the hazard analysis portion of the plan development by looking at each hazard and determining how it will be controlled. For instance, if an ingredient were found to have a possibility of a high micro level, it would be assumed that somewhere in the process a kill step would need to be completed. Furthermore, if an ingredient spec sheet showed the possibility of a toxin and the supplier agreed to send a "toxin-free" certificate of analysis along with every lot shipment, then it would be assumed that a parameter to evaluate would be a review of the COA to verify its cleanliness. The team then looks at each of the processing steps and determines what parameter will be analyzed. During this process, it will be determined that some of the processing steps will not have parameters to evaluate and some will. The ones that do not have parameters that contain milk should be left empty.

Upon completion of the first two columns, it must be determined if the parameter that is being evaluated is to control a chemical, physical, microbiological, or allergen risk—or none of them—and the correct box filled in. In the last column, whether it is a critical control point or just a control point should be indicated. To determine whether it is or not, the following decision tree is helpful:

Q_1 Could the raw material contain the hazard at dangerous levels?
   No—not a critical control point, possibly a control point
   Yes—proceed to Q_2
Q_2 Will further processing or handling, including correct customer use, remove it or reduce it to an acceptable level?
   Yes—not a critical control point, possibly a control point.
   No—a critical control point; a program needs to be in place to eliminate and monitor the elimination of the hazard
Q_3 Is the formulation or composition of the intermediate or final product essential to prevent an unacceptable increase of the hazard?
   No—not a critical control point, possibly a control point
   Yes—formula or composition is a critical control point
Q_4 Is contamination or recontamination possible or is the increase of the hazard possible?
   Yes—proceed to Q_5
   No—not a critical control point
Q_5 Will further processing or handling, including correct consumer handling, remove the hazard?
   Yes—not a critical control point
   No—a critical control point; a program needs to be in place to eliminate and monitor the elimination of the hazard
Q_6 Is the process stage intended to reduce the hazard to safe levels?
   No—not a critical control point
   Yes—a critical control point; a program needs to be in place to eliminate and monitor the elimination of the hazard

The natural tendency for many HACCP plan developers is to overclassify control or eliminates points as critical. This is not necessary because critical control points are only those that involve the measurement of a processing step that controls a hazard. Most HACCP plans may have as few as one CCP but may have several. The key is to make sure that when a CCP is chosen, the choice is based on whether the potential hazard can be controlled or eliminated.

All completed product/process control point evaluation worksheets are to be filed in section 5, "Critical Control Worksheets," of the HACCP Book 1. A final administrative step to be completed after determination of the critical control points is to return to the process flows and place the designated CCP numbers in the left-hand column.

Example 1

A typical critical control point evaluation worksheet for a bakery is shown on the CD (Book 1_HACCP Program Examples\Section 5:Critical Control Point Worksheet Complete 1) and at the end of this chapter. It shows the processing steps that occur and each of the parameters that are evaluated along the way. For each step, it is determined whether there is a hazard present or not. First, the ingredients are received, the truck is inspected, and the certificates of analysis for the ingredients that require them are checked against their respective specifications. When the packaging and ingredients are received, their ingredient statements are compared against approved statements to verify that they have not changed. Next, the ingredients are mixed and extruded before going into the oven. The baking process is monitored for its time and temperature to ensure that a complete microbiological kill occurs. After the cookies are cooled, they are packed and run through a metal detector. This ensures that all of the extraneous metal that may have gotten into the cookies is removed. Finally, all of the cookies are coded, cased, and palletized before freezing. Although there are many control points within this process, there are only three critical control points: one to prevent an allergen hazard, one to eliminate a microbiological hazard, and one to monitor a potential physical hazard.

Example 2

The control point evaluation for a jelly-making application is shown on the CD (Book 1_HACCP Program Examples\Section 5:Critical Control Point Worksheet Complete 2) and at the end of the chapter. First, ingredients are received and the needed COAs are verified. Next, the ingredient statements on the ingredients and packaging are verified to ensure that no new ingredients have been added or changed. Then, the ingredients are mixed, heated, and passed through an in-line metal detector before packaging. The containers are labeled, coded, cased, palletized, and warehoused. In this process lots of points are measured to maintain a consistent quality product; however, there are only three critical control points: one to prevent an allergen hazard, one to prevent a microbiological hazard, and one to prevent a potential physical hazard.

Example 3

The control point evaluation for a chopped-candy factory is shown on the CD (Book 1_HACCP Program Examples\Section 5:Critical Control Point Worksheet Complete 3) and at the end of this chapter. Ingredients and packaging are received and the ingredients are checked to make sure that no extraneous allergens are added. Next, the ingredients are stored in a dry warehouse or freezer, and then staged for production as needed. After the product is put on a conveyor, a free-flowing ingredient is added. The candy is then chopped and run into an automatic scale system in preparation for bagging. After the chopped candy is weighed, it passes through a metal detector and is bagged, cased, taped, coded, palletized, and put into the freezer for sale. In this process, there are many control points, but only two critical control points. The first one, the ingredient statement inspection of the raw materials and packaging, is an allergen hazard control. The second critical control point is the metal detector, which acts as a physical hazard control.

BOOK 1: SECTION 5a

CONTROL LIMIT ESTABLISHMENT

Once the critical control points are determined, the next step for the HACCP team is to institute control measures and establish criteria to measure control at them. Establishing the measures and criteria usually takes the form of setting control limits. This is the range of acceptability of the hazard that is allowed in the process. Typically, these are specified with a target and/or upper and lower limits, outside of which range the operator would deem that the product is out of specification. Sometimes there is no range. The critical control limit is set by being present or not—in other words, positive or negative. Because of the nature of critical control points, the point at which the limits should be set can be found by a search of the literature, common sense, or discussing it with other industry professionals. As a general rule, the control point should effectively eliminate the hazard altogether, so the limit should be zero, negative, or set at the effective limit of the test. Some examples of typical critical control point control limits include:

Control Point	Typical Limit
Incoming ingredient verification	No changes
Incoming packaging verification	No changes
Pasteurization	Time and temperature
Metal detection	Reject at tested sensitivity
Baking	No micro growth
Positive release program	No micro growth
Allergen test	Negative or limit of test
Magnets	Tested and operating

Although it may seem that setting the control limits is difficult, remember that each of these control points is designed to detect whether a physical, microbiological, chemical, or allergenic hazard remains in the finished product. Always err on the side of caution and make the limits tight but manageable within the scope of the capabilities within the plant.

The second part of establishing the control limits is developing a method of documenting that the control limit is within the defined limits. Known as monitoring forms, they can be made using any software program, such as Microsoft Word or Excel. The forms will be plant, product, and line specific. In whatever form they are made, they should always contain the following information: date, line, product, parameter measured, control limits of the parameter measured, person checking the control point, time the control point was checked, a space for the measurement, a space for the action taken if the control point is outside the limits, and a space for a second measurement of the control point after any action. Like all other forms, they should have an effective date and a replaces date for document tracking. When these are in development, they may serve a dual purpose of also documenting the observation or measurement of quality control parameters, GMPs, or production data. After the monitoring forms are completed, they are to be reviewed for completeness and comparison against the critical control limits, initialed, and filed. Copies of the monitoring forms for each critical control point are to be put in tab section 7, "Record Keeping."

CORRECTIVE ACTIONS

After the hazards have been determined, critical control points identified, and control limits established, the next step in the development of the HACCP plan is to develop methods to monitor the critical control points. These are programs or systems developed to determine and implement appropriate corrective action when the critical control parameter is outside the critical control limit; they usually involve rejecting the out-of-specification product in some manner, isolating the rejected product, destroying or reworking the rejected product to bring it into compliance, documenting the problem, and communicating the issue to others within the company. All of these specific programs need to have separate designated programs for the management of the measurement and validation of each of these critical control points. These program documents are tabbed sections within the quality control manual and will be dealt with separately later in this book.

BOOK 1: SECTION 6

HAACP PLAN SUMMARY SHEET

As a means of documenting these required actions, a HACCP plan summary chart should be filled out (found on the CD under Book 1_HACCP Program\Section 6_ Plan Documentation:HACCP Plan Summary Chart and at the end of this chapter). In the "CCP Description" column, the type of hazard being controlled (i.e., physical, microbiological, chemical, or allergen) should be filled in. Next, the critical limits that were identified in the preceding section are filled in.

In the "Monitoring Program" column the name of the program that covers where the critical control point monitoring details and documentation are referred is filled in. Below this space, the title of the tab section from the quality control manual is indicated. For example, the critical control point of packaging inspection for ingredient

Book 1: HACCP Program

changes is traditionally completed at the receiving dock and the program and documents are stored in the quality manual tab section titled "Receiving Program."

The next column, "Frequency," lists how often the critical limit is checked. For example, is every product checked, or the sleeve, or the case, or the pound, or the package, or some other unit of measure? The second to last column is where the corrective action or "what you do if the product is out of specification for the hazard" is listed. Here, what happens if the critical limit is exceeded is listed step by step. When this column is being filled in, it is necessary to be specific, reference any other programs that pertain to the monitoring of the hazard, and outline exactly what is done with the out-of-spec product. To complete the chart, a CCP number is assigned to each critical control point identified in the previous section. A convenient convention is to use "P" for physical hazards, "A" for allergen hazards, "C" for chemical hazards, and "M" for microbiological hazards, along with a numerical qualifier. This identifier is indicated in the left column of the chart. The final column is where the responsible party is listed. The tendency when filling in this column is to specify the line operator. This is only a partial solution; in actuality, dual responsibility exists between a quality control employee and a production employee, a hierarchal responsibility involving a technician and manager or an operator and manager, or some combination of the two. It is important to make sure that when responsibility is considered, someone from the technical and operational sides of the company's upper management is held accountable for the enforcement of critical control points.

After the HACCP plan summary chart is completed, it is placed in the HACCP book in tab section 6, "HACCP Plan Summary Sheet."

Example

An illustration of how an HACCP plan summary chart should be completed is shown on the CD (Book 1_HACCP Program Examples\Section 6_Control Limits and Corrective Actions:HACCP Plan Summary Chart) and at the end of this chapter. In the left-hand column, the CCP numbers are listed: P-1 for the first physical hazard critical control point, A-1 for the first allergen critical control point, and M-1 for the first microbiological critical control point. The next column describes the critical control point by the type of hazard controlled and the point or procedure at which the critical control point monitoring occurs. In the case of this example, these points are an allergen check at the receiving inspection, metal check at the metal detector, and bacteria control at the pasteurizer. Next, the critical limits are filled in for each critical control point based on the critical control point determination section and the section of the quality manual that details the program. The frequency at which the critical control point is monitored, what is to be done when it is or becomes out of specification, and, finally, who is responsible for monitoring the critical control point are indicated in the final three columns.

BOOK 1: SECTION 7

VERIFICATION

The verification step of the HACCP plan development requires a periodic review of the program to ensure that the plan is functioning effectively, the Critical Control

Points (CCP's) that were established are satisfactory, and that the plan is being audited on a regular basis. To verify these aspects of the HACCP program, the HACCP team meets on a regularly scheduled basis, when new information comes to light regarding an ingredient or process that may introduce risk, and when a critical control point is found to be out of conformance or unannounced to review critical control point monitoring. This meeting should be documented.

First, the HACCP team evaluates the number and type of non-conformances that have occurred during the previous period. The team should classify them into two groups, control point deviations and critical control point deviations. Control point deviations are the responsibility of the quality department while critical control point deviations are to be reviewed under the second section of the verification, determining if the CCP's are satisfactory.

Second, the committee evaluates each of the critical control point deviations and verifies that the control point monitoring plan and limits are effective and appropriate. The deviations should be evaluated against the hazard analysis for the particular product or process flow with a view toward implementing new or improved control measures. Some things the committee may consider during this evaluation are:

1. Were critical control points monitored on the designated schedule?
2. Were critical control points monitored at their respective control limits?
3. Did all records indicate correct monitoring procedures?
4. Are control records reviewed on a timely basis?
5. Are control records maintained adequately?

Any changes made to a control point should be reflected on all the hazard worksheets including the process flows, risk analysis sheets, critical control worksheets, and HACCP plan summary sheet. When this occurs, make sure that the effective date and replaces date in the footer of the document are updated.

Third, the committee verifies that the entire HACCP plan has been audited on a regular basis and the audit is documented. Normally, this is not a problem since the quality control manager and plant manager conduct the audits. The committee should expect a HACCP audit to occur on a regular basis and be documented.

A template to record who attended the meeting and the topics reviewed can be found on the CD (Book 1_HACCP Program\Section 8_Verification Procedures:HACCP Team Yearly Auditing Meeting Attendance) and at the end of this chapter. At the beginning of the team meeting, each member should sign the attendance sheet and then, at the end of the meeting, the areas reviewed should be checked. This should be filed in the HACCP book under Section 7.

BOOK 1: SECTION 8

AUDITING

A daily function of the plant manager and quality control manager is that of auditing the HACCP program. This is done by reviewing the daily critical control point monitoring documentation. The measurements that were taken should be carefully reviewed and compared against the specifications as set forth during the control point measurement determination process of HACCP program development. If the measurements are out of spec, then appropriate action must be taken according to the individual measurement program. These actions might include reworking the product or destroying the violating product.

When the control point data has been checked, these should be initialed by either the plant manager or the quality control manager, and then filed according to the designated record keeping policy.

Auditing the records is a critical yet often overlooked component of the HACCP program. Too often, especially in smaller companies with limited labor, the daily audit gets assigned to non-trained personnel, pushed to the back burner, or just not completed. Care must be taken to ensure that this does not happen as the audit is an important component of a positive release program and is the last line of defense in eliminating risk prior to the product getting to the customer.

BOOK 1: SECTION 9

Training

A primary responsibility of the HACCP committee is to share the knowledge of HACCP to all those within the company. The purpose of this is to make everyone aware of the types of risks associated with the products being manufactured and the control steps that are in place to control them. Training is an important component in making HACCP successful in any food company. Like all programs, HACCP and what it represents work best when it is integrated into the normal daily activities and duties of each employee rather than added as an extra responsibility.

How much and the type of training given to each employee depends on the particular employee's responsibility within the company. Management employees will need a greater understanding of the principles of HACCP more so than sales or accounting. Likewise, production and warehouse employees need more information than marketing or administrative personnel.

The training regimen should provide an overview of the HACCP principles and how they are applied within the scope of the manufacturing environment. It should be routine and include details regarding the specific control points that are involved within the scope of their responsibility and/or the line they work on. For all employees, the goal is to make them proficient in completing the tasks that the HACCP plan requires them to perform and make educated decisions in a timely manner when situations arise. As a means of documenting this training, a HACCP training verification sheet may be found on the CD (Book 1_HACCP Program\Section 9_Training:HACCP Training Verification) and at the end of this chapter. This should be completed and filed in the HACCP manual under Section 9.

SUPPLEMENTAL MATERIALS

Book 1: HACCP Program　　　　　　　　　　　　　　　　　　　　　　　　　　67

HACCP PROGRAM

General Overview

Section A

_____ hazard analysis critical control point (HACCP) program is a systematic approach to ensuring that all products produced are free from microbiological, chemical, physical, and allergenic contaminants. It is based on the requirements promulgated by the federal government as a means to develop and ensure a safe food supply. The program is divided into several parts:

1. Hazard analysis and risk assessment: raw materials and finished goods
   21CFR123.6c(1)　　9CFR417.2c(1)
2. Process analysis
3. Critical control point identification
   21CFR123.6c(2)　　9CFR417.2c(2)
4. Establishing critical limits
   21CFR123.6c(3)　　9CFR417.2c(3)
5. Monitoring procedure development
   21CFR123.6c(4)　　9CFR417.2c(4)
6. Corrective action development
   21CFR123.6c(5)　　9CFR417.2c(5)
7. Verification procedure development
   21CFR123.6c(6)　　9CFR417.2c(7)
8. Record keeping procedures
   21CFR123.6c(7)　　9CFR417.2c(6)
9. Program auditing

The management of _____ is committed to providing the resources necessary to complete a thorough hazard analysis and supports the monitoring and enforcement of all critical control points.

Section B

1. Hazard analysis and risk assessment.
   a. For each finished good, the product description form is filled out. This gives a written description of the product, where it is used, the conditions under which it is packaged, and how it is packaged and sold. It also gives a concise summary of how the raw materials contained within the product are received.
   b. All raw materials and finished goods are subjected to a hazard analysis and risk assessment using the microbiological risk analysis, the chemical risk analysis, the physical risk analysis, and the allergen risk analysis forms.

Section C

2. Process analysis
   a. For each process used, a processing chart will be drawn that identifies the flow of the process and codes each potential area for physical, chemical, microbiological, and allergenic hazards.

Section D

3. Critical control point identification
   a. From the processing chart, the critical control points are identified. These are the points in the process where a loss of control may result in an unacceptable health risk. These may be within the ingredient, growing or procurement of raw materials, ingredient receiving and handling, processing, packaging, distribution, or handling at the retail, food service, or home level. The critical control points are placed on the processing charts, where an M represents a microbiological hazard, a C represents a chemical hazard, a P represents a physical hazard, and an A represents an allergenic hazard.

Section E

4. Establishing critical limits
   a. For each critical control point a critical limit must be established. This is the minimum or maximum level that is acceptable for health and regulatory standards. A control sheet is filled out summarizing the critical control point, the critical limit, frequency of audit, action plan if it is out of specification, and who is responsible.

Section F

5. Monitoring procedures
   a. Each critical control point that is established will have a method developed to monitor its compliance to the set specification. Monitoring methods shall be based on current proper scientific methodology. Control points shall be audited on a timely and routine basis.

Section G

6. Corrective action
   a. For each critical control point that is monitored, an action plan will be developed for the operator, the quality control technician, and the company to use as a guide for their response when the control point is outside the control limits.
   b. Corrective actions shall include the action, retesting protocols, and accountability.

Section H

7. Program verification
   a. Within 24 hours of when a critical control point is monitored, this record shall be reviewed by a member of management.
   b. After review, the document shall be signed and dated.

Section I

8. Record keeping system
   a. All HACCP program development documents shall be kept in HACCP Book 1.
   b. CCP monitoring documents shall be filed with the daily production and quality records.

Section J

9. Program auditing
   a. On a yearly basis, the control point record is audited to verify that the specification has been met and that no product or ingredient was produced or received out of specification.

Approved by: _____ Date: _____

HACCP PROGRAM

Development and Implementation Team Roster

Date: _____

The following employees have been designated as members of the _____ _____ HACCP development and implementation team. The team will meet on a minimum of a monthly basis to develop and implement a formal hazard analysis critical control point program. This includes performing a hazard analysis on raw materials, packaging, and processing; determining critical control points; establishing critical limits; determining corrective actions; reviewing records; and verifying and auditing the program. It is the responsibility of the team leader to confirm that all steps in the development of the program have been executed properly and have been completed in a timely manner.

Employee Name	Department Represented
_____	_____
_____	_____
_____	_____
_____	_____
_____	_____
_____	_____
_____	_____
_____	_____

The team leader is: _____

HACCP PROGRAM

HACCP Team Meeting Attendance

Date: _____

The following employees were in attendance at the HACCP development and implementation team meeting on the date above.

Employee Name	Department Represented

SUPPLIER DATA SHEET

Item Number: _____

Ingredient Name: _____

Supplier Name: _____ Vendor Item Number _____

Address: _____

Sales Contact: _____
1 _____ Telephone Number: _____
2 _____ Fax Number: _____
                           Email: _____
                           Cell Number: _____

Emergency Contact:
1 _____ Telephone/cell Number: _____
2 _____ Telephone/cell Number: _____
3 _____ Telephone/cell Number: _____

Items to be in folder: Effective Date
Specification Sheet   Storage _____
Ingredients   Shelf Life _____
Nutritional

	QA Verified	Certification	Expires Date		
Kosher Certification					
Organic Certification			Lot Code Example		
Vegan Certification					
Halal Certification			Pack Size to be Purchased		
Continuing Guaranty					
MSDS			Pallet Configuration		
Non-GMO Verification					
Biosecurity Act Verification					

Allergen	Contains (Y/N)	QA Verified		Contains (Y/N)	QA Verified
Milk			Celery Sds		
Egg			Mustard Sds		
Peanuts			Sesame Sds		
Wheat			Colors:		
Cereal Gluten					
Soy					
Fish					
Shellfish					
Tree Nut					
Sulfites					
MSG					
Molusks					

PRODUCT DESCRIPTION WORKSHEET

Finished Product Number:

Finished Product Name:

Raw Material Item Number	Ingredient Name	How Received or Prepared

HACCP Process Flow Master List

Finished Good Number	Product Name	Process Flow Diagram

HACCP PROGRAM
HACCP Principle 1

Process Flow For: _____

Step 1 _____

Step 2 _____

Step 3 _____

Step 4 _____

Step 5 _____

Step 6 _____

Step 7 _____

Step 8 _____

Step 9 _____

Step 10 _____

Step 11 _____

Step 12 _____

Step 13 _____

Step 14 _____

Step 15 _____

Step 16 _____

Step 17 _____

Step 18 _____

Step 19 _____

Step 20 _____

HACCP Principle 1

RISK ASSESSMENT WORKSHEET FOR **CHEMICAL** FOOD HAZARDS

Product Name: _____ Date: _____

Product Item Number: _____

Finished Product

	How Stored			Potential Introduction of Hazard Y/N	If yes, can the hazard be detected and removed by the customer
	Frozen	Refrigerated	Room Temp		
Company Warehouse					
Distributor					
Retailer					
Customer					

Ingredients

	Raw Materials / Ingredients	Ingredient might/does contain hazard	Hazard removed during mfg	Can be recontaminated between mfg and pkg	Is hazard detectable or removable by the customer	If Ingred. contains hazard, can supplier remove it?	COA available
1							
2							
3							
4							
5							
6							
7							
8							
9							
10							
11							
12							
13							
14							
15							

Book 1: HACCP Program

HACCP Principle 1

RISK ASSESSMENT WORK-SHEET FOR **MICROBIOLOGICAL** FOOD HAZARDS

Product Name: _____   Date: _____

Product Item Number: _____

Finished Product

	How Stored			Potential Introduction of Hazard Y/N	If yes, is the product heated by the customer
	Frozen	Refrigerated	Room Temp		
Company Warehouse					
Distributor					
Retailer					
Customer					

Ingredients

	Raw Materials / Ingredients	Ingredient contains hazard	Hazard removed during mfg	Can be recontaminated between mfg and pkg	Can be recontaminated during storage	If ing. contains hazard, can supplier remove it?	COA available
1							
2							
3							
4							
5							
6							
7							
8							
9							
10							
11							
12							
13							
14							
15							

HACCP Principle 1

RISK ASSESSMENT WORK-SHEET FOR **PHYSICAL** FOOD HAZARDS

Product Name: _____ Date: _____

Product Item Number: _____

Finished Product

	How Stored			Potential Introduction of Hazard Y/N	If yes, is the hazard detected or removed by the customer
	Frozen	Refrigerated	Room Temp		
Company Warehouse					
Distributor					
Retailer					
Customer					

Ingredients

	Raw Materials / Ingredients	Ingredient might/does contain hazard	Hazard removed during mfg	Can be recontaminated between mfg and pkg	Is hazard detectable or removed by the customer	If ingredient contains hazard, can supplier remove it?	COA available
1							
2							
3							
4							
5							
6							
7							
8							
9							
10							
11							
12							
13							
14							
15							

HACCP Principle 1

RISK ASSESSMENT WORK-SHEET FOR **ALLERGEN** FOOD HAZARDS

Product name: _____   Date: _____

Product Item Number: _____

Finished Product

	How Stored			Potential Introduction of Hazard Y/N	If yes, is the hazard detected or removed by the customer
	Frozen	Refrigerated	Room Temp		
Company Warehouse					
Distributor					
Retailer					
Customer					

Raw Material Analysis

Item #	Ingredient	Milk	Eggs	Fish	Shellfish	Tree Nuts	Peanuts
1							
2							
3							
4							
5							
6							
7							
8							
9							
10							
11							
12							
13							
14							
15							

Item #	Ingredient	Wheat	Soybeans	Seeds	MSG	Sulfites	Colors
1							
2							
3							
4							
5							
6							
7							
8							
9							
10							
11							
12							
13							
14							
15							

HACCP Principle 2
PRODUCT/PROCESS CONTROL POINT EVALUATION WORKSHEET

Process Flow: Baking 1
Evaluation Date: 15-Jul-07

Processing Step	Parameter to Evaluate	Chemical Hazard	Physical Hazard	Micro Hazard	Allergen Hazard	Critical Control Point or Control Point
Incoming Ingredients	container cleanliness					Control Point
Incoming Ingredients	COA		yes-pecan shells			Control Point
Incoming Ingredients	ingredient statements				yes	Critical Control Point (1)
Packaging receipt	ingredient statements				yes	Critical Control Point
Ingredient blending	consistancy					Control Point
Extruding	consistancy					Control Point
Wire cut	cookie weight					Control Point
Baking	time and temperature			yes		Critical Control Point (2)
Cooling	temperature					Control Point
Packaging	seal					Control Point
Metal detection	metallic foreign objects		yes-metal			Critical Control Point (3)
Coding - indvidual	accuracy					Control Point
Casing	weight					Control Point
Coding - case	accuracy					Control Point
Palletizing	accuracy					Control Point
Pallet label	accuracy					Control Point
Freezing	temperature					Control Point

Products ran on this process
- chocolate chip 1630
- lemon 1231
- macadamea 3345
- peanut butter 6549
- pecan shortbread 1134

HACCP Plan Summary Chart

CCP Number	CCP Description		Critical Limit(s) Description	Monitoring Program		Frequency	Action Plan If Out Of Specification	Responsible Person
A-01	Hazard Controlled	Allergen	No ingredient changes on labels	Name	Receiving	All packaging and ingredients received	1. Place on hold 2. Notify QC	1. Receiving Inspector 2. QC Manager
	Point or Procedure	Receiving Inspection		Quality Manual Section	Receiving Program			
P-01	Hazard Controlled	Metal	< 1.5 mm Ferrous < 2.0 mm Non-Ferrous < 2.5 mm 316 Stainless Steel	Name		Prior to startup Every two hours End of shift	1. Notify operations and QA 2. Isolate product back to previous documented working time 3. Fix unit 4. Retest held product	1. Line Operator 2. Plant Manager 3. QC Manager
	Point or Procedure	Metal Detector		Quality Manual Section	Metal Detection Program			
M-01	Hazard Controlled	Bacteria	185 °F minimum 29 seconds	Name		All products	1. Place on hold 2. Fix pasteurizer and document working condition 3. repasteurize product if possible 4. destroy product on hold if not possible	1. Pasteurizer Operator 2. Plant Manager 3. QC Manager
	Point or Procedure	Pasteurizer		Quality Manual Section	HTST Program			
	Hazard Controlled			Name				
	Point or Procedure			Quality Manual Section				
	Hazard Controlled			Name				
	Point or Procedure			Quality Manual Section				

HACCP PROGRAM

HACCP Team Yearly Auditing Meeting Attendance

Date: _____

The following employees were in attendance at the HACCP auditing team meeting on the date above:

Employee Name   **Department Represented**

_____   _____

_____   _____

_____   _____

_____   _____

_____   _____

_____   _____

_____   _____

Areas Reviewed:

- ☐ Program Document
- ☐ Process Flows
- ☐ Chemical Hazard Analysis
- ☐ Physical Hazard Analysis
- ☐ Critical Control Points
- ☐ HACCP Summary Chart
- ☐ Supplier Information
- ☐ Product Descriptions
- ☐ Microbiological Hazard Analysis
- ☐ Allergen Hazard Analysis
- ☐ Monitoring Forms
- ☐ Verification Process

VERIFICATION DOCUMENTATION
HACCP Training

    I _____ have been trained on the seven principals of HACCP and understand how they are to be applied at _____. Furthermore I will notify my supervisor if I see any risk being introduced into the products or process or if I witness a critical control point either out of compliance for any reason.

    I will share my knowledge with others employees I work including temporary help. If I have questions regarding any quality or production system or control point I will immediately contact my supervisor.

Date: _____

Name: _____

Instructor: _____

EXAMPLES

SUPPLIER DATA SHEET

Item Number:	1601
Ingredient Name:	Apple Flavor

Supplier Name:	Mission Flavors	Vendor Item Number:	AF-23-516
Address:	8500 Condio St.		
	Mission, CA 95218		
Sales Contact:		Telephone Number:	503-705-3128
1	Dave Mattson	Fax Number:	503-256-9947
2		Email:	davemattson@aol.com
		Cell Number:	503-461-2234

Emergency Contact:

1	Paul Williams	Telephone/cell Number:	803-995-2654
2	Sally Hartley	Telephone/cell Number:	803-995-2659
3	Bill Wagoner	Telephone/cell Number:	801-926-4432

Items to be in folder:

		Effective Date		
Specification Sheet	x	9/1/2001	Storage	6 months
Ingredients	x		Shelf Life	refrigerated temp
Nutritional	x			

	QA Verified	Certification	Expires Date	
Kosher Certification	x	OU	6/1/2008	
Organic Certification	no		Lot Code Example	
Vegan Certification	no			6237 julian date yddd
Halal Certification	no		Pack Size to be Purchased	
Continuing Guaranty	yes			4 x 1 gallon
MSDS	yes		Pallet Configuration	
Non-GMO verification	yes			15 x 8
Biosecurity Act Verification		yes		

Allergen	Contains (Y/N)	QA Verified		Contains (Y/N)	QA Verified
Milk	n	y	Celery Sds	n	y
Egg	n	y	Mustard Sds	n	y
Peanuts	n	y	Sesame Sds	n	y
Wheat	n	y	Colors: ↓	n	y
Cereal Gluten	n	y			
Soy	n	y			
Fish	n	y			
Shellfish	n	y			
Tree Nut	n	y			
Sulfites	n	y			
MSG	n	y			
Molusks	n	y			

HACCP PROGRAM

HACCP Principle 1

Process Flow for Chopped Product A

Incoming inspection and adherence to specifications
Receipt of materials
    Refrigerated ingredients: ≤60°F
    Dry goods: Room temperature
Storage: Freezer
Unwrapped
Storage: Freezer
Staging
Conveyor
Starch added
Chopping machine
Conveyor
Weighed
Conveyor
Metal detector
Conveyor
Casing
Conveyor
Palletized
Storage: Freezer

HACCP PROGRAM

HACCP Principle 1

Process Flow for Baked Product A

Incoming inspection and adherence to specifications
Receipt of materials
    Refrigerated ingredients: ≤60°F
    Dry goods: Room temperature
Storage: Freezer
    Warehouse
    Cooler
Staging
Blending
Extruder
Conveyor
Oven
Conveyor
Weighing
Conveyor
Metal detector
Conveyor
Packaging
Casing
Conveyor
Palletized
Storage: Freezer

PRODUCT DESCRIPTION WORKSHEET

Finished Product Number: C100-320

Finished Product Name: Chocolate Chip Cookies, 3 oz IW

C100 is an individually wrapped 3.2 oz chocolate chip cookie. It is ready to use by the customer at room temperature. This product is manufactured by combining ingredients in a mixer, extruding them onto cookie sheets, then baking them in an oven at 375 F. They are then cooled and individually wrapped on a vertical formfill machine then hand packed 24 to a case. The finished good is stored at room temperature until distributed.

Raw Material Item Number	Ingredient Name	How Received or Prepared
D43061	Dry Blend - Chocolate Chip Cookie	Room Temperature
B41192	Butter - low fat	Refrigerated
M42714	Milk 2%	Refrigerated
E42513	Egg Substitute	Refrigerated
C41247	Cream Cheese - low fat	Refrigerated
S45164	Liquid Sugar	Bulk - Room Temp.

Book 1: HACCP Program

PRODUCT DESCRIPTION WORKSHEET

Finished Product Number: 1385-12.5

Finished Product Name: Steak Sauce, hot

Gordon's hot steak sauce is a combination of spices and other prepared ingredients. It is processed by combining these ingredients in a steam jacketed kettle untill it is attains a temperature of at least 190 F. It is then pumped to the filler where it is packaged into glass jars. These are lidded, labeled, and packed into corrugated cases. The cases are stored in an ambient temperature warehouse.

Raw Material Item Number	Ingredient Name	How Received or Prepared
261	Tomato Puree-asceptic pack	Room Temperature
311	Pineapple Juice	Frozen
349	Corn Syrup	Room Temperature
441	Distilled Vinegar	Room Temperature
513	Raisin Paste	Refrigerated
523	Spice Blend	Room Temperature
165	Orange Peel, diced	Frozen
53	Garlic Powder	Room Temperature
57	Onion Powder	Room Temperature
71	Potassium Sorbate	Room Temperature
136	Caramel Color	Refrigerated
512	Xanthan Gum	Room Temperature

HACCP Principle 1

RISK ASSESSMENT WORKSHEET FOR **CHEMICAL** FOOD HAZARDS

Product Name: Steak Sauce, hot   Date: 15-Jul-07

Product Item Number: 1385-12.5

Finished Product

	How Stored			Potential Introduction of Hazard Y/N	If yes, can the hazard be detected and removed by the customer
	Frozen	Refrigerated	Room Temp		
Company Warehouse			yes	no	
Distributor			yes	no	
Retailer			yes	no	
Customer		yes	yes	no	

Ingredients

Raw Materials / Ingredients	Ingredient might/does contain hazard	Hazard removed during mfg	Can be recontaminated between mfg and pkg	Is hazard detectable or removable by the customer	If Ingred. contains hazard, can supplier remove it?	COA available
Tomato Puree-aseptic pack	yes*	no	no	no	yes	yes
Pineapple Juice	yes*	no	no	no	yes	yes
Corn Syrup	no					
Distilled Vinegar	no					
Raisin Paste	no					
Spice Blend	no					
Orange Peel, diced	no					
Garlic Powder	no					
Onion Powder	no					
Potassium Sorbate	no					
Caramel Color	no					
Xanthan Gum	no					
	* possible pesticides					

Book 1: HACCP Program

HACCP Principle 1

RISK ASSESSMENT WORKSHEET FOR **MICROBIOLOGICAL** FOOD HAZARDS

Produc Name: Steak Sauce, hot   Date: 15-Jul-07

Product Item Number: 1385-12.5

Finished Product

	How Stored			Potential Introduction of Hazard Y/N	If yes, is the product heated by the customer
	Frozen	Refrigerated	Room Temp		
Company Warehouse			yes	no	
Distributor			yes	no	
Retailer			yes	no	
Customer		yes	yes	yes	no

Ingredients

Raw Materials / Ingredients	Ingredient contains hazard	Hazard removed during mfg	Can be recontaminated between mfg and pkg	Can be recontaminated during storage	If Ingred. contains hazard, can supplier remove it?	COA available
Tomato Puree-asceptic pack	yes	yes	no	no	yes	yes
Pineapple Juice	no					
Corn Syrup	no					
Distilled Vinegar	no					
Raisin Paste	yes	yes	no	no	yes	yes
Spice Blend	yes	yes	no	no	yes	yes
Orange Peel, diced	yes	yes	no	no	yes	yes
Garlic Powder	no					
Onion Powder	no					
Potassium Sorbate	no					
Caramel Color	no					
Xanthan Gum	no					

HACCP Principle 1

RISK ASSESSMENT WORKSHEET FOR **PHYSICAL** FOOD HAZARDS

Product Name: Steak Sauce, hot   Date: 15-Jul-07

Product Item Number: 1385-12.5

Finished Product

	How Stored			Potential Introduction of Hazard Y/N	If yes, can the hazard be detected and removed by the customer
	Frozen	Refrigerated	Room Temp		
Company Warehouse			yes	no	
Distributor			yes	no	
Retailer			yes	no	
Customer		yes	yes	yes	yes

Ingredients

Raw Materials / Ingredients	Ingredient might/does contain hazard	Hazard removed during mfg	Can be recontaminated between mfg and pkg	Is hazard detectable or removable during the process	If Ingred. contains hazard, can supplier remove it?	COA available
Tomato Puree-asceptic pack	yes	yes	no			
Pineapple Juice	no					
Corn Syrup	no					
Distilled Vinegar	no					
Raisin Paste	yes	yes				
Spice Blend	yes	yes	yes	yes	yes	yes
Orange Peel, diced	no					
Garlic Powder	no					
Onion Powder	no					
Potassium Sorbate	no					
Caramel Color	no					
Xanthan Gum	no					

Book 1: HACCP Program

HACCP Principle 1

RISK ASSESSMENT WORKSHEET FOR **ALLERGENS** FOOD HAZARDS

Product: Baking Mix   Date: 15-Jul-07

Product Item Number: BM42170.40

Finished Product

	How Stored			Potential Introduction of Hazard Y/N	If yes, can the hazard be detected and removed by the customer
	Frozen	Refrigerated	Room Temp		
Company Warehouse			yes	no	
Distributor			yes	no	
Retailer			yes	no	
Customer			yes	no	

Raw Material Analysis

	Item #	Ingredient	Milk	Eggs	Fish	Shellfish	Tree Nuts	Peanuts
1	FW1890	Flour, Wheat, enrich	no	plant	no	no	no	no
2	OS1234	Oil, Canola	no	no	no	no	no	no
3	DS1145	Dextrose	no	no	no	no	no	no
4	SC1163	Sugar	no	no	no	no	no	no
5	BS9820	Baking Soda	no	no	no	no	no	no
6	PS2983	Sodium Al. Phos	no	no	no	no	no	no
7	SC7581	Salt	no	no	no	no	no	no
8	PT2981	Trical Phos	no	no	no	no	no	no
9	PM2987	Monocal Phos	no	no	no	no	no	no
10	DM0182	Datem	no	no	no	no	no	no
11								
12								
13								
14								
15								

	Item #	Ingredient	Wheat	Soybeans	Seeds	MSG	Sulfites	Colors
1	FW1890	Flour, Wheat, enr	yes	no	no	no	no	no
2	OS1234	Oil, Canola	no	no	no	no	no	no
3	DS1145	Dextrose	no	no	no	no	no	no
4	SC1163	Sugar	no	no	no	no	no	no
5	BS9820	Baking Soda	no	no	no	no	no	no
6	PS2983	Sodium Al. Phos	no	no	no	no	no	no
7	SC7581	Salt	no	no	no	no	no	no
8	PT2981	Trical Phos	no	no	no	no	no	no
9	PM2987	Monocal Phos	no	no	no	no	no	no
10	DM0182	Datem	no	no	no	no	no	no
11								
12								
13								
14								
15								

HACCP Principle 2

PRODUCT/PROCESS CONTROL POINT EVALUATION WORKSHEET

Process Flow: Baking 1　　　　　　　Evaluation Date: 15-Jul-07

Processing Step	Parameter to Evaluate	Chemical Hazard	Physical Hazard	Micro Hazard	Allergen Hazard	Critical Control Point or Control Point
Incoming Ingredients	container cleanliness					Control Point
Incoming Ingredients	COA		yes-pecan shells			Control Point
Incoming Ingredients	ingredient statements				yes	Critical Control Point (1)
Packaging receipt	ingredient statements				yes	Critical Control Point
Ingredient blending	consistancy					Control Point
Extruding	consistancy					Control Point
Wire cut	cookie weight					Control Point
Baking	time and temperature			yes		Critical Control Point (2)
Cooling	temperature					Control Point
Packaging	seal					Control Point
Metal detection	metalic foreign objects		yes-metal			Critical Control Point (3)
Coding - individual	accuracy					Control Point
Casing	weight					Control Point
Coding - case	accuracy					Control Point
Palletizing	accuracy					Control Point
Pallet label	accuracy					Control Point
Freezing	temperature					Control Point

Products run on this process:
- chocolate chip 1630
- lemon 1231
- macadamea 3345
- peanut butter 6549
- pecan shortbread 1134

Book 1: HACCP Program

HACCP Principle 2
PRODUCT/PROCESS CONTROL POINT EVALUATION WORKSHEET

Process Flow: Jelly 1
Evaluation Date: 15-Jul-07

Processing Step	Parameter to Evaluate	Chemical Hazard	Physical Hazard	Micro Hazard	Allergen Hazard	Critical Control Point or Control Point
Incoming Ingredients	container cleanliness					Control Point
Incoming Ingredients	COA					Control Point
Incoming Ingredients	comparison to spec					Control Point
Incoming Ingredients	ingredient statements				yes	Critical Control Point (1)
Packaging receipt	ingredient statements				yes	Critical Control Point
Ingredient blending	consistancy					Control Point
Heating	time and temperature			yes		Critical Control Point (2)
Metal detection	metallic foreign objects		yes-metal			Critical Control Point (3)
Packaging	fill					Control Point
Labeling	accuracy					Control Point
Coding	accuracy					Control Point
Casing						Control Point
Coding - case	accuracy					Control Point
Palletizing	weight					Control Point
Pallet label	accuracy					Control Point
Warehouse	accuracy					Control Point

Products run on this process

raspberry 890
straw/rasp 456
strawberry 362
marionberry 786
current 091

blueberry 778

HACCP Principle 2
PRODUCT/PROCESS CONTROL POINT EVALUATION WORKSHEET

Process Flow: Chopped Candy
Evaluation Date: 15-Jul-07

Processing Step	Parameter to Evaluate	Chemical Hazard	Physical Hazard	Micro Hazard	Allergen Hazard	Critical Control Point or Control Point
Incoming Ingredients	container cleanliness					Control Point
Incoming Ingredients	COA					Control Point
Packaging receipt	ingredient statements				yes	Critical Control Point (1)
Storage						
Staging						
Free Flowing Addition						
Chopping	granulation					Control Point
Scale						
Metal detector	metallic foreign material		yes-metal			Critical Control Point (2)
Bagger	weight					Control Point
Casing						
Taper						
Coder	accuracy					Control Point
Palletizing						
Freezing						

Products run on this process:
- Snickers
- Peanut butter cups
- M&M's
- Baby Ruth

HACCP Plan Summary Chart

CCP Number	CCP Description		Critical Limit(s) Description	Monitoring Program		Frequency	Action Plan If out of Specification	Responsible Person
	Hazard Controlled	Point or Procedure		Name	Quality Manual Section			
	Hazard Controlled	Point or Procedure		Name	Quality Manual Section			
	Hazard Controlled	Point or Procedure		Name	Quality Manual Section			
	Hazard Controlled	Point or Procedure		Name	Quality Manual Section			
	Hazard Controlled	Point or Procedure		Name	Quality Manual Section			

5 Quality Control Program
Overview

THEORY

No single program within a food manufacturing company can be as polarizing yet as important as that of quality control (QC). Its triple focus is to develop a system that protects the brand from enduring regulatory intervention, protects the customer from undue contamination, and protects the product from undue variation. Due to its importance, it is critical that the quality control manager understand what a quality control system is and of what it is composed.

At its fundamental core, quality control is defined as a procedure or set of procedures designed to make certain that the end product conforms to a designated set of criteria as set forth by either the company or the customer. The sum of the procedures that the company establishes is considered its quality control system. In modern food manufacturing environments, quality control systems are the supporting programs that are outcrops of the hazard analysis critical control point (HACCP) program. The information gained through measurements, observation, and documentation from these programs provides a constant picture of the conformance of the product to a specification. In order to implement an effective QC program, a company must first decide the specification that the product or service must meet. Then the type of measuring methods are developed and applied. After analyzing the data, the technician determines if the measurements are within the designated specification. If they are not, then corrective action is implemented and a remeasurement is done. These results are reported to management, who reviews them and determines if process or product improvement is needed.

The components of the quality control system are programs that are required by the federal government, by the customer, or internally. Each program is documented in a program general overview and verified on a monitoring form. Copies of each program's documents are placed in the appropriate tab section in the Quality Control Manual. As a means to document the entire quality control system, a quality monitoring scheme form is completed. This document offers a clear and succinct glimpse of all the quality related programs established by the company, what they monitor, the specification, action to be taken, and whether it is a critical control point.

APPLICATION

The development of the quality control system is begun by drawing up a quality control program general overview document, as found on the CD (Book 2_Quality Control Manual\Section 1_Quality Overview:General Overview) and at the end of this chapter. This document can be customized to outline the exact programs that the company will develop. Programs specific to the operation can be added and those that do not fit the current programs of the company can be deleted. When this is complete, it should be signed by the quality control manager and placed in the control section of the quality manual.

As quality control programs are set up, the required information should be placed on the quality monitoring scheme (QMS) form, which can be found on the CD (Book 2_Quality Control Manual\Section 1_Quality Overview:Quality Monitoring Scheme) and at the end of this chapter. The form is provided as a template for the company to utilize as it develops its own QMS. This form is a unique way to record what is happening in the quality control program, where it is happening, why it is happening, who is responsible, and what happens if it is out of compliance.

EXAMPLES

A quality monitoring scheme document completed for a typical grinding operation can be found on the CD (Book 2_Quality Control Manual\Section 1 _ Quality Overview:QMS Example 1) and at the end of this chapter. It outlines each quality control program within the system and its purpose. The next three columns are related to any samples that are taken in relation to the specified program. This section explains where the samples are taken, how often they are taken, and who is responsible for taking the samples. The next section relates to any examination that takes place related to the samples and specific quality program. It denotes what analysis is to be completed, who is responsible, and the specified method to be used for analysis. The ninth column lists any specification to which the sample analysis must adhere. This is followed by how the results are to be reported, what action is to be taken if the results of the sample analysis are out of specification, and, finally, whether the item is considered a critical control point. This particular QMS is split into four pages and presents 18 separate control points, including two critical control points.

SUPPLEMENTAL MATERIAL

QUALITY CONTROL PROGRAM

GENERAL OVERVIEW

Section A: General

The _____ quality program is based on a company-wide desire to provide a perfect product for our customers each and every time. To meet this goal, _____ has established programs, procedures, and policies to ensure regulatory compliance, consistent products, and customer satisfaction. The cornerstone of the quality assurance program is a hazard analysis critical control point (HACCP) plan for food safety, a quality control program, and a rodent/pest control program. The underlying regulatory statute that drives many of these programs is found in 21CFR110, good manufacturing principles.

Section B: Hazard Analysis Critical Control Point (HACCP) Plan

_____'s HACCP plan is broken down into several parts, including hazard analysis on all ingredients and finished goods, as well as identifying and establishing critical control points, critical control point monitoring programs, corrective action plans for each critical control point, and a verification system. The HACCP program is documented in Book 1: HACCP.

Section C: Quality Control

The quality control program augments the HACCP program to ensure that all products produced are of consistent quality. It includes all of the precursor programs that are dictated by 21CFR110.

Section D: Rodent/Pest Control

_____'s rodent/pest control program involves an outside pest control contractor, as well as in-house monitoring. It is based on the principle of excluding the pests. All pest control devices are checked on a routine basis.

Section E: Lot Coding

Each incoming raw material and finished good is tracked through a lot coding system. _____'s "lot code" on the finished good is a Julian day code followed by the production period (e.g., 02573 would be a product produced on January 25, 2007, during the third period of 12:00–14:30).

Section F: Sanitation

_____ utilizes a complete and documented sanitation program to clean the production room and all other areas inside and outside the facility.

Quality Control Program

Section G: Pre-operational Check

_____ inspects the production room's, equipment's, and employees' good manufacturing practices (GMPs) before startup each day. Any discrepancies found are cleaned and sanitized before any production is run. All discrepancies are documented.

Section H: Environmental Testing

On a weekly basis, samples are drawn from random areas of production and the warehouse for microbiological testing for various pathogens. No food contact areas are tested for pathogens.

Section I: Allergens

All ingredients processed and stored at _____ are segregated by their allergen-containing components. Segregation is done both vertically and horizontally. Nonallergen items are processed before allergen items.

Section J: GMPs

All employees who are employed at _____ review and sign documented GMPs. These are reviewed on a yearly basis. All visitors are required to read, acknowledge, and adhere to the GMP policy. Any construction or building maintenance that occurs shall be done in accordance with written standards dictated by good manufacturing practice and shall be designed and completed with food safety as the basis.

Section K: Metal Detection

_____ passes all products through an in-line metal detector located directly before the packaging step. The sensitivity is ____ mm nonferrous, ____ mm ferrous, and ____ mm stainless. The metal detector is checked every 2 hours.

Section L: Receiving

All trucks are inspected for cleanliness and integrity before unloading. All receipts are documented.

Section M: Shipping

All trucks are inspected for cleanliness prior to loading. All shipments are documented.

Section N: Documentation

_____ maintains accurate and clear records of all data taken to support programs contained within the scope of the quality, HACCP, and food safety. Separate books shall be maintained for the HACCP, quality control, pest control, and other programs.

Section O: Customer Complaint

_____ tracks and responds to all customer complaints in an orderly manner. This is part of the continuous improvement process.

Section P: Recall

A formal recall program is established that documents clear, concise procedures to be followed in the event of a product withdrawal or mandatory product recall. This program includes a designated point person for communications inside and outside the company.

Section Q: Regulatory Inspection

A program that outlines the responsibilities of the company toward regulatory inspections is established. This shall explain what should and should not be done in order for the company to handle federal, state, and local inspectors, whether the inspection is the annual inspection or a complaint-dictated one.

Section R: Weight Control

The company shall establish a program that documents compliance to federal weights and measures requirements. This shall be accomplished either manually or electronically and will apply to all products manufactured.

Section S: Defective Material

A program shall be established to document and monitor the handling of product that is out of compliance with company-set specifications. This involves three parts:

- holds: setting aside those ingredients, packaging, or finished goods that are out of specification; designating them so that they are clearly marked; and in many cases, isolating them from all other products;
- releases: the formal process of allowing the out-of-specification item to be reworked or destroyed; and
- destructions: the documented process of removing products from stock, either through garbage or donation.

Section T: Loose Tool/Material

A documented program that keeps track of all loose tools—those used for maintenance or during processing, or other separate materials such as knives, scoops, and tape guns—shall be in place.

Section U: Supplier Certification

A detailed written program shall be in place that evaluates all suppliers of raw materials to determine if they comply with food safety requirements, remove all hazards, and comply with federal, state, and local food laws.

Section V: Glass/Hard Plastic/Wood

A written program that outlines the use, or lack thereof, of glass, wood, and hard plastic within the plant and provides the means to contain it shall be implemented.

Section W: Responsibility

It is the responsibility of the quality control department to establish food safety programs. Quality control and operations share enforcement responsibility.

Approved by: _____ Date: _____

Quality Monitoring Scheme

ITEM	PURPOSE	SAMPLING			EXAMINATION				REPORT	ACTION IN CASE OF DEVIATION	CCP
		WHERE	FREQUENCY	BY WHOM	ANALYSIS	BY WHOM	METHOD	SPECIFICATION			

EXAMPLES

Quality Monitoring Scheme

ITEM	PURPOSE	SAMPLING			EXAMINATION				REPORT	ACTION IN CASE OF DEVIATION	CCP
		WHERE	FREQUENCY	BY WHOM	ANALYSIS	BY WHOM	METHOD	SPECIFICATION			
Environment	Pathogen Monitoring	Various Random Areas	Weekly	QC Tech	Salmonella, Listeria	Diebel or Warren Labs	BAM	See Environmental Monitoring Program	See Environmental Testing Log	Notify TSD of Any Positive Results in Any Zone	
GMP Audit	Regulatory Compliance	Operations	Hourly	QC Tech	Typical GMP Requirements	QC Tech	Visual	See GMP Rules	Daily Line Check Sheet	Fix Problem with Supervisor, Record Offense	
Inspection	Quality, Safety and Maintenance Audit	Entire Facility	Monthly	Plant Manager, TSD, Warehouse Manager	Look for Cleanliness, Hazards, Maintenance Issues	Plant Manager, TSD, Warehouse Manager	Visual	See Inspection Policies and Procedures	Inspection Form	Fix Problem, Document	
Sanitation	To Monitor Daily Sanitation	Entire Facility	Daily		Verify Thoroughness and Dryness	QC Tech, Prod. Sup., TSD, Sanitation Lead	Visual	See Daily Plant Sanitation Procedures	Daily Inspection Form	Notify TSD	
	To Monitor General Sanitation	Entire Facility	As Scheduled		Work Accomplished	Plant Manager, TSD	Visual	See Master Sanitation Schedule	Master Cleaning Schedule	Notify TSD	
Pest Control	To Monitor Pest Activity	Outside and Inside	Monthly Contract	Terminex	Ketch-alls and Bait Stations	Licensed Pest Applicator	Visual	See Rodent & Pest Control Program	Inspection Form, Terminex Form	Notify TSD	
		Outside and Inside	Twice Weekly	QC Tech	Ketch-alls, Bait Stations, Insect Attractors	QC Tech	Visual	See Rodent & Pest Control Program	Inspection Form	Notify TSD	

Quality Control Program

Quality Monitoring Scheme

ITEM	PURPOSE	SAMPLING			EXAMINATION				REPORT	ACTION IN CASE OF DEVIATION	CCP
		WHERE	FREQUENCY	BY WHOM	ANALYSIS	BY WHOM	METHOD	SPECIFICATION			
Staging area	Ensure quality of RM					Prod. Room Emp./forklift operator	Visual & product screening	Clean and not damaged	Integrity of containers and cleanliness of pallets	Notify QA	
									Proper weight of ingredients	Notify QA	
	Ingredient Traceability				Ingredient lot identification	prod. Room employee	Record ingredient lot # & cs used	All lot # & cs must be recorded	lot sheet	Fill in all blanks, all lot # & quantities recorded	
	Rework				Ingredient lot identification	prod. Room employee	Visual	All lot # & cs must be recorded	lot sheet	Fill in all blanks, all lot # & quantities recorded	
Grinder	Ensure unit is setup and functioning properly				Set up according to proc. Spec.	qc	visual before startup	matches spec	daily line check sheet	notify prod. Sup.	
Weight control	Ensure proper fill weights	bagger	5 bags every hour minimum	QC Tech.	weights	qc	hand weigh	in control limits for ingredient	weight control sheet	notify operator/prod sup.	
Ink Jet Coder	Verify that proper code is applied				Proper & legible code	Stacker, QC	Visual	TR Coding system	daily line check sheet	Hold product. Notify prod sup/QC. Correct code. Recode. if nec.	
Palletizer	Verify correct pallet pattern & number of cases	Palletizer			Stacking, product condition	Stacker, QC	Visual & Manual	Correct pattern & number of cases per pallet;	daily line check sheet	Hold product. Notify prod sup/QC. Correct stack. Restack, if nec.	
Granulation	To verify partical size	after grinder	minimum once per lot	QC Tech.	rotap	qc	rotap methods and proc	must be in spe	Rotap inspection form	notify prod. Sup. put on hold. notify TSD	

Quality Monitoring Scheme

ITEM	PURPOSE	SAMPLING			EXAMINATION				REPORT	ACTION IN CASE OF DEVIATION	CCP
		WHERE	FREQUENCY	BY WHOM	ANALYSIS	BY WHOM	METHOD	SPECIFICATION			
Staging area	Ensure quality of RM					Prod. Room Emp./forklift operator	Visual & product screening	Clean and not damaged	Integrity of containers and cleanliness of pallets	Notify QA	
									Proper weight of ingredients	Notify QA	
	Ingredient Traceability				Ingredient lot identification	prod. Room employee	Record ingredient lot # & cs used	All lot # & cs must be recorded	lot sheet	Fill in all blanks, all lot # & quantities recorded	
	Rework				Ingredient lot identification	prod. Room employee	Visual	All lot # & cs must be recorded	lot sheet	Fill in all blanks, all lot # & quantities recorded	
Grinder	Ensure unit is setup and functioning properly				Set up according to proc. Spec.	qc	visual before startup	matches spec	daily line check sheet	notify prod. Sup.	
Weight control	Ensure proper fill weights	bagger	5 bags every hour minimum	QC Tech.	weights	qc	hand weigh	in control limits for ingredient	weight control sheet	notify operator/prod sup.	
Ink Jet Coder	Verify that proper code is applied				Proper & legible code	Stacker, QC	Visual	TR Coding system	daily line check sheet	Hold product. Notify prod sup/QC. Correct code. Recode, if nec.	
Palletizer	Verify correct pallet pattern & number of cases	Palletizer			Stacking, product condition	Stacker, QC	Visual & Manual	Correct pattern & number of cases per pallet	daily line check sheet	Hold product. Notify prod sup/QC. Correct stack. Restack, if nec.	
Granulation	To verify partical size	after grinder	minimum once per lot	QC Tech.	rotap	qc	rotap methods and proc	must be in spe	Rotap inspection form	notify prod. Sup. put on hold, notify TSD	

Quality Control Program

Quality Monitoring Scheme

ITEM	PURPOSE	SAMPLING			EXAMINATION				REPORT	ACTION IN CASE OF DEVIATION	CCP
		WHERE	FREQUENCY	BY WHOM	ANALYSIS	BY WHOM	METHOD	SPECIFICATION			
Environment	Pathogen Monitoring	Various Random Areas	Weekly	QC Tech	Salmonella, Listeria	Diebel or Warren Labs	Bacteriological Analytical Manual BAM)	Environmental Monitoring Program	Environmental Testing Log	Notify QC Manager of Any Positive Results in Any Zone	
GMP Audit	Regulatory Compliance	Operations	Hourly	QC Tech	Typical GMP Requirements	QC Tech	Visual	See GMP Rules	Daily Line Check Sheet	Fix Problem with Supervisor, Record Offense	
Inspection	Quality, Safety and Maintenance Audit	Entire Facility	Monthly	Plant Manager, QC Manager, Warehouse Manager	Look for Cleanliness, Hazards, Maintenance Issues	Plant Manager, QC Manager, Warehouse Manager	Visual	See Inspection Policies and Procedures	Inspection Form	Fix Problem, Document	
Sanitation	To Monitor Daily Sanitation	Entire Facility	Daily	QC Tech	Verify Thoroughness and Dryness	QC Tech, Production Supervisor, QC Manager, Sanitation Lead	Visual	See Daily Plant Sanitation Procedures	Daily Inspection Form	Notify QC Manager	
	To Monitor General Sanitation	Entire Facility	As Scheduled	QC Tech	Work Accomplished	Plant Manager, QC Manager	Visual	See Master Sanitation Schedule	Master Cleaning Schedule	Notify QC Manager	
Pest Control	To Monitor Pest Activity	Outside and Inside	Monthly Contract	Terminex	Ketch-alls and Bait Stations	Licensed Pest Applicator	Visual	See Rodent & Pest Control Program	Inspection Form, Terminex Form	Notify QC Manager	
		Outside and Inside	Twice Weekly	QC Tech	Ketch-alls, Bait Stations, Insect Attractors	QC Tech	Visual	See Rodent & Pest Control Program	Inspection Form	Notify QC Manager	

6 Organizational Chart

PROGRAM TYPE: REQUIRED

THEORY

Who is responsible? Or, more commonly in some companies, who is at fault? This question or concern crops up constantly within a manufacturing facility as the two basic functional parts, production and quality control, strive to produce a product that is of the highest quality and yet attain the highest efficiency possible. In many manufacturing environments there is a struggle between production and quality control personnel as each department strives to fulfill its assigned responsibility. Production wants to make as many products as it can within a given period of time; this is acceptable until quality control institutes measures to control the consistency of the product and ensure the safety of the food. This causes conflict because, in most cases, it requires additional time and energy. Do not worry because this is a normal departmental relationship that must be managed correctly by management.

One tool that a company can utilize to help manage this relationship is the organizational chart. This chart is a visual representation of how management pictures the lines of authority and communication in all departments of the company—an authority tree per se that dictates the responsibility of each person and to whom he or she reports problems, concerns, and specification deviations. It creates a formal chain of command for the solving of problems and supervision of employees. When created, the organizational chart can be used as a training tool for new employees and as reinforcement for existing ones.

An added advantage of having a current working organizational chart is to provide customers with an easy reference for communication when problems occur. This also indicates to the customer that the company is organized and utilizes specific areas of responsibility. Large customers and third-party auditors will request to see, and sometimes request, a copy of the organizational chart when they audit the facility.

APPLICATION

The development of an organizational chart can be completed in many ways. It can be hand drawn or computer generated. A template generated using Microsoft Word can be found on the CD (Book 2_Quality Control Manual\Section 2_Organizational Chart:Organizational Chart) and at the end of this chapter. It is designed for a typical food manufacturing company that is split into four basic functional parts: finance, operations, technical, and sales. Each department has a designated leader who reports to a top officer. *Note:* in any company, the operations and technical lines of responsibility cannot have a common leader. They must be separate due to their competing responsibilities.

To customize this chart, the drawing tool bar from Word can be utilized. Additional help can be found by looking up "organization chart" in the help index. Once completed, this chart should be placed in the organization chart tab section of the quality control manual.

SUPPLEMENTAL MATERIALS

ORGANIZATIONAL CHART

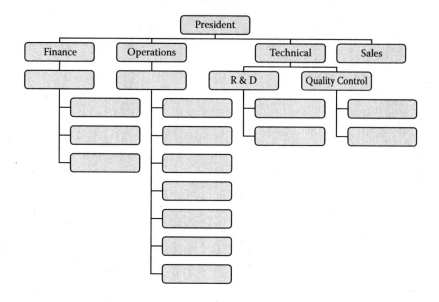

7 Good Manufacturing Practices Program

PROGRAM TYPE: REQUIRED

THEORY

In addition to the United States, for every food manufacturer in the world, the single core program that underlies most other food safety programs contained within the quality control manual is the Federal Food Drug and Cosmetic Act. Predicated on the Food and Drug Administration's early landmark 1938 Food Drug and Cosmetic Act and subsequent amendments, it is written specifically to deal with the serious issues of food adulteration and food misbranding. In it, 21USC342 SEC. 402 states:

> A food shall be deemed to be adulterated (a)(1) If it bears or contains any poisonous or deleterious substance which may render it injurious to health; but in case the substance is not an added substance such food shall not be considered adulterated under this clause if the quantity of such substance in such food does not ordinarily render it injurious to health; 2 (2)(A) 3 if it bears or contains any added poisonous or added deleterious substance (other than a substance that is a pesticide chemical residue in or on a raw agricultural commodity or processed food, a food additive, a color additive, or a new animal drug) that is unsafe within the meaning of section 406; or (B) if it bears or contains a pesticide chemical residue that is unsafe within the meaning of section 408(a); or (C) if it is or if it bears or contains (i) any food additive that is unsafe within the meaning of section 409; or (ii) a new animal drug (or conversion product thereof) that is unsafe within the meaning of section 512; or (3) if it consists in whole or in part of any filthy, putrid, or decomposed substance, or if it is otherwise unfit for food; or (4) if it has been prepared, packed, or held under unsanitary conditions whereby it may have become contaminated with filth, or whereby it may have been rendered injurious to health; or (5) if it is, in whole or in part, the product of a diseased animal or of an animal which has died otherwise than by slaughter; or (6) if its container is composed, in whole or in part, of any poisonous or deleterious substance which may render the contents injurious to health.
>
> (b)(1) If any valuable constituent has been in whole or in part omitted or abstracted there from; or (2) if any substance has been substituted wholly or in part therefore; or (3) if damage or inferiority has been concealed in any manner; or (4) if any substance has been added thereto or mixed or packed therewith so as to increase its bulk or weight, or reduce its quality or strength, or make it appear better or of greater value than it is.

This section and the subsequent sections specifically state that it is the responsibility of the food manufacturer to produce a product that is unadulterated and not misbranded.

In support of the preceding directive, the FDA has developed rules and guidelines that inform manufacturers of specific requirements to aid them in the manufacture and distribution of unadulterated and nonmisbranded foods. These guidelines are located in the Code of Federal Regulations 21CFR part 110 as Current Good Manufacturing Practice in Manufacturing, Packing or Holding Human Food. Inside this tome are the rules and regulations for food manufacturers and distributors in producing clean, safe, and wholesome foods; therefore, it is the standard by which the government determines if the food is adulterated. As such, it is critical that every food manufacturer and distributor understand the regulations and develop systems and programs that prove adherence to the letter of this law. The regulations are split into several sections, including 110.5 Current Good Manufacturing Practice, 110.10 Personnel, 110.20 Plant and Grounds, 110.35 Sanitary Operations, 110.37 Sanitary Facilities and Controls, 110.40 Equipment and Utensils, 110.80 Processes and Controls, 110.93 Warehousing and Distribution, and 110.110 Natural or Unavoidable Defects in Food for Human Use That Present No Health Hazards. To understand them, it is helpful to see each section in its entirety.

21CFR110.5 Current Good Manufacturing Practice is the legal tie-in to the Federal Food, Drug and Cosmetic Act and states:

(a) The criteria and definitions in this part shall apply in determining whether a food is adulterated (1) within the meaning of section 402(a)(3) of the act in that the food has been manufactured under such conditions that it is unfit for food; or (2) within the meaning of section 402(a)(4) of the act in that the food has been prepared, packed, or held under unsanitary conditions whereby it may have become contaminated with filth, or whereby it may have been rendered injurious to health. The criteria and definitions in this part also apply in determining whether a food is in violation of section 361 of the Public Health Service Act (42 U.S.C. 264). (b) Food covered by specific current good manufacturing practice regulations also is subject to the requirements of those regulations.

Thus, it is required that each and every food manufacturer of distributor adhere strictly to the regulations as a means to maintaining a safe food supply.

110.10 Personnel states:

The plant management shall take all reasonable measures and precautions to ensure the following: (a) Disease control. Any person who, by medical examination or supervisory observation, is shown to have, or appears to have, an illness, open lesion, including boils, sores, or infected wounds, or any other abnormal source of microbial contamination by which there is a reasonable possibility of food, food-contact surfaces, or food-packaging materials becoming contaminated, shall be excluded from any operations which may be expected to result in such contamination until the condition is corrected. Personnel shall be instructed to report such health conditions to their supervisors. (b) Cleanliness. All persons working in direct contact with food, food-contact surfaces, and food-packaging materials shall conform to hygienic practices while on duty to the extent necessary to protect against contamination of food. The methods for maintaining cleanliness include, but are not limited to: (1) Wearing outer garments suitable to the operation in a manner that protects against the contamination of food, food-contact surfaces, or food-packaging materials. (2) Maintaining adequate personal cleanliness. (3) Washing hands thoroughly (and sanitizing if necessary to protect against contamination with undesirable microorganisms) in an adequate hand-washing facility before starting work, after each absence from the work station, and at any other time when the hands may have become soiled or contaminated. (4) Removing all unsecured jewelry and other objects that might fall into food, equipment, or containers, and removing hand jewelry that cannot be adequately sanitized during periods in which food is manipulated by hand. If such hand jewelry cannot be removed, it may be covered by material which can be maintained in an intact, clean, and sanitary condition and which effectively protects against the contamination by these objects of the food, food-contact surfaces, or food-packaging materials. (5) Maintaining gloves, if they are used in food handling, in an intact, clean, and sanitary condition. The gloves should be of an impermeable material. (6) Wearing, where appropriate, in an effective manner, hair nets, headbands, caps, beard covers, or other effective hair restraints. (7) Storing clothing or other personal belongings in areas other than where food is exposed or where equipment or utensils are washed. (8) Confining the following to areas other than where food may be exposed or where equipment or utensils are washed: eating food, chewing gum, drinking beverages, or using tobacco. (9) Taking any other necessary precautions to protect against contamination of food, food-contact surfaces, or food-packaging materials with microorganisms or foreign substances including, but not limited to, perspiration, hair, cosmetics, tobacco, chemicals, and medicines applied to the skin. (c) Education and training. Personnel responsible for identifying sanitation failures or food contamination should have a background of education or experience, or a combination thereof, to provide a level of competency necessary for production of clean and safe food. Food handlers and supervisors should receive appropriate training in proper food handling techniques and food-protection principles and should be informed of the danger of poor personal hygiene and unsanitary practices. (d) Supervision. Responsibility for assuring compliance by all personnel with all requirements of this part shall be clearly assigned to competent supervisory personnel.

110.20 Plant and Grounds states:

(a) Grounds. The grounds about a food plant under the control of the operator shall be kept in a condition that will protect against the contamination of food. The methods for adequate maintenance of grounds include, but are not limited to: (1) Properly storing equipment, removing litter and waste, and cutting weeds or grass within the immediate vicinity of the plant buildings or structures that may constitute an attractant, breeding place, or harborage for pests. (2) Maintaining roads, yards, and parking lots so that they do not constitute a source of contamination in areas where food is exposed. (3) Adequately draining areas that may contribute contamination to food by seepage, foot-borne filth, or providing a breeding place for pests. (4) Operating systems for waste treatment and disposal in an adequate manner so that they do not constitute a source of contamination in areas where food is exposed. If the plant grounds are bordered by grounds not under the operator's control and not maintained in the manner described in paragraph (a) (1) through (3) of this section, care shall be exercised in the plant by inspection, extermination, or other means to exclude pests, dirt, and filth that may be a source of food contamination. (b) Plant construction and design. Plant buildings and structures shall be suitable in size, construction, and design to facilitate maintenance and sanitary operations for food-manufacturing purposes. The plant and facilities shall: (1) Provide sufficient space for such placement of equipment and storage of materials as is necessary for the maintenance of sanitary operations and the production of safe food. (2) Permit the taking of proper precautions to reduce the potential for contamination of food, food-contact surfaces, or food-packaging materials with microorganisms, chemicals, filth, or other extraneous material. The potential for contamination may be reduced by adequate food safety controls and operating practices or effective design, including the separation of operations in which contamination is likely to occur, by one or more of the following means: location, time, partition, air flow, enclosed systems, or other effective means. (3) Permit the taking of proper precautions to protect food in outdoor bulk fermentation vessels by any effective means, including: (i) Using protective coverings. (ii) Controlling areas over and around the vessels to eliminate harborages for pests. (iii) Checking on a regular basis for pests and pest infestation. (iv) Skimming the fermentation vessels, as necessary. (4) Be constructed in such a manner that floors, walls, and ceilings may be adequately cleaned and kept clean and kept in good repair; that drip or condensate from fixtures, ducts, and pipes does not contaminate food, food-contact surfaces, or food-packaging materials; and that aisles or working spaces are provided between equipment and walls and are adequately unobstructed and of adequate width to permit employees to perform their duties and to protect against contaminating food or food-contact surfaces with clothing or personal contact. (5) Provide adequate lighting in hand-washing areas, dressing and locker rooms, and toilet rooms and in all areas where food is examined, processed, or stored and where equipment or utensils are cleaned; and provide safety-type light bulbs, fixtures, skylights, or other glass suspended over exposed food in any step of preparation or otherwise protect against food contamination in case of glass breakage. (6) Provide adequate ventilation or control equipment to minimize odors and vapors (including steam and noxious fumes) in areas where they may contaminate food; and locate and operate fans and other air-blowing equipment in a manner that minimizes the potential for contaminating food, food-packaging materials, and food-contact surfaces. (7) Provide, where necessary, adequate screening or other protection against pests.

110.35 Sanitary Operations states:

(a) General maintenance. Buildings, fixtures, and other physical facilities of the plant shall be maintained in a sanitary condition and shall be kept in repair sufficient to prevent food from becoming adulterated within the meaning of the act. Cleaning and sanitizing of utensils and equipment shall be conducted in a manner that protects against contamination of food, food-contact surfaces, or food-packaging materials. (b) Substances used in cleaning and sanitizing; storage of toxic materials. (1) Cleaning compounds and sanitizing agents used in cleaning and sanitizing procedures shall be free from undesirable microorganisms and shall be safe and adequate under the conditions of use. Compliance with this requirement may be verified by any effective means including purchase of these substances under a supplier's guarantee or certification, or examination of these substances for contamination. Only the following toxic materials may be used or stored in a plant where food is processed or exposed: (i) Those required to maintain clean and sanitary conditions; (ii) Those necessary for use in laboratory testing procedures; (iii) Those necessary for plant and equipment maintenance and operation; and (iv) Those necessary for use in the plant's operations. (2) Toxic cleaning compounds, sanitizing agents, and pesticide chemicals shall be identified, held, and stored in a manner that protects against contamination of food, food-contact surfaces, or food-packaging materials. All relevant regulations promulgated by other federal, state, and local government agencies for the application, use, holding of these products should be followed. (c) Pest control. No pests shall be allowed in any area of a food plant. Guard or guide dogs may be allowed in some areas of a plant if the presence of the dogs is unlikely to result in contamination of food, food-contact surfaces, or food-packaging materials. Effective measures shall be taken to exclude pests from the processing areas and to protect against the contamination of food on the premises by pests. The use of insecticides or rodenticides is permitted only under precautions and restrictions that will protect against the contamination of food, food-contact surfaces, and food-packaging materials. (d) Sanitation of food-contact surfaces. All food-contact surfaces, including utensils and food-contact surfaces of equipment, shall be cleaned as frequently as necessary to protect against contamination of food. (1) Food-contact surfaces used for manufacturing or holding low-moisture food shall be in a dry, sanitary condition at the time of use. When the surfaces are wet-cleaned, they shall, when necessary, be sanitized and thoroughly dried before subsequent use. (2) In wet processing, when cleaning is necessary to protect against the introduction of microorganisms into food, all food-contact surfaces shall be cleaned and sanitized before use and after any interruption during which the food-contact surfaces may have become contaminated. Where equipment and utensils are used in a continuous production operation, the utensils and food-contact surfaces of the equipment shall be cleaned and sanitized as necessary. (3) Non-food-contact surfaces of equipment used in the operation of food plants should be cleaned as frequently as necessary to protect against contamination of food. (4) Single-service articles (such as utensils intended for one-time use, paper cups, and paper towels) should be stored in appropriate containers and shall be handled, dispensed, used, and disposed of in a manner that protects against contamination of food or food-contact surfaces. (5) Sanitizing agents shall be adequate and safe under conditions of use. Any facility, procedure, or machine is acceptable for cleaning and sanitizing equipment and utensils if it is established that the facility, procedure, or machine will routinely render equipment and utensils clean and provide adequate cleaning and sanitizing treatment. (e) Storage and handling of cleaned portable equipment and utensils. Cleaned and sanitized portable equipment with food-contact surfaces and utensils should be stored in a location and manner that protects food-contact surfaces from contamination.

110.37 Sanitary Facilities and Controls states:

Each plant shall be equipped with adequate sanitary facilities and accommodations including, but not limited to: (a) Water supply. The water supply shall be sufficient for the operations intended and shall be derived from an adequate source. Any water that contacts food or food-contact surfaces shall be safe and of adequate sanitary quality. Running water at a suitable temperature, and under pressure as needed, shall be provided in all areas where required for the processing of food, for the cleaning of equipment, utensils, and food-packaging materials, or for employee sanitary facilities. (b) Plumbing. Plumbing shall be of adequate size and design and adequately installed and maintained to: (1) Carry sufficient quantities of water to required locations throughout the plant. (2) Properly convey sewage and liquid disposable waste from the plant. (3) Avoid constituting a source of contamination to food, water supplies, equipment, or utensils or creating an unsanitary condition. (4) Provide adequate floor drainage in all areas where floors are subject to flooding-type cleaning or where normal operations release or discharge water or other liquid waste on the floor. (5) Provide that there is not backflow from, or cross-connection between, piping systems that discharge waste water or sewage and piping systems that carry water for food or food manufacturing. (c) Sewage disposal. Sewage disposal shall be made into an adequate sewerage system or disposed of through other adequate means. (d) Toilet facilities.

Each plant shall provide its employees with adequate, readily accessible toilet facilities. Compliance with this requirement may be accomplished by: (1) Maintaining the facilities in a sanitary condition. (2) Keeping the facilities in good repair at all times. (3) Providing self-closing doors. (4) Providing doors that do not open into areas where food is exposed to airborne contamination, except where alternate means have been taken to protect against such contamination (such as double doors or positive air-flow systems). (e) Hand-washing facilities. Hand-washing facilities shall be adequate and convenient and be furnished with running water at a suitable temperature. Compliance with this requirement may be accomplished by providing: (1) Hand-washing and, where appropriate, hand-sanitizing facilities at each location in the plant where good sanitary practices require employees to wash and/or sanitize their hands. (2) Effective hand-cleaning and sanitizing preparations. (3) Sanitary towel service or suitable drying devices. (4) Devices or fixtures, such as water control valves, so designed and constructed to protect against recontamination of clean, sanitized hands. (5) Readily understandable signs directing employees handling unprotected food, unprotected food-packaging materials, or food-contact surfaces to wash and, where appropriate, sanitize their hands before they start work, after each absence from post of duty, and when their hands may have become soiled or contaminated. These signs may be posted in the processing room(s) and in all other areas where employees may handle such food, materials, or surfaces. (6) Refuse receptacles that are constructed and maintained in a manner that protects against contamination of food. (f) Rubbish and offal disposal. Rubbish and any offal shall be so conveyed, stored, and disposed of as to minimize the development of odor, minimize the potential for the waste becoming an attractant and harborage or breeding place for pests, and protect against contamination of food, food-contact surfaces, water supplies, and ground surfaces.

110.40 Equipment and Utensils states:

(a) All plant equipment and utensils shall be so designed and of such material and workmanship as to be adequately cleanable, and shall be properly maintained. The design, construction, and use of equipment and utensils shall preclude the adulteration of food with lubricants, fuel, metal fragments, contaminated water, or any other contaminants. All equipment should be so installed and maintained as to facilitate the cleaning of the equipment and of all adjacent spaces. Food-contact surfaces shall be corrosion-resistant when in contact with food. They shall be made of nontoxic materials and designed to withstand the environment of their intended use and the action of food, and, if applicable, cleaning compounds and sanitizing agents. Food-contact surfaces shall be maintained to protect food from being contaminated by any source, including unlawful indirect food additives. (b) Seams on food-contact surfaces shall be smoothly bonded or maintained so as to minimize accumulation of food particles, dirt, and organic matter and thus minimize the opportunity for growth of microorganisms. (c) Equipment that is in the manufacturing or food-handling area and that does not come into contact with food shall be so constructed that it can be kept in a clean condition. (d) Holding, conveying, and manufacturing systems, including gravimetric, pneumatic, closed, and automated systems, shall be of a design and construction that enables them to be maintained in an appropriate sanitary condition. (e) Each freezer and cold storage compartment used to store and hold food capable of supporting growth of microorganisms shall be fitted with an indicating thermometer, temperature-measuring device, or temperature-recording device so installed as to show the temperature accurately within the compartment, and should be fitted with an automatic control for regulating

Good Manufacturing Practices Program

temperature or with an automatic alarm system to indicate a significant temperature change in a manual operation. (f) Instruments and controls used for measuring, regulating, or recording temperatures, pH, acidity, water activity, or other conditions that control or prevent the growth of undesirable microorganisms in food shall be accurate and adequately maintained, and adequate in number for their designated uses. (g) Compressed air or other gases mechanically introduced into food or used to clean food-contact surfaces or equipment shall be treated in such a way that food is not contaminated with unlawful indirect food additives.

110.80 Processes and Controls states:

All operations in the receiving, inspecting, transporting, segregating, preparing, manufacturing, packaging, and storing of food shall be conducted in accordance with adequate sanitation principles. Appropriate quality control operations shall be employed to ensure that food is suitable for human consumption and that food-packaging materials are safe and suitable. Overall sanitation of the plant shall be under the supervision of one or more competent individuals assigned responsibility for this function. All reasonable precautions shall be taken to ensure that production procedures do not contribute contamination from any source. Chemical, microbial, or extraneous material testing procedures shall be used where necessary to identify sanitation failures or possible food contamination. All food that has become contaminated to the extent that it is adulterated within the meaning of the act shall be rejected, or if permissible, treated or processed to eliminate the contamination. (a) Raw materials and other ingredients. (1) Raw materials and other ingredients shall be inspected and segregated or otherwise handled as necessary to ascertain that they are clean and suitable for processing into food and shall be stored under conditions that will protect against contamination and minimize deterioration. Raw materials shall be washed or cleaned as necessary to remove soil or other contamination. Water used for washing, rinsing, or conveying food shall be safe and of adequate sanitary quality. Water may be reused for washing, rinsing, or conveying food if it does not increase the level of contamination of the food. Containers and carriers of raw materials should be inspected on receipt to ensure that their condition has not contributed to the contamination or deterioration of food. (2) Raw materials and other ingredients shall either not contain levels of microorganisms that may produce food poisoning or other disease in humans, or they shall be pasteurized or otherwise treated during manufacturing operations so that they no longer contain levels that would cause the product to be adulterated within the meaning of the act. Compliance with this requirement may be verified by any effective means, including purchasing raw materials and other ingredients under a supplier's guarantee or certification. (3) Raw materials and other ingredients susceptible to contamination with aflatoxin or other natural toxins shall comply with current Food and Drug Administration regulations and action levels for poisonous or deleterious substances before these materials or ingredients are incorporated into finished food. Compliance with this requirement may be accomplished by purchasing raw materials and other ingredients under a supplier's guarantee or certification, or may be verified by analyzing these materials and ingredients for aflatoxins and other natural toxins. (4) Raw materials, other ingredients, and rework susceptible to contamination with pests, undesirable microorganisms, or extraneous material shall comply with applicable Food and Drug Administration regulations and defect action levels for natural or unavoidable defects if a manufacturer wishes to use the materials in manufacturing food. Compliance with this requirement may be verified by any effective means, including purchasing the materials under a supplier's guarantee or certification, or examination of these materials for contamination. (5) Raw materials, other ingredients, and rework shall be held in bulk, or in containers designed and constructed so as to protect against contamination and shall be held at such temperature and relative humidity and in such a manner as to prevent the food from becoming adulterated within the meaning of the act. Material scheduled for rework shall be identified as such. (6) Frozen raw materials and other ingredients shall be kept frozen. If thawing is required prior to use, it shall be done in a manner that prevents the raw materials and other ingredients from becoming adulterated within the meaning of the act. (7) Liquid or dry raw materials and other ingredients received and stored in bulk form shall be held in a manner that protects against contamination. (b) Manufacturing operations. (1) Equipment and utensils and finished food containers shall be maintained in an acceptable condition through appropriate cleaning and sanitizing, as necessary. Insofar as necessary, equipment shall be taken apart for thorough cleaning. (2) All food manufacturing, including packaging and storage, shall be conducted under such conditions and controls as are necessary to minimize the potential for the growth of microorganisms, or for the contamination of food. One way to comply with this requirement is careful monitoring of physical factors such as time, temperature, humidity, AW, pH, pressure, flow rate, and manufacturing operations such as freezing, dehydration, heat processing, acidification, and refrigeration to ensure that mechanical breakdowns, time delays, temperature fluctuations, and other factors do not contribute to the decomposition or contamination of food. (3) Food that can support the rapid growth of undesirable microorganisms, particularly those of public health significance, shall be held in a manner that prevents the food from becoming adulterated within the meaning of the act. Compliance with this requirement may be accomplished by any effective means, including: (i) Maintaining refrigerated foods at 45°F (7.2°C) or below as appropriate for the particular food involved. (ii) Maintaining frozen foods in a frozen state. (iii) Maintaining hot foods at 140°F (60°C) or above. (iv) Heat treating acid or acidified foods to destroy mesophilic microorganisms when those foods are to be held in hermetically sealed containers at ambient temperatures. (4) Measures such as sterilizing, irradiating, pasteurizing, freezing, refrigerating, controlling pH, or controlling AW that are taken to destroy or prevent the growth of undesirable microorganisms, particularly those of public health significance, shall be adequate under the conditions of manufacture, handling, and distribution to prevent food from being adulterated within the meaning of the act. (5) Work-in-process shall be handled in a manner that protects against contamination. (6) Effective measures shall be taken to protect finished food from contamination by raw materials, other ingredients, or refuse. When raw materials, other ingredients, or refuse are unprotected, they shall not be handled simultaneously in a receiving, loading, or shipping area if that handling could result in contaminated food. Food transported by conveyor shall be protected against contamination as necessary. (7) Equipment, containers, and utensils used to convey, hold, or store raw materials, work-in-process, rework, or food shall be constructed, handled, and maintained during manufacturing or storage in a manner that protects against contamination. (8) Effective measures shall be taken to protect against the inclusion of metal or other extraneous material in food. Compliance with this requirement may be accomplished by using sieves, traps, magnets, electronic metal detectors, or other suitable effective means. (9) Food, raw materials, and other ingredients that are adulterated within the meaning of the act shall be disposed of in a manner that protects against the contamination of other food. If the adulterated food is capable of being reconditioned, it shall be reconditioned using a method that has been proven to be effective or it shall be reexamined and found not to be adulterated within the meaning of the act before being incorporated into other food. (10) Mechanical manufacturing steps such as washing, peeling, trimming, cutting, sorting and inspecting, mashing, dewatering, cooling, shredding, extruding, drying, whipping, defatting, and forming shall be performed so as to protect food against contamination. Compliance with this requirement may be accomplished by providing adequate physical protection of food from contaminants that may drip, drain, or be drawn into the food. Protection may be provided by adequate cleaning and sanitizing of all food-contact surfaces, and by using time and temperature controls at and between each manufacturing step. (11) Heat blanching, when required in the preparation of food, should be effected by heating the food to the required temperature, holding it at this temperature for the required time, and then either rapidly cooling the food or passing it to subsequent manufacturing without delay. Thermophilic growth and contamination in blanchers should be minimized by the use of adequate operating temperatures and by periodic cleaning. Where the blanched food is washed prior to filling, water used shall be safe and of adequate sanitary quality. (12) Batters, breading, sauces, gravies, dressings, and other similar preparations shall be treated or maintained in such a manner that they are protected against contamination. Compliance with this requirement may be accomplished by any effective means, including one or more of the following: (i) Using ingredients free of contamination. (ii) Employing adequate heat processes where applicable. (iii) Using adequate time and temperature controls. (iv) Providing adequate physical protection of components from contaminants that may drip, drain, or be drawn into them. (v) Cooling to an adequate temperature during manufacturing. (vi) Disposing of batters at appropriate intervals to protect against the growth of microorganisms. (13) Filling, assembling, packaging, and other operations shall be performed in such a way that the food is protected against contamination. Compliance with this requirement may be accomplished by any effective means, including: (i) Use of a quality control operation in which the critical control points are identified and controlled during manufacturing. (ii) Adequate cleaning and sanitizing of all food-contact surfaces and food containers. (iii) Using materials for food containers and food-packaging materials that are safe and suitable, as defined in Sec. 130.3(d) of this chapter. (iv) Providing physical protection from contamination, particularly airborne contamination. (v) Using sanitary handling procedures. (14) Food such as, but not limited to, dry mixes, nuts, intermediate moisture food, and dehydrated food, that relies on the control of AW for preventing the growth of undesirable microorganisms shall be processed to and maintained at a safe moisture level. Compliance with this requirement may be accomplished by any effective means, including employment of one or more of the following practices: (i) Monitoring the AW of food. (ii) Controlling the soluble solids–water ratio in finished food. (iii) Protecting finished food from moisture pickup, by use of a moisture barrier or by other means, so that the AW of the food does not increase to an unsafe level. (15) Food such as, but not limited to, acid and acidified food, that relies principally on the control of pH for preventing the growth of undesirable microorganisms shall be monitored and maintained at a pH of 4.6 or below. Compliance with this requirement may be accomplished by any effective means, including employment of one or more of the following practices: (i) Monitoring the pH of raw materials, food in process, and finished food. (ii) Controlling the amount of acid or acidified food added to low-acid food. (16) When ice is used in contact with food, it shall be made from water that is safe and of adequate sanitary quality, and shall be used only if it has been manufactured in accordance with current good manufacturing practice as outlined in this part. (17) Food-manufacturing areas and equipment used for manufacturing human food should not be used to manufacture nonhuman food-grade animal feed or inedible products, unless there is no reasonable possibility for the contamination of the human food.

110.93 Warehousing and Distribution states

Storage and transportation of finished food shall be under conditions that will protect food against physical, chemical, and microbial contamination as well as against deterioration of the food and the container.

Finally, 110.110 Natural or Unavoidable Defects in Food for Human Use That Present No Health Hazards states

(a) Some foods, even when produced under current good manufacturing practice, contain natural or unavoidable defects that at low levels are not hazardous to health. The Food and Drug Administration establishes maximum levels for these defects in foods produced under current good manufacturing practice and uses these levels in deciding whether to recommend regulatory action. (b) Defect action levels are established for foods whenever it is necessary and feasible to do so. These levels are subject to change upon the development of new technology or the availability of new information. (c) Compliance with defect action levels does not excuse violation of the requirement in section 402(a)(4) of the act that food not be prepared, packed, or held under unsanitary conditions or the requirements in this part that food manufacturers, distributors, and holders shall observe current good manufacturing practice. Evidence indicating that such a violation exists causes the food to be adulterated within the meaning of the act, even though the amounts of natural or unavoidable defects are lower than the currently established defect action levels. The manufacturer, distributor, and holder of food shall at all times utilize quality control operations that reduce natural or unavoidable defects to the lowest level currently feasible. (d) The mixing of a food containing defects above the current defect action level with another lot of food is not permitted and renders the final food adulterated within the meaning of the act, regardless of the defect level of the final food. (e) A compilation of the current defect action levels for natural or unavoidable defects in food for human use that present no health hazard may be obtained upon request from the Center for Food Safety and Applied Nutrition (HFS-565), Food and Drug Administration, 5100 Paint Branch Pkwy., College Park, MD 20740.

Because the preceding regulatory sections dictate the need for full compliance, the onus is on the manufacturer to interpret them and develop programs and documents that establish fulfillment. To assist the manufacturer in knowing how to fulfill these regulations, it is helpful to correlate the requirement with the program to be developed. The following table establishes this correlation:

Regulation	Description	Program
Section 110.10 part a	Disease control	GMP—employee; GMP—visitor
Section 110.10 part b (1–9)	Cleanliness	GMP—employee
Section 110.10 part c	Education and training	GMP—training
Section 110.10 part d	Supervision	GMP—overview
Section 110.20 part a (1–4)	Grounds	GMP—standards inspection
Section 110.20 part b (1–7)	Plant construction and design	GMP—standards inspection
Section 110.35 part a	General maintenance	GMP—standards inspection
Section 110.35 part b (1–2)	Substances used in cleaning and sanitizing	Sanitation inspection
Section 10.35 part d (5)		

Good Manufacturing Practices Program

Regulation	Description	Program
Section 110.35 part c	Pest control	Pest control
Section 110.35 part d (1–5)	Sanitation	Sanitation inspection
Section 110.35 part e	Storage and handling of cleaned portable equipment and utensils	Inspection
Section 110.37 part a	Water supply	HACCP specification
Section 110.37 part b (1–5)	Plumbing	GMP—standards
Section 110.37 part c	Sewage disposal	GMP—standards
Section 110.37 part d (1–4)	Toilet facilities	GMP—standards
Section 110.37 part e (1–6)	Hand-washing facilities	GMP—standards
Section 110.37 part f	Rubbish and offal disposal	GMP—standards
Section 110.40 parts a–g	Equipment and utensils	GMP—standards inspection
Section 110.40 part f	Instruments and controls	Calibration
Section 110.80 part a (1)	Raw materials and other ingredients	Receiving
Section 110.80 part a (2–3)	Raw materials and other ingredients	Receiving HACCP specification
Section 110.80 part a (4)	Raw materials and other ingredients	HACCP specification
Section 110.80 part a (5–7)	Raw materials and other ingredients	Receiving
Section 110.80 part b (1–5)	Manufacturing operations	Specification
Section 110.80 part b (5–7)	Manufacturing operations	Specification
Section 110.80 part b (8)	Manufacturing operations	Metal detection
Section 110.80 part b (9–14)	Manufacturing operations	Specification GMP—standards
Section 110.10 part b (15–16)	Manufacturing operations	Specification
Section 110.10 part b (17)	Manufacturing operations	Inspection
Section 110.93	Warehousing and distribution	GMP—standards inspection; shipping
Section 110.110	Defect action levels	Specification

The primary program for compliance with the regulations as put forth by 21CFR110 is that of the good manufacturing practices (GMPs) program. Consisting of three fundamental parts—GMP standards, employee GMPs, and visitor GMPs—it forms a foundation for the construction and maintenance of the building and equipment, the determination of how the process equipment is built and maintained, the behavior of employees, and the behavior of visitors. Adherence to and enforcement of these GMPs is an essential component of a company's overall food safety program and must be supported by management in its entirety.

Good manufacturing practices standards are succinct statements that expound on how the company will build and maintain its facility. They deal with topics such as walls, electrical boxes, windows, ceilings, floors, doors, ventilation, unessential items, equipment maintenance, design and installation, stock arrangement, objectionable odors, equipment cleanliness, product protection, hand-washing facilities, toilet facilities, and surroundings. They state specifically what the company does to follow the GMPs in these areas. Because every company is different and manufacturing situations change, the standards are guidelines for employees to use during day-to-day operations and inspections. These standards need to be reviewed by all operational and quality employees prior to signing a document stating that they understand them and that they will bring any issues that are not in compliance to the immediate attention of their supervisor.

The second part of a company's adherence to the federal good manufacturing practices involves the establishment of a written set of standards for employees to follow while they are on company premises. This is a contract with employees that specifies what they will wear; what they will not wear; how they handle the product; their cleanliness, their health, and their personal items; and how they conduct themselves. It should be reviewed with an employee as part of the hiring process with the quality manager and a signed copy should be placed in the employee's personnel file. During this initial review, it should be made clear to the employee that these are federal rules and that failure to adhere to them will result in disciplinary action up to and including possible termination. On an annual basis, these GMPs should be reviewed with the employees so that they are reinforced. This training should be documented and placed in the employee's file.

The final component of an effective GMP strategy involves how the company deals with visitors to the facility. Although most visitors will probably have entered a manufacturing facility previously, it is imperative that, each time a visitor enters, that the GMPs are reviewed with him or her and that the visitor signifies his or her full compliance via signature. Visitors are required to follow a less encompassing set of GMPs, so the company should develop a visitor-specific document for them to review and sign prior to entry each time. These should be kept in a file by date for a period of 1 year.

APPLICATION

GMP Standards

To develop a strong GMP program, the company first needs to establish a set of written standards that outlines the company's commitment to the federal GMP requirements. The forms for sanitation standards 1–14 on the CD (Book 2_Quality Control Manual\Section 3_GMP:Sanitation Standards 1–14) and at the end of this chapter offer the company standards that in total cover all of the designated requirements regarding facility construction and maintenance. These standards include:

Standard 1: Walls/Electrical Boxes/Windows
Standard 2: Ceilings
Standard 3: Floors
Standard 4: Doors
Standard 5: Ventilation
Standard 6: Unessential Items
Standard 7: Equipment Maintenance, Design, and Installation

Good Manufacturing Practices Program

Standard 8: Stock Arrangement
Standard 9: Objectionable Odors
Standard 10: Equipment Cleanliness
Standard 11: Product Protection
Standard 12: Hand-Washing Facilities
Standard 13: Toilet Facilities
Standard 14: Surroundings

Once the standards are established, all employees need to read, understand, and agree to comply with them at all times during their daily duties. The form found on the CD (Book 2_Quality Control Manual\Section 3_GMP:GMP Standards Training Verification) and at the end of this chapter can be used to document this training. After the employee completes the form, it should be placed in the employee's file. When a GMP program is started or an existing one is brought into compliance, it is recommended that management conduct a thorough walk-through using the monthly inspection form to determine compliance with the preceding standards. The initial list of changes to be made will tend to be long, cumbersome, and sometimes expensive, but this should not be a deterrent: The facility must be brought into compliance as the first step to meeting the letter of the law and manufactured food that can be adequately protected and deemed safe for human consumption.

The next step in development of the GMP program is to establish a set of good manufacturing practices for *all* employees to follow. These rules must apply to all employees at the company, including the owner, top management, and office, technical, operations, and sanitation personnel. Management must set the example of adherence to the GMPs for other employees to see as a fundamental part of enforcing the rules and achieving employee buy off. A document for employees to review and sign can be found on the CD (Book 2_Quality Control Manual\Section 3_GMP:Employee GMPs) and at the end of this chapter. It covers the legal basis for the rules and the employee's involvement with disease control, cleanliness, and the plant and grounds with regard to good manufacturing practices. This form should be utilized for the initial employee GMP training; the employee should complete it and then be given a copy and the original placed in the employee's personnel file.

On an annual basis, all employees need to review the GMPs. The document on the CD (Book 2_Quality Control Manual\Section 3_GMP:Employee GMP Annual Review) and at the end of this chapter can be used to document that the annual GMP review has taken place. This should be completed by the employee and placed in the his or her personnel file.

The third step in the development of the GMP program is to develop a set of visitor GMPs. These are usually employee GMPs that have been modified to remove just the portion that applies to the plant and grounds. A template for a set of visitor GMPs to be used for every visitor and nonemployee to document that he or she understands what can and cannot be done when visiting the facility can be found on the CD (Book 2_Quality Control Manual\Section 3_GMP:Visitor GMPs) and at the end of this chapter.

To support the implementation of the GMP standards, the employee GMPs, and the visitor GMPs, it is important to establish an effective GMP auditing program. A typical GMP auditing program has two parts: daily auditing and monthly auditing. Daily auditing is conducted as part of the regular quality control line checks and adherence to the program documented on a line sheet such as the daily line check form. (CD supplemental forms: line check sheet and at the end of this chapter.) Responsibility for daily auditing usually is left to the quality control staff. Monthly GMP auditing is conducted as part of the inspection program and responsibility falls on the designated inspection team.

Finally, when all the components of the good manufacturing practices program are in place, the next step is to write the program document. This usually takes the form of a general statement referring to the regulatory statutes that pertain to GMPs and how the program applies. This is followed by sections on employee GMPs, the employee GMP yearly review, visitor GMPs, sanitation standards, GMP auditing, and program responsibility.

SUPPLEMENTAL MATERIALS

DAILY LINE CHECK SHEET

Julian Date: _____
Date:

TIME

GMP'S
Hairnets, gloves, jewelry, clothing

ITEM
Type

Brand

LINE
Metal
  product protected
Coverings
  product covered
Floor
  no excessive waste
  garbage taken out
Metal detector in line
Product Insp. Position 1
Product Insp. Position 2
Product Insp. Position 3
Product Insp. Position 4
Product Insp. Position 5

PRODUCT
Packaging

Correct case
Correct bag/film
Kosher sticker
Vendor codes marked out

Correct label
Free Flow rate
Code

  side 1

  side 2
Pallet
  configuration
  stacked straight

Code

INSPECTOR
SUPERVISOR

SANITATION STANDARD 1

WALLS/ELECTRICAL BOXES/WINDOWS

Section A: General

1. All walls should be constructed of a smooth, nontoxic, easily cleanable surface that is impervious to water, sanitizers, etc.
2. Walls should be free of cracks and holes, which may harbor pests and reduce cleaning efficiencies.
3. Walls should be free of dust, dirt, product accumulation, and flaking paint.
4. Walls should be sealed at the wall–floor juncture.
5. All walls should be sealed around the openings through which equipment, pipes, or other items pass.
6. All walls should be free of dust, dirt, and food accumulation.
7. All windows should be in good repair, with no broken panes of glass.
8. All windows should be clean and free from dust, dirt, and mold.
9. All electrical boxes mounted to walls should be in good repair, free of rust and flaking paint, and mounted flush and caulked.
10. Electrical boxes should be free of dust, dirt, and food accumulation inside and out.
11. Electrical boxes should be in good repair with no rust or flaking paint and should be tightly sealed to prevent insects from entering.

Section B: Responsibility

Monitoring: Operations and Quality Control
Enforcement: Operations

SANITATION STANDARD 2

CEILINGS

Section A: General

1. All ceilings should be constructed of a smooth, nonabsorbent, easily cleanable material.
2. All ceiling surfaces, as well as overhead installations or structures including ventilation units, light fixtures, electrical raceways, piping, etc., should be clean and free of product buildup, dust, mold, webbing, rust, and peeling or flaking paint.
3. All ceiling surfaces and overhead installations should be free of condensation.
4. There should be no evidence of or active water leaks on ceilings.
5. Insulation material used on overhead lines should be in good repair, smooth, nonabsorbent, and easily cleanable.
6. Nails, staples, or screws should not be used to secure ceiling material in processing, ingredient, or packaging areas.
7. No strings, ropes, wires, or tapes are to be used as pipe or line supports.

Section B: Responsibility

Monitoring: Operations and Quality Control
Enforcement: Operations

SANITATION STANDARD 3

FLOORS

Section A: General

1. Floors should be well drained, smooth, and clean, with no open cracks, holes, or broken areas.
2. All perimeters should have an uncluttered 18-inch wide strip to assure cleanliness, allow proper inspections, and comply with fire regulations.
3. Floors should be maintained in a clean condition with no excessive accumulations or static buildup of product.
4. Concrete floors should not be cracked or eroded, which causes absorption and unclean surfaces.
5. There should be no holes, missing concrete, or low spots causing standing water.
6. Drains should have traps and drain covers and should be maintained in clean condition, with care taken to keep water in the trap and prevent objectionable odors.
7. All cracks at the floor–wall junctures must be sealed to prevent product buildup and possible insect harborage.

Section B: Responsibility

Monitoring: Operations and Quality Control
Enforcement: Operations

SANITATION STANDARD 4

Doors

Section A: General

1. All doors should be tight fitting, clean, and in good repair.
2. Doors should be self-closing if open to the outside.
3. If the door is open, it should be equipped with an effective air curtain or strip door.
4. Doors should be tight fitting at the bottom to preclude the entrance of insects and rodents.
5. All holes in hollow wood or metal doors must be sealed to prevent harborage inside the door.

Section B: Responsibility

Monitoring: Operations and Quality Control
Enforcement: Operations

SANITATION STANDARD 5

VENTILATION

Section A: General

1. Adequate ventilation in the processing area should be provided to maintain proper environmental and sanitary conditions for equipment, ingredients, finished goods, and packaging materials.
2. All systems should be cleanable, function properly, and be designed in a manner to prevent product contamination from condensation, mold, bacteria, insects, dust, and odors.
3. Objectionable odors, fumes, or vapors should not be present.
4. All fans, fan guards, duct work, louvers, and heat/air conditioning units should be clean and in good repair.
5. All ceiling-mounted fans or vents should have properly functioning, self-closing louvers and should be screened to prevent insect entry.
6. All critical processing areas should be maintained under positive air pressure to prevent dust, flying insect entry, and cross-contamination.
7. All filters and screens used in ventilation systems should be routinely inspected and replaced or cleaned as necessary.
8. All toilet and locker room facilities should be vented mechanically to the outer air.
9. Condensation should not be present on walls or ceilings.

Section B: Responsibility

Monitoring: Operations and Quality Control
Enforcement: Operations

SANITATION STANDARD 6

UNESSENTIAL ITEMS

Section A: General

1. Unessential items are items that, when not stored in their proper places, contribute to poor housekeeping and pest control problems. These should be properly managed and stored.
2. Brooms, squeegees, and mops should have hangers and should be kept off the floor.
3. Small articles such as pencils, knives, fuses, and tools should not be left in or on electrical panels, switchboards, processing equipment, etc.
4. Idle equipment should be removed from processing areas and stored in a clean fashion.
5. Obsolete equipment should not be allowed to accumulate to the point of causing housekeeping or pest control problems.

Section B: Responsibility

Monitoring: Operations and Quality Control
Enforcement: Operations

Good Manufacturing Practices Program   139

SANITATION STANDARD 7

EQUIPMENT MAINTENANCE, DESIGN, AND INSTALLATION

Section A: General

1. All equipment should be designed in a manner to fulfill its use effectively and efficiently while protecting the product from contamination.
2. All equipment should be of food-grade, smooth, impervious, nontoxic, nonabsorbent, and corrosion-resistant material where it has direct product contact.
3. All non-food-grade materials, such as wire, tape, string, wood, and cardboard, should not be used as a temporary design or repair.
4. All equipment should be free of rough surfaces and cracked walls where products may become static and make cleaning difficult.
5. All equipment should be free of oil leaks and excessive grease buildup on bearings and motor housing where it may contaminate product.
6. Equipment should be constructed in a manner to preclude metal-to-metal contact.
7. Appropriate lids/covers should be provided to protect the product from contamination.
8. All food equipment should be installed in a manner to provide easy access for cleaning and inspections.
9. Equipment should be free of flaking paint and rust.
10. Conveyor belts should be free of loose string, metal clips, and loose pieces of rubber.
11. All processing equipment should be free of leaks at valves, gaskets, fittings, etc.
12. Thermometers, recording charts, and pressure gauges should be provided where applicable. They must be in good working order and easily accessible.
13. Equipment should be designed to preclude or divert condensation away from products and product contact surfaces.
14. All equipment should be designed to eliminate areas where products may become static.
15. All grease, oil, other lubricants, and cleaning supplies must be acceptable for food use (food grade) by the FDA/USDA.

Section B: Responsibility

Monitoring: Operations and Quality Control
Enforcement: Operations

SANITATION STANDARD 8

STOCK ARRANGEMENT

Section A: General

1. All stock should be stored 18 inches from the wall and 5 inches off the floor.
2. All stock should be stored in a clean and dry environment.
3. Finished goods, raw materials, and quarantined damaged goods should be stored separately.
4. All stock should be stored in an orderly manner and properly stacked to prevent damage.
5. All partially used ingredient or packaging containers should be properly resealed or covered to preclude contamination.
6. All pallets, racks, and shelving should be clean and in good repair.
7. All damaged finished products should be disposed of in a timely manner.
8. All stock should be rotated on a first in, first out (FIFO) basis.
9. Finished products, packaging material, equipment, or ingredients should not be stored in close proximity to any chemicals, cleaning compounds, pesticides, or odorous materials.

Section B: Responsibility

Monitoring: Operations and Quality Control
Enforcement: Operations

SANITATION STANDARD 9

OBJECTIONABLE ODORS

Section A: General

1. No objectionable or abnormal odors should be present.
2. Rotten or spoiled food should be removed as they can result in objectionable odors and create other hazards.
3. All strong chemical odors, which could contaminate food, should be prevented.

Section B: Responsibility

Monitoring: Operations and Quality Control
Enforcement: Operations

SANITATION STANDARD 10

EQUIPMENT CLEANLINESS

Section A: General

1. All processing equipment should be maintained in a clean and sanitary manner.
2. Each piece of equipment should have written procedures for complete breakdown and cleaning.
3. All equipment/utensils should be stored in a sanitary manner when not in use.
4. All equipment should be cleaned and sanitized on a regular schedule that ensures the wholesomeness of the product by reducing bacterial loads.
5. All cleaning procedures should be followed as outlined in the cleaning manual.
6. Valves, gaskets, and pumps should be maintained in a clean and sanitary manner.
7. All reusable ingredient containers should be thoroughly cleaned before reuse.
8. No steel wool or metal sponges should be used in the cleaning of equipment.

Section B: Responsibility

Monitoring: Operations and Quality Control
Enforcement: Operations

SANITATION STANDARD 11

PRODUCT PROTECTION

Section A: General

1. All ingredients, packaging material, and finished products should be handled, stored, or processed in such a manner as to assure a safe, wholesome, and unadulterated product.
2. All pesticides should be stored in a locked area and separated from all ingredients, cleaning material, equipment/utensils, and sanitizers.
3. All sanitizers, cleaning compounds, and chemicals should be stored separately from all ingredients, packaging material, and finished products in a manner to prevent any contamination.
4. There should be no flaking or peeling paint, static product, soil buildup, or rust on or above product zones.
5. All products or product containers should be adequately protected to preclude contamination.
6. There should be no condensation above product zones.
7. All conveyor belts in direct product contact should be constructed of a nontoxic and nonabsorbent material and should be clean and in good repair.
8. Lights should be provided with adequate shielding to protect from breakage.
9. Wood, cardboard, or other absorbent materials should not be used for product contact surfaces.
10. Air used for conveying product or packaged materials should be filtered, clean, and oil free.
11. Glass should not be used in, above, or near processing or ingredient areas.
12. All gasket material should be nontoxic, nonabsorbent, and in good condition (not torn, frayed, or deteriorated).
13. Pressure cleaning/sanitizing should not be conducted near, on, or above product zones during processing or packaging.
14. All tanks, vats, and blenders should be covered.
15. All product piping, pumps, and other equipment should be capped or closed when not in use and stored in a sanitary manner.
16. All partially used or damaged ingredient containers should be properly resealed in a manner to protect against contamination.
17. Product containers should not be used for purposes other than their intended use.
18. All product and ingredient containers should be covered and properly identified.
19. All tanks, vats, and kettles should be self-draining.
20. All bearings should be properly protected, designed, and/or constructed so that no lubricant can leak, drip, or be forced into the product zone.
21. A program should be developed and maintained that provides adequate means of eliminating foreign material, metal, etc. in finished products (e.g., metal detectors, in-line filters, screens, magnets).

Section B: Responsibility

Monitoring: Operations and Quality Control
Enforcement: Operations

SANITATION STANDARD 12

Hand-Washing Facilities

Section A: General

1. An adequate number of hand-washing stations should be provided within the processing area.
2. All hand-washing stations should be maintained in a clean condition.
3. Single-service towels, tempered water, dispenser soap, and proper disposal should be provided at each hand-washing station.
4. A sign at each hand-washing station that instructs employees to wash their hands prior to returning to work should be conspicuously posted.
5. Hand-sanitizing stations should be provided where deemed necessary.

Section B: Responsibility

Monitoring: Operations and Quality Control
Enforcement: Operations

SANITATION STANDARD 13

TOILET FACILITIES

Section A: General

1. Employees should be provided with clean, sanitary, properly functioning toilet and hand-washing facilities.
2. All toilets, urinals, and hand-washing facilities should be clean and properly functioning.
3. Hand-washing stations should have sufficient pressures of both hot and cold water.
4. Hand-washing soap, in a suitable dispenser, must be provided at each basin.
5. An adequate supply of toilet tissue and single-service towels should be provided.
6. Doors to toilet facilities should be self-closing and should not open directly into processing, ingredient, or packaging areas.
7. There should be no clogged drains or overflowing toilets.
8. Legible signs should be posted conspicuously in all toilet facilities directing employees to wash their hands before returning to work.
9. All walls, ceilings, and floors should be clean and in good repair.
10. Waste containers should be provided for used towels or other waste and should have self-closing covers.
11. All toilet facilities should be mechanically ventilated to outer air.

Section B: Responsibility

Monitoring: Operations and Quality Control
Enforcement: Operations

Good Manufacturing Practices Program

SANITATION STANDARD 14

SURROUNDINGS

Section A: General

1. The growth of weeds and tall grass along the building perimeter should be arrested.
2. The surrounding premises should be free of standing water or other drainage problems. All storm sewers and catch basins should be kept clear and operative.
3. The area should be free of droppings, tunneling, or any other signs of rodent activity. Rodent traps should be maintained in an orderly fashion.
4. There should be no apparent signs of pest activity immediately surrounding the building.
5. All roads, lots, and yards should be adequately paved and maintained.
6. All scrap, pallet, equipment, etc., should be stored off the ground and away from building walls.
7. Excessive amounts of scrap or idle equipment should be disposed of to reduce pest activity.
8. Any outside waste compactor/disposal area should be clean and properly maintained.
9. Suitable covers should be provided for outside waste containers. The surrounding area should be free of paper, trash, and litter.
10. The areas under and around loading docks are to be kept clean and uncluttered.
11. The building roof should be free of standing water, product dust, accumulated filth, and unessential materials.
12. There should be no signs of pest or rodent activity on the building roof.

Section B: Responsibility

Monitoring: Operations and Quality Control
Enforcement: Operations

VERIFICATION DOCUMENTATION OF GMP STANDARDS TRAINING

I, _____, have read and understand the 14 good manufacturing standards for plant construction and maintenance. I will comply with the standards in the course of my daily duties and will notify my immediate supervisor if I see an area of noncompliance inside or outside the plant or with any machinery. If I have questions, I will discuss them with my supervisor.

Date: _____

Name: _____

Instructor: _____

GOOD MANUFACTURING RULES
21CFR110

Section A: General

The federal government requires that all manufacturers of food products comply with the good manufacturing practices (GMPs) (21CRF110). The purpose of these rules is to ensure that all products that are produced at _____ are safe for human consumption. Therefore, as a condition of employment, all company employees should abide by the following sanitation rules.

Section B: 110.10 Personnel

110.10.a Disease Control

Employees should take all reasonable measures and precautions to prevent the spread of disease and to ensure general personal cleanliness.
1. If you have a boil, open sore, or an infected area on your hands, arms, or face, notify your supervisor **before** starting work.
2. If you feel ill, notify your supervisor **before** starting work.

Section C

110.10.b Cleanliness

Employees should keep their bodies, clothes, and areas clean.
1. No sleeveless shirts, shorts, cutoffs, tank tops, or open-toed shoes should be worn in the facility. When appropriate or required, outer garments, such as aprons, should be worn.
2. Good personal hygiene should be practiced. Hands must be washed after contact with any part of the face, nose, or mouth.
3. Hands should be washed before starting work and before returning to work from breaks, meal periods, and using the restroom. The sink in the production area is to be used in addition to the restroom sink.
4. All jewelry, including watches, necklaces, chains, bracelets, exposed piercings, earrings, and rings, should be removed before entering the production area.
5. All employees that come in contact with the product or food contact surfaces should wear gloves. They should keep these clean and change them when necessary.
6. A hair net must be worn at all times, by all personnel in the processing area, and should cover all exposed hair. Hats, bandannas, shirts, or other types of hair restraints are not allowed. Beard nets should be worn to cover beards and mustaches of any length.
7. All personal belongings, food for consumption, and beverages should be stored away from the production area (i.e., in the locker room or lunch room).
8. No food, drink, gum, or candy is allowed in the processing area.

9. Smoking is permitted only in designated areas and never in the processing area. Chewing tobacco is not permitted in the production facility.
10. Containers are not to be used for anything other than the product.
11. No employee should store pencils, pens, tools, cigarettes, gum, etc. in his or her breast pockets.
12. No fake fingernails or fingernail polish should be worn in any processing area.

Section D

110.20 Plant and Grounds

1. All employees should keep their areas and the plant clean and free from clutter.
2. All products should be covered when not being processed (e.g., during breaks, meal periods, in storage).
3. Scoops, utensils, and parts of equipment are not to be placed on the floor. When not in use, they should be cleaned thoroughly and stored in their proper place.

I have read the above rules and understand them fully. I also understand that failure to follow them will result in disciplinary action up to and including immediate termination.

_____   _____
Employee Signature   Date

_____   _____
Supervisor Signature   Date

VERIFICATION DOCUMENTATION OF
EMPLOYEE GMP ANNUAL REVIEW

I, _____, have read and understand the good manufacturing standards for employees at _____. I will comply with the standards in the course of my daily duties and will notify my immediate supervisor if I see an area of noncompliance inside or outside the plant, with any machinery, or with any other employee. If I have questions, I will discuss them with my supervisor.

Date: _____

Name: _____

Instructor: _____

GOOD MANUFACTURING RULES: VISITORS
21CFR110

Section A: General

The federal government requires that all manufacturers of food products comply with the good manufacturing practices (GMPs) (21CRF110). The purpose of these rules is to ensure that all products that are produced at _____ are safe for human consumption. Therefore, as a condition of visitation, all visitors should abide by the following sanitation rules.

Section B: 110.10 Personnel

110.10.a Disease Control

Visitors should take all reasonable measures and precautions to prevent the spread of disease and to ensure general personal cleanliness.
1. If you have a boil, open sore, or an infected area on your hands, arms, or face, notify your company guide **before** entering the processing area.
2. If you feel ill, notify your company guide **before** entering the processing area.

Section C

110.10.b Cleanliness

Visitors should enter the facility only if their clothes and bodies are clean.
1. No sleeveless shirts, shorts, cutoffs, tank tops, or open-toed shoes should be worn in the facility. When appropriate or required, outer garments, such as aprons, should be worn.
2. Good personal hygiene should be practiced. Hands must be washed after contact with any part of the face, nose, or mouth.
3. Hands are to be washed and sanitized before entering the facility.
4. All jewelry, including watches, necklaces, chains, bracelets, exposed piercings, earrings, and rings, should be removed before entering the production area.
5. All visitors that come in contact with the product or food contact surfaces should wear gloves. They should keep these clean and change them when necessary.
6. A hair net must be worn at all times, by all visitors in the processing area, and should cover all exposed hair. Hats, bandannas, shirts, or other types of hair restraints are not allowed. Beard nets should be worn to cover beards and mustaches of any length.
7. All personal belongings, food for consumption, and beverages should be stored away from the production area (i.e., in the locker room or lunch room).
8. No food, drink, gum, or candy is allowed in the processing area.

9. Smoking is permitted only in designated areas and never in the processing area. Chewing tobacco is not permitted in the production facility.
10. Containers are not to be used for anything other than the product.
11. No visitor should store pencils, pens, tools, cigarettes, gum, etc. in his or her breast pockets.

I have read the above rules and understand them fully. I also understand that failure to follow them will result in refusal to allow entrance into the production facility.

_____   _____
Visitor Signature   Date

_____   _____
Company Guide Signature   Date

GMP AUDIT PROCEDURE/SCHEDULE

Section A   **General**

As a means of verifying that all GMP's are being followed at all times during manufacturing, Quality Control will do an hourly GMP audit.

B   **Procedure**

Using the Daily Line Check form, the quality control technician will observe all production employees and their surrounding production areas to verify that there is complete compliance to the Good Manufacturing Practices.

    Non-conformance shall be detailed on the inspection form and the employees' supervisor notified.

C   **Corrective Action**

Employees that are found to be out of compliance will be notified and the non-compliance corrected. Repeat offenses will result in disciplinary action up to and including termination.

D   **Responsibility**

    Monitoring: Quality Control
    Enforcement: All Management

Approved by: _____ Date: _____

GOOD MANUFACTURING PROGRAM
General Overview

Section A: General

_____'s good manufacturing program is based on the federal government's food manufacturing guidelines as set forth in 21CFR110. This program entails three parts: employee GMPs, visitor GMPs, and manufacturing sanitation standards.

Section B: Employee GMPs

Each employee of _____ should read and understand the good manufacturing rules. As a part of _____'s new employee training, the quality control manager will review each section with the employee and answer all questions. Furthermore, the employee should understand that complete adherence to _____'s GMP policy is required and failure to do so will lead to termination. As a condition of employment, the employee will sign and date the GMP rules and a copy will be given to him or her and one placed in the employee's file.

Section C: Employee GMP Yearly Review

_____ will conduct yearly GMP reviews with all employees. This will be documented and placed in the employee files.

Section D: Visitor GMPs

Upon entrance to _____, each visitor should read, understand, and sign _____'s GMP policy. These forms will be placed in a file for a period of 30 days. Each time the visitor enters the facility, he or she will be required to sign a new document.

Section E: Sanitation Standards

All facility construction, equipment construction, maintenance, and facility management should be conducted in a clean and sanitary manner. Sanitation guidelines for various parts of the company should be set forth in documented sanitation standards.

Section F: GMP Auditing

Employee GMPs should be audited as part of the operational quality checks and sanitation standards should be audited as part of the internal inspection program.

Section G: Responsibility

Responsibility for enforcement of this program lies equally between quality control and operations.

Approved by: _____ Date: _____

8 Pest Control Program

PROGRAM TYPE: REQUIRED

THEORY

Black flies, fruit flies, silverfish, gnats, mosquitoes, meal moths, cockroaches, mice, rats, snakes, birds, and numerous other vermin and insects are just some of the pest control challenges facing the modern quality manager. Each creature may pose a serious risk to the safety of the food, either by direct infestation or through indirect contamination. The Food and Drug Administration deems the risk of contamination from pests to be so important that it has dedicated several sections of 21CFR110 to the current good manufacturing practice in manufacturing, packing, or holding human food part of the Code of Federal Regulations. For example, 21CFR110.20(a) states:

> The grounds about a food plant under the control of the operator shall be kept in a condition that will protect against the contamination of food. The methods for adequate maintenance of grounds include, but are not limited to (1) properly storing equipment, removing litter and waste, and cutting weeds or grass within the immediate vicinity of the plant buildings or structures that may constitute an attractant, breeding place or harborage for pests.

Later, that same section states:

> If the plant grounds are bordered by grounds not under the operator's control and not maintained in the manner described in paragraph (a) (1) through (3) of this section, care shall be exercised in the plant by inspection, extermination, or other means to exclude pests, dirt, and filth that may be a source of food contamination.

Section 110.35 (c) specifically addresses pest control in a food facility:

> Pest control. No pests shall be allowed in any area of a food plant.... Effective measures shall be taken to exclude pests from the processing areas and to protect against contamination of food on the premises by pests. The use of insecticides or rodenticides is permitted only under precautions and restrictions that will protect against the contamination of food, food-contact surfaces, and food-packaging materials.

Although there are many tomes related to the identification of and specific methods used to control pests, proper plant pest control relies on the establishment of a complete systematic approach through the implementation of monitoring points, monitoring these control points, and the periodic verification of effectiveness. Due to the importance of this program and the scientific and specific nature of it, an outside pest control provider *must be* consulted and hired before implementing monitoring points. These contractors provide an understanding of the biology and ecology of the available pests, as well as methods of exclusion and eradication, and are licensed and trained. They should always seek a permanent solution to pest problems rather than a series of temporary solutions; they should use principles of integrated pest prevention and pest control methods that have the least impact on the environment and nontarget organisms. An integrated pest prevention program employs multiple control measures, including sanitation, mechanical control, cultural control, biological control, and chemical control.

Sanitation

Not only is proper sanitation necessary for prevention, but it is also an effective way to control and eliminate infestations. The removal of food and water sources stresses populations, making traps and baits more effective. Residual oils and greases also render many insecticides ineffective. This goes hand in hand with the plant sanitation, training, and inspection programs. It is important to design the plant sanitation program so that all cracks, joints, crevices, and hidden surfaces are cleaned and inspected routinely. If this is not done, then buildup occurs and pests are attracted.

Mechanical Control

These methods involve the use of traps (mechanical and nonmechanical), barriers (caulking, strips, seals, foam), and mechanical exclusion (screens, bird wire, strip curtains), as well as using air currents and manipulation of environmental factors (temperature, humidity). Typical traps are designed to allow the pest to enter and be detained or to feed on a tainted food substance and return to its habitat. The most common mechanical rodent catching traps are the low-profile Tin Cat and Ketch-alls. Nonmechanical insect control involves using either VECTORs (wall stations that utilize a blue light to attract insects to a glue board) or Insectecutors (wall units that utilize a blue light to attract insects and then electrocute them when they come in contact with an electrically charged set of metal rods). Barriers, usually of a pliable material, foam, or strip material, fill door–floor, wall–floor, and other joints. Rubber strips fill other oddly shaped holes and cracks. Canned foam is a unique addition to the pest control applicator's arsenal. Found at most local hardware stores, canned foam can be sprayed into any opening and, when released from the pressure, expands to many times its initial size, thus filling the hole or crack. This product should not be used as a permanent maintenance tool or as a temporary tool for filling cracks or holes leading to the outside. Another barrier that can be used it that of plastic strip curtains to create "compartments" within the facility. They help to prevent the unfettered movement of pests throughout the facility because they act as semipermanent doors.

In conjunction with the Ketch-alls located in the interior of the facility, the company must establish an 18-inch barrier between the wall and any racking or other materials. This barrier should be painted white to allow the pest control inspector to see rodent droppings easily. At 18 inches, the barrier is wide enough for inspectors to have easy access for traveling between the traps as well as access to the back of pallets for inspection and inventory purposes. All exterior and interior walls should have this barrier.

CULTURAL CONTROL

This involves changing the habits or behaviors of employees and visitors. Too often, visitors who come to the plant—whether seasoned food products professionals, first time visitors or, even company employees—have a varying degree of pest control knowledge. This imbalance can often lead to routes of pest entry through open doors, dropped food, or poor cleaning practices.

One way to prevent this contamination is to institute programs and procedures for visitors and employees to follow that assist in the exclusion of pests and change the cleaning schedules to prevent waste accumulations. These procedures should be part of the good manufacturing process (GMP) rules that employees and visitors sign.

BIOLOGICAL CONTROL

These methods utilize biological organisms or their byproducts to control pests. This includes the use of parasitic wasps for fly control, bacteria-based products for fruit fly control, and pheromone traps for insect collection and identification. These are complicated strategies for control of pests and generally should be left to licensed professionals.

CHEMICAL CONTROL

This involves the correct, effective, and safe use of pesticides for controlling insects and pests. Only chemicals that are not harmful to the environment and application techniques that are target specific should be used. In most states, only a licensed professional may apply pesticides.

Depending on how large the facility is, what pests are in the area, and the knowledge level of the quality control manager, the contract pest control provider should provide any and/or all of the following services:

plan development;
plan implementation;
periodic evaluation;
accurate record keeping; and
pest elimination.

Hiring a contract pest control provider involves requesting proposals from local providers, typically found in the local phone book, and national providers such as Steritech, Sprague, and Terminex, which are found in the phone book or online. Before hiring the contract pest control provider, each prospective company should be interviewed and a proposal received that outlines the type of program, its validation procedures and timelines, and initial and ongoing costs associated with the program. In anticipation of the proposal, an initial meeting to discuss the proposed program and evaluate the facility should take place. This involves inspecting the facility and any specific areas of concern; preparing a basic diagram of the facility for use in preparing the proposal; determining the state of the current pest control program, if any; and interviewing the quality control manager to determine the level of involvement the company will have during the development and implementation phase. Suggested topics for this initial meeting are past or present pest problems, methods of control, responsibility for the program, and company expectations.

It is important that during this initial meeting the company representative discloses any knowledge of prior or current pest problems. This helps to direct the potential pest control provider in developing the program. For instance, in Colorado during May and June the Miller moths migrate from the plains states to the Rocky Mountains to mate. They are not prevalent during any other time of year, so it is important to disclose this at off times of the year to the potential pest control provider so that the provider may plan accordingly.

The methods of control to be used are important to discuss to ensure that the program established is comprehensive, effective, and does not pose any food safety issues. Some questions that should be considered are: (1) Will the program be strictly organic or nonorganic? (2) Will all three areas of basic pest control—rodents, insects, and birds—be addressed within the program? (3) What types of equipment will be used (e.g., Tin Cats, Ketch-alls, or both? VECTORs or electrocutors?) (4) Will there be spraying or no spraying? (5) Will there be targeted control applications, what chemicals if any will be used, and will this be applied inside or outside? Whether the program will be organic or nonorganic is important because this fact determines the types of chemicals and bait to be used. As these questions are considered and answered in conjunction with the contract pest control company, food safety must be considered. For example, periodic external spraying for ants or spiders may be necessary in some areas but periodic internal spraying may pose food contamination issues. Care must be taken never to sacrifice food safety when choosing the proper pest control measures.

Finally, responsibility for the program's development, implementation, and enforcement is important. Typically, the company and provider will cooperatively establish the system; the provider will implement the program and enforcement will be split between both parties, with the provider offering a periodic service (normally, monthly) and the company conducting bimonthly or weekly inspections. Sometimes the provider will offer to provide a periodic service and the company will be tempted to allow it to enforce the program unilaterally. This is like the fox watching the hen house. It is important for the company to participate in program enforcement to ensure the provider's compliance to the initial contract and as a method of verifying that the provider is performing the service was hired to provide in its entirety.

The basis of developing a strong and effective integrated pest control system relies on pest exclusion. This takes two forms: perimeter exclusion and internal exclusion. Perimeter exclusion involves inspecting the outside building perimeter specifically looking for entrance points. There are many pests that can invade the facility; once inside, they use a variety of channels to remain undetected. Common entrance points are cracks between walls, floors, and ceilings; unscreened vents on the roof; outside doors; dock doors; and cracks and seams in the floor and ceiling.

Pest Control Program

Internal exclusion includes eliminating food buildup and clutter; proper maintenance of doors, walls, and other exclusion devices; and proper sanitation procedures. The fundamentals of pest control should be considered during the design and implementation of the preventative maintenance program, the employee training program, and the facility inspection program.

Once the plan is developed, a contract is signed with the pest control provider and the implementation phase begins. There are three parts to this phase: placement, document development, and assignment of responsibility. During the previous phase, a map of the facility is drawn that includes the basic wall structure, doors, warehouse racking, and equipment. This does not have to be a very detailed drawing, but it must be clear enough to act as a guide for trap placement and program verification. Next, based on the most effective method of control, determine where the mechanical control units are to be placed. It is recommended that a Ketch-all or Tin Cat be placed on both sides of the interior of any exterior door, both faces and both sides of interior nonoffice doors, both sides of freezer and cooler access, corners of interior warehouse walls, and every 50 feet on interior-facing walls. Because mice move along walls, this offers the best chance to catch them if they are present.

Bait stations should be placed on the outside of the facility within 2 feet of each exterior door and every 50 feet on exterior walls. The bait stations are to be attached directly to the ground or attached to a block and then attached to the ground. One method commonly used is to use liquid nails to attach the bait station to a 1-inch block of cement and then attach the block to the ground. The bait used inside the stations will depend on the type of program installed. If the program is generic, then chemical- or poison-laced bait will be used. If it is an organic program, then the bait should not be laced with poison and used as just an indicator of feeding. The specific type of bait is dependent on the contract pest control provider and can only be monitored and replaced by this provider because the bait stations are locked and the provider has the key. Also, laws and regulations dictate that only licensed applicators may apply bait.

Insect-attracting units should be placed strategically in the facility; they should be near exterior doors but not facing them so that they do not draw insects into the facility. Insectecutors should be located in areas where there are no open food products or packaging; VECTORs should be located in the more sensitive areas. This prevents insect parts from contaminating food or packaging.

Both the contract pest control provider and the company should be involved in determining the location of the rodent traps, bait stations, and insect-attracting units. Once the locations of these are determined and diagrammed on the map of the facility, the contract pest control provider will begin installing the units. For each unit, the contract pest control provider should install a numbered sign above its location. This number should match the numbered location on the map and also the number written on the top of the respective bait station, trap, or insect attractor.

Once the placement is complete, the documentation of the program is developed and organized. This involves the creation of two notebooks: Pest Control 1 and Pest Control 2. These notebooks should be clearly labeled on both the cover and the spine. Pest Control 1 houses the static documents and should always be stored in the quality control office with the other quality control manuals. The Pest Control 1 notebook is separated into the following seven clearly labeled sections: contract, company license, company insurance, labels, MSDSs (material safety data sheets), map, and applicator's license. In section 1 a copy of the contract between the pest control provider and the company is placed. This acts as the basis for the entire pest control program; it outlines the physical and monetary agreement between the two companies. In the next section, a copy of the contract pest control company's pest control applicator's license is placed. It is important the expiration of this document is verified and that it is current. The contract pest control provider should also provide a copy of its current insurance policy statement that indicates the level of coverage in case someone is hurt or products are damaged by any action its employees make take. This is to be placed in the company insurance section. For both the company license and the company insurance, the expiration date should be highlighted for quick reference.

The next two sections are specific to the agreed upon pesticides to be used for insect, bird, or rodent control. This includes the bait for the bait stations and any chemicals to be sprayed. For each of these, a copy of the label from the container and an MSDS should be obtained. Place the chemical labels in their section and the MSDSs in their section. It is critical that labels and MSDSs for all chemicals used be obtained and an agreement made between the company and the contract pest control provider stating that no undocumented chemicals will be used.

The next section in Pest Control 1 is a current copy of the facility map that has been designed in conjunction with the contract pest control provider during the first part of the implementation phase. This shows by number what traps, bait stations, and insect attractors are installed and located throughout the facility. Finally, a copy of the pest control applicator's license for the contract pest control provider representative who validates the program on the agreed basis is provided. He or she should have this on his or her person and a copy should be obtained and placed in the final section after it is verified to be current. The expiration date should be highlighted for quick reference.

Pest Control 2 acts as the "working" notebook and has only two sections: a current blank map and completed validation documents. The current blank map in the first section is the same map that was placed in the sixth section of Pest Control 1 and acts as a guide for validation persons. Finally, the bulk of Pest Control 2 is the documents created during the pest control program validation process and includes in-house completed check sheets, contract pest control provider completed check sheets, and the pest control service reports.

The final part to the implementation phase of the pest control program is the validation program. This usually involves the contract pest control provider inspecting on a monthly basis and company personnel inspecting on a weekly basis. The contract pest control provider's inspection frequency is written into the contract and the company involvement is covered in the program document. A typical contract pest control provider inspection would follow these steps:

1. Announce presence at front desk and sign in.
2. Obtain blank check sheet from Pest Control 2.

3. Proceed to first unit, open, inspect, and sign and date the card contained inside.
 4. Initial contractor validation for unit on lower part of check sheet.
 5. After all units are checked, signed, and dated, meet with quality manager and discuss findings.
 6. Quality control manager signs and dates pest control service report.
 7. Pest control service report and completed check sheet are placed in section 2 of the Pest Control 2 notebook.

For some, step 3 might be argued against by the pest control provider because a signature card has to be placed in each unit and that takes a little more time. Some pest control service providers, especially national companies, maintain that they have a scanning device that reads bar codes. Although these devices might work, they pose no guarantees that the inspector actually has inspected workings of the trap. Having to open each unit forces the inspector to examine the integrity and cleanliness of the trap and verify that the unit is baited correctly, wound correctly, and working properly. This step is also a critical component of the program because it provides a means for the company to determine if the contract pest control provider is doing the job each month agreed upon via contract. Many times the nature of a small to medium size company involves most people wearing several hats and having numerous responsibilities. This makes it even more essential to be able to verify that outside contractors are correctly completing the agreed upon tasks.

A typical in-house inspection involves the following steps:

 1. Obtain a blank check sheet from the Pest Control 2 notebook.
 2. Open and inspect each unit. If the unit is damaged, make a note on the check sheet and notify the quality control manager at the end of the inspection. Note on the check sheet the presence of any rodents or other pests. For bait station units, do not open them. Just verify that they are correctly placed and still attached to the ground. Verify that all units are located directly under their respective numbered signs.
 3. a. If the inspection is a company inspection directly following a contract pest control provider inspection, remove the signature card and verify that the inspector signed and dated the card. If he or she did, check off the unit number on the contractor half of the check sheet. If he or she did not, make a note on the sheet and notify the quality control manager at the end of the inspection.
    b. If the inspection does not follow a contract pest control provider inspection, then inspect and reset the unit, and complete the check sheet.
 4. Meet with the quality control manager and discuss the findings.
 5. Sign and date the completed check sheet. Place it in section 2 of the Pest Control 2 notebook.

When all three parts of the implementation phase are complete, the final step in development of the pest control program is to write the program document. This should include a general statement of the program's compliance with 21CFR110.35c; a section on how rodents, insects, and birds will be controlled inside and outside the facility; details on how and when the facility will be inspected and who will be responsible for inspection; and, finally, how the program will be validated and documented. This program document should be signed, dated, and placed in the Quality Control Manual.

APPLICATION

On the CD (Book 2_Quality Control Manual:Section 4_Pest Control Program:Pest Control Overview) and at the end of this chapter is a template for a rodent and pest control program overview. The name of the company should be filled in and the form modified as needed to fit the specific company application.

EXAMPLES

On the CD (Book 3_Pest Control Examples) and at the end of this chapter are specific examples of each of the components of Pest Control 1. These are separated into each of the specific tab sections to be created: contract, company license, company insurance, labels, MSDSs, map, and applicator's license. The contract section is separated into pest control contract example, and pest control contract example attachment. The company license section contains a copy of a commercial pest control company license and the company insurance section contains a copy of an insurance policy coversheet for a commercial pest control company. The label section provides several examples of commercial pesticides used by commercial applicators; the MSDS section contains several examples of material safety data sheets for commercial pest control applicators' pesticides. The map section contains the pest control inspection form front map and pest control inspection form side B. The front map is a facility drawing that shows the walls and doors of the facility as well as the placement of the pest control units by number. When it is printed, the map should be placed on the opposite side of the pest control inspection form side B. The applicator's license section contains a copy of an individual applicator's license to apply pesticides. A sample commercial pest control service report that would typically be given at the end of a contract pest control provider inspection can also be found on the CD (Book 4_Pest Control-2) and at the end of this chapter.

SUPPLEMENTAL MATERIALS

RODENT AND PEST CONTROL PROGRAM

GENERAL OVERVIEW

Section A: General

In accordance with 21CFR110.35c, _____ will employ and maintain adequate means to exclude rodents, insects, and birds from all processing and storage areas. These means will include, but not be limited to, mechanical traps, glue boards, and Insectecutors and will involve both in-house and contract inspections. Inspections will be made by an outside pest control contractor and a quality control technician on a timely basis and will be properly documented. Rodent burrows, rodent runs, and any conditions consistent with the attraction of rodents, both inside and outside the facility, will be eliminated. It is the responsibility of the quality control manager to establish and monitor this program.

Section B: Rodent

Internal types
> Mechanical traps, such as Ketch-alls, shall be placed on either side of all overhead and pedestrian doors, every 20 linear feet along interior walls, and at least one per wall on walls of less than 20 feet in storage areas. Traps will be commercial in nature, be numbered, and have a posted sign indicating the location and number of the trap below. This sign will be placed at eye level and will be maintained with the traps. Each trap will have a label or card attached to the inside of the removable lid for the pest control contractor to sign and date upon inspection.

External types
> Surrounding the perimeter of the facility will be bait stations. These will be placed at intervals of between 30 and 40 feet on running walls and on either side of all outside doors. They will be numbered and have a posted sign indicating the location and number of the trap below. This sign will be placed at eye level and will be maintained with the traps.

Section C: Insect

Electric flying insect control units (Insectecutors or VECTORs) will be placed as needed inside the facility. They will be located at least 10 feet from all production lines and should face away from the outside of the building so as not to attract insects. VECTORs should be used where insect fragments might come in contact with packaged finished goods or ingredients. Control units should be clearly marked.

Section D: Bird

Birds should be controlled by exclusion. Netting, screening, or mechanical traps should only be used as an interim measure. All food sources should be removed from the exterior of the facility or properly contained. Access to the inside of the facility

Pest Control Program

should be denied by keeping all exterior doors closed, proper curtaining of loading facilities, and scheduled maintenance of the outside walls.

Section E: Inspection

In house

All traps, bait stations, and Insectecutors will be inspected by Quality Control personnel at least once per week on a scheduled basis. A floor plan of the facility indicating all of the traps and Insectecutors will be used. Rodent, insect, and bird activity will be noted as well as any action taken.

Contract

A certified pest control service will inspect all traps, bait stations, and Insectecutors on a monthly basis. The service should be licensed to apply pesticides and rodenticides and will make recommendations as needed. It will be required to provide the following documents prior to any service:

a contract describing specific services to be rendered, including chemicals, methods, precautions, and material safety data sheets;

sample labels of all pesticides used;

service reports and pesticide usage logs; and

copies of insurance, company license, and applicators' licenses.

Section F: Documentation

A separate inspection log book that houses all inspection records will be kept. These will include the in-house inspection sheets and the contract inspection sheets. A template floor plan indicating the location of traps and Insectecutors will be placed at the front of the book for reference.

Approved by: _____ Date: _____

COMPENSATION WORKSHEET
FOR
SMITH'S CANDY COMPANY
1997 Sinclair Lane
Boise, Idaho 83707

Service Cost

Initial month investment	$120.00
Service once per month at $55.00 per visit	$55.00

Note: All traps are property of Smith's Candy Company. If damaged, they may necessitate replacement at Smith's Candy Company's expense.

Material Cost

Ketch-alls: 39 at $12.50 each	$487.50
Exterior weatherproof, tamper-resistant bait stations: 16 at $16.00 each	$256.00
Total material cost	$743.50
Total cost	**$863.50**

Attachments

Company license
Proof of insurance
Rodenticide labels
MSDS sheets
Service report examples

CONTRACT

July 15, 2007

Smith's Candy Company
Attn: Mr. Smith
PO Box 6307
Boise, ID 83707

RE: Pest Control Proposal

Dear Mr. Smith:

A survey of the plant and warehouse was done April 3, 2007. The following represents my determination of the rodent control layout and insect treatment of the plant.

I. Rodent control program
   A. Exterior treatment
      Set up weatherproof, tamper-resistant bait stations along exterior perimeters of the building. Stations will be anchored and each station will be numbered as shown on the floor plan. Rodent bait will be tied to the inside of each station to avoid any possibility of bait being carried into production areas. All rodenticides will be approved in the meat and poultry manual subpart 8-6 (regs: M-318; P subpart H) p. 40 8.49 rodenticides paragraph A.
   B. Interior treatment
      Ketch-all multiple catch traps will be placed as shown on the floor plan. Ketch-alls will be numbered on top and on the wall. Ketch-alls will have date stickers placed on the inside of the lid. No rodenticides will be used.
   C. Treatment frequency
      Each interior Ketch-all and exterior bait station will be inspected, dated, and cleaned once a month by our professional pest control technicians.

II. Insect control program
   No insect treatments will be done inside the facility unless pre-approved in writing by the quality control manager. All nonresidual and residual materials will meet the guidelines specified in the meat and poultry inspection manual 8.48, p. 38. Proposed material to be used is PT 3-6-10 aerosol 565-XLO. We will inspect and service fly machines in the plant once a month. Smith's Candy Company is to provide replacement parts as needed.

III. Fly control equipment

Two Insectecutors and one VECTOR are recommended. The VECTOR will be on the path of ingredients entering the production area. All equipment will meet poultry and meat specifications.

IV. Insurance

BugBusters will furnish a certificate of insurance with this proposal.

V. Compensation for service

See attached compensation work sheet.

VI. Terms of agreement

This agreement shall be effective for an initial period of one year and shall renew itself annually for a period of one year unless thirty (30) days written notice is given prior to the anniversary date of this agreement.

Submitted for BugBusters

By: _____ Date: _____

Accepted for Smith's Candy Company

By: _____ Date: _____

Colorado Department of Agriculture
Division of Plant Industry
700 Kipling Street, Suite 4000
Lakewood, Colorado 80215-5894
(303) 239-4146

COMMERCIAL APPLICATOR
CERTIFICATE OF LICENSE

License Number: 00223
Date Issued: 12/30/2003

Effective Date: 01/01/2004
Good Through: 12/31/2004

TERMINIX INTERNATIONAL COMPANY L.P.(THE)
2215 EXECUTIVE CR
COLORADO SPRINGS, CO 80906

DBA: TERMINIX INTERNATIONAL AKA: TERMINIX
ECONOMY PEST CONTROL

This certificate is evidence that a license has been issued to the person(s) listed above to do business in the name(s) listed above under the provisions of the Pesticide Applicators' Act and may be revoked, suspended, or have other lawful discipline imposed for cause.

COMMISSIONER OF AGRICULTURE
Donald D. Ament

CERTIFICATE OF INSURANCE

CERTIFICATE NUMBER: CHI-001092908-01

PRODUCER:
The ServiceMaster Certificate Team
Fax - 877-732-7799
MARSH USA
500 West Monroe Street
Chicago, IL 60661
2289

THIS CERTIFICATE IS ISSUED AS A MATTER OF INFORMATION ONLY AND CONFERS NO RIGHTS UPON THE CERTIFICATE HOLDER OTHER THAN THOSE PROVIDED IN THE POLICY. THIS CERTIFICATE DOES NOT AMEND, EXTEND OR ALTER THE COVERAGE AFFORDED BY THE POLICIES DESCRIBED HEREIN.

COMPANIES AFFORDING COVERAGE

- COMPANY A: ZURICH AMERICAN INS. CO.
- COMPANY B: AMERICAN-ZURICH INSURANCE COMPANY
- COMPANY C: ILLINOIS NATIONAL INSURANCE COMPANY
- COMPANY D:

INSURED:
THE TERMINIX INTERNATIONAL COMPANY
LIMITED PARTNERSHIP
PO BOX 17167
MEMPHIS, TN 38187

COVERAGES

THIS IS TO CERTIFY THAT POLICIES OF INSURANCE DESCRIBED HEREIN HAVE BEEN ISSUED TO THE INSURED NAMED HEREIN FOR THE POLICY PERIOD INDICATED. NOTWITHSTANDING ANY REQUIREMENT, TERM OR CONDITION OF ANY CONTRACT OR OTHER DOCUMENT WITH RESPECT TO WHICH THE CERTIFICATE MAY BE ISSUED OR MAY PERTAIN, THE INSURANCE AFFORDED BY THE POLICIES DESCRIBED HEREIN IS SUBJECT TO ALL THE TERMS, CONDITIONS AND EXCLUSIONS OF SUCH POLICIES. AGGREGATE LIMITS SHOWN MAY HAVE BEEN REDUCED BY PAID CLAIMS.

CO LTR	TYPE OF INSURANCE	POLICY NUMBER	POLICY EFFECTIVE DATE (MM/DD/YY)	POLICY EXPIRATION DATE (MM/DD/YY)	LIMITS	
A	GENERAL LIABILITY [X] COMMERCIAL GENERAL LIABILITY [ ] CLAIMS MADE [X] OCCUR [ ] OWNER'S & CONTRACTOR'S PROT	GLO 8343850-05	01/01/03	01/01/06	GENERAL AGGREGATE	$5,000,000
					PRODUCTS - COMP/OP AGG	$1,000,000
					PERSONAL & ADV INJURY	$1,000,000
					EACH OCCURRENCE	$1,000,000
					FIRE DAMAGE (Any one fire)	$1,000,000
					MED EXP (Any one person)	$5,000
A	AUTOMOBILE LIABILITY [X] ANY AUTO [X] ALL OWNED AUTOS [ ] SCHEDULED AUTOS [ ] HIRED AUTOS [ ] NON-OWNED AUTOS	BAP 8343861-05 (AOS) TAP 8343854-05 (TX) BAP 8343859-05 (VA)	01/01/03 01/01/03 01/01/03	01/01/06 01/01/06 01/01/06	COMBINED SINGLE LIMIT	$1,000,000
					BODILY INJURY (Per person)	$
					BODILY INJURY (Per accident)	$
					PROPERTY DAMAGE	$
	GARAGE LIABILITY [ ] ANY AUTO				AUTO ONLY - EA ACCIDENT	$
					OTHER THAN AUTO ONLY:	
					EACH ACCIDENT	$
					AGGREGATE	$
C	EXCESS LIABILITY [X] UMBRELLA FORM [ ] OTHER THAN UMBRELLA FORM	BE 309-79-07	04/01/01	04/01/04	EACH OCCURRENCE	$5,000,000
					AGGREGATE	$5,000,000
						$
A A B	WORKERS COMPENSATION AND EMPLOYERS' LIABILITY THE PROPRIETOR/ PARTNERS/EXECUTIVE OFFICERS ARE: [X] INCL [ ] EXCL OTHER	WC 8343847-05 (AOS) WC 8343805-05 (WI) WC 8343833-05 (IL)	01/01/03 01/01/03 01/01/03	01/01/06 01/01/06 01/01/06	[X] WC STATU-TORY LIMITS [ ] OTHER	
					EL EACH ACCIDENT	$1,000,000
					EL DISEASE-POLICY LIMIT	$1,000,000
					EL DISEASE-EACH EMPLOYEE	$1,000,000

DESCRIPTION OF OPERATIONS/LOCATIONS/VEHICLES/SPECIAL ITEMS

CERTIFICATE HOLDER

Tr Toppers
Attn: Mark
P.O. Box 6308
Boise, ID 83707

CANCELLATION

SHOULD ANY OF THE POLICIES DESCRIBED HEREIN BE CANCELLED BEFORE THE EXPIRATION DATE THEREOF, THE INSURER AFFORDING COVERAGE WILL ENDEAVOR TO MAIL __30__ DAYS WRITTEN NOTICE TO THE CERTIFICATE HOLDER NAMED HEREIN, BUT FAILURE TO MAIL SUCH NOTICE SHALL IMPOSE NO OBLIGATION OR LIABILITY OF ANY KIND UPON THE INSURER AFFORDING COVERAGE, ITS AGENTS OR REPRESENTATIVES, OR THE ISSUER OF THIS CERTIFICATE.

MARSH USA INC.
BY: Christy N. Phoebus *(signature)*

VALID AS OF: 12/03/03

Pest Control Program

Prescription Treatment® brand
565 PLUS XLO®
Formula 2

Contact Insecticide

KILLS:
Angoumois Grain Moths, Ants, Bed Bugs, Booklice, Carpet Beetles, Centipedes, Chocolate Moths, Cigarette Beetles, Clothes Moths, Clover Mites, Cluster Flies, Cockroaches, Confused Flour Beetles, Crickets, Drugstore Beetles, Fleas, Flies, Fruit Flies, Gnats, Grain Mites, Granary Weevils, Horn Flies, House Flies, Indianmeal Moths, Mediterranean Flour Moths, Millipedes, Mosquitoes, Mud Daubers, Red Flour Beetles, Rice Weevils, Sawtoothed Grain Beetles, Silverfish, Small Flying Moths, Sowbugs, Spiders, Stable Flies, Ticks and Wasps

FOR USE IN AND AROUND:
Apartments, Campgrounds, Food Storage Areas, Homes, Hospitals†, Hotels, Motels, Nursing Homes†, Resorts, Restaurants and other Food Handling Establishments†, Schools†, Supermarkets, Transportation Equipment (Buses, Boats, Ships, Trains, Trucks, Planes†), Utilities, Warehouses, and other Commercial and Industrial Buildings
† See special instructions for these sites under Directions for Use.

ACTIVE INGREDIENTS:
Pyrethrins, a botanical insecticide 0.5%
Piperonyl Butoxide, Technical* 1.0%
n-Octyl Bicycloheptene Dicarboximide 1.0%
OTHER INGREDIENTS: 97.5%
* Equivalent to 0.8% (butylcarbityl)(6-propylpiperonyl) ether and 0.2% related compounds.
TOTAL: 100.0%

EPA Reg. No. 499-290

KEEP OUT OF REACH OF CHILDREN
CAUTION

FIRST AID

IF SWALLOWED: Immediately call a poison control center or doctor. Do not induce vomiting unless told to do so by a poison control center or doctor. Do not give ANY liquid to the person. Do not give anything by mouth to an unconscious person.
IF ON SKIN OR CLOTHING: Take off contaminated clothing. Rinse skin immediately with plenty of water for 15 - 20 minutes. Call a poison control center or doctor for treatment advice.
IF IN EYES: Hold eyes open and rinse slowly and gently with water for 15-20 minutes. Remove contact lenses, if present, after the first 5 minutes, then continue rinsing eyes. Call a poison control center or doctor for treatment advice.
IF INHALED: Move person to fresh air. If person is not breathing, call 911 or an ambulance, then give artificial respiration, preferably by mouth-to-mouth, if possible. Call a poison control center or doctor for further treatment advice.
Have the product container or label with you when calling a poison control center or doctor, or going for treatment. You may also contact 1-800-225-3320 for emergency medical treatment information.

PRECAUTIONARY STATEMENTS
HAZARDS TO HUMANS AND DOMESTIC ANIMALS
CAUTION: Harmful if swallowed or absorbed through skin. Avoid contact with eyes, skin or clothing. Wash thoroughly with soap and water after handling.
NOTICE TO APPLICATOR: In order to avoid any unpleasant drying of the nose and throat when dispensing this product in confined areas or when prolonged exposure to spray mist occurs, wear a NIOSH approved respirator with an organic vapor (OV) cartridge or canister with any R, P or HE prefilter.

ENVIRONMENTAL HAZARDS
This product is toxic to fish. Do not apply directly to water, or to areas where surface water is present or to intertidal areas below the mean high water mark. Do not contaminate water by cleaning of equipment or disposal of wastes.
PHYSICAL OR CHEMICAL HAZARDS
Extremely Flammable. Contents under pressure. Keep away from heat, sparks and heated surfaces. Do not puncture or incinerate container. Exposure to temperatures above 130°F may cause bursting.

DIRECTIONS FOR USE
IT IS A VIOLATION OF FEDERAL LAW TO USE THIS PRODUCT IN A MANNER INCONSISTENT WITH ITS LABELING.
† **IN HOSPITALS AND NURSING HOMES:** Do not apply this product in hospital rooms while occupied or in any rooms occupied by the elderly or infirm. Apply according to specific directions for pest(s) to be treated. Patients should be removed from room prior to treatment. Room should be ventilated for 2 hr after spraying. Do not return patients to room until after ventilation.
† **FOOD AREAS OF FOOD HANDLING ESTABLISHMENTS:** Apply according to specific directions for pest(s) to be treated. When applying a space treatment in food handling areas, the food handling operation should be shut down. Foods should be removed or covered during treatment. In food processing areas, all utensils, shelving, etc., where food will be handled should be covered or removed before treatment or thoroughly washed with an effective cleaning compound followed by a potable water rinse prior to use. Careful Crack & Crevice® application for inspection purposes is permissible at all times except in federally inspected meat and poultry plants, as provided below.
† **FEDERALLY INSPECTED MEAT AND POULTRY PLANTS:** Apply according to specific directions for pest(s) to be treated. When using this product, apply only when processing area is not in operation. Foods should be removed or covered during treatment. After a space treatment, all food processing surfaces and equipment should be thoroughly washed with an effective cleaning compound followed by a potable water rinse before using. To prevent food product contamination, the treated area should be free of odor before the food product is placed in the area.
† **SCHOOLS:** Do not apply to classrooms when in use.
† **PLANES:** Do not apply in aircraft cabins.
Do not spray on plastic, painted or varnished surfaces or directly into any electronic equipment such as radios, televisions, computers, etc. Do not use this product in electrical equipment (where there is exposed wiring or contact points) due to possibility of short circuits and shock hazard. Product should only be used when can temperature is above 60°F. If can temperature is below 60°F, store at room temperature until a temperature above 60°F is reached.

HOW TO USE WHITMIRE MICRO-GEN INJECTION SYSTEM
CRACK & CREVICE APPLICATION: Use PT 565 PLUS XLO with the supplied actuator and injection tubes or other Whitmire Micro-Gen equipment. Inject PT 565 PLUS XLO into cracks and crevices or void spaces where insects may be harboring, living and breeding. Place injector tip into cracks, crevices, holes and other small openings. Apply product for 1 sec. For light infestations, move injector tip along cracks while treating at the rate of 3 linear ft per sec. For heavy infestations, move injector tip along at 1 linear ft per sec.
For wall and equipment voids calculate the void's cubic area and treat at the rate of 5 - 10 sec per 3 cu ft. Several holes may be required in long-running voids.
SPACE TREATMENT: Calculate the cubic footage to be treated and apply at the dosage rate specified below for the pest(s) to be treated. Ventilate treated areas before reoccupying.

WHITMIRE MICRO-GEN
RESEARCH LABORATORIES, INC.

NOTE: This specimen label is for informational purposes only. All uses may not be approved in all states. See labeling which accompanied product for Directions for Use or call 800-777-8570 for more information. **For automatic specimen label updates, register at www.wmmg.com.**

565 PLUS XLO® Formula 2 Contact Insecticide

INDOOR TREATMENTS

FLYING INSECTS (Cluster Flies, Flies, Fruit Flies, Gnats, Mosquitoes, Small Flying Moths): Apply as a space treatment. Close all windows and doors and direct fog upward at a rate of 1 - 3 sec per 1,000 cu ft. Disperse in all locations. Keep area closed for 15 min. Open and ventilate before reoccupying. Treatment may be repeated daily if necessary.

CLUSTER FLIES, CLOVER MITES, WASPS AND BEES INSIDE STRUCTURAL VOIDS: For treatment of these pests, locate all insect entrances and close all but one opening. Determine size of void. Inject 5 - 10 sec per 3 cu ft of void into remaining opening. For best results, treat at dusk or later. Treatment may be repeated every 3 days if necessary.

ANTS, BOOKLICE, CENTIPEDES, CLOVER MITES, CRICKETS, MILLIPEDES, ROACHES, SILVERFISH, SOWBUGS AND SPIDERS: Apply as a Crack & Crevice treatment into cracks and crevices in all hiding places behind baseboards, sinks, cabinets, meter boxes, door frames, windows or as a space treatment at a rate of 20 sec per 1,000 cu ft. Open cabinets and doors in area to be treated. Turn off air conditioners and fans and close doors and windows before treating. Disperse toward area suspected of harboring the greatest insect infestations. Disperse in all locations contacting as many insects as possible. Keep area closed for 15 min. Open and ventilate the treated area before reoccupying. Treatment may be repeated every 3 days if necessary. Inspections may be made on a daily basis.

For treatment of void areas such as attics, false ceilings and crawl spaces, calculate the volume of the void and apply as a space treatment at a rate of 20 sec per 1,000 cu ft. Treatment may be repeated 3 days if necessary.

BED BUGS: Take bed apart and treat cracks, joints and interior of framework. Treat mattresses and box springs; especially tufts, folds and edges. Also treat picture frames, moldings and all cracks and crevices in the room. Do not use in patient rooms in hospitals and nursing homes for treatment of bed bugs.

CARPET BEETLES (In Residences): Hold product 36" above floor and direct spray toward floor and lower walls treating at a rate of 10 sec per 100 sq ft. Also apply as a Crack & Crevice treatment to cracks and crevices near the source of the infestation.

CLOTHES MOTHS: Locate source of infestation. Treat nearby cracks and crevices and perform a space treatment of the infested area at a rate of 10 - 20 sec per 1,000 cu ft. To protect woolens and other keratin containing materials, brush clean, then air out. Before treating entire article, treat a small hidden area of fabric to determine if staining will occur. Treat thoroughly by holding dispenser at least 18" from garment. Treated articles should be put into a darkened closet or other darkened storage such as cedar chests, storage drawers or sealable plastic fabric bags. For best protection clean and treat inside of closets or storage areas by treating 1 sec per 10 sq ft before putting treated materials away. Preventative treatment should be repeated every 6 months. To control a present infestation, treatment may be repeated every 3 days if necessary. Dry clean treated clothes before wearing.

FOR EXPOSED ADULT AND LARVAL STAGES OF CARPET BEETLES, CIGARETTE BEETLES, CONFUSED FLOUR BEETLES, DRUGSTORE BEETLES, GRAIN MITES, GRANARY WEEVILS, RED FLOUR BEETLES, RICE WEEVILS AND SAWTOOTHED GRAIN BEETLES: Apply as a Crack & Crevice treatment into cracks and crevices of pantries, cabinets, food processing and handling equipment and other places where insects harbor or as a space treatment at a rate of 20 sec per 1,000 cu ft. Apply around cartons and containers, and other areas where these insects tend to congregate. Treatment may be repeated every 3 days if necessary. Inspections may be made on a daily basis. Infested stored products should be fumigated or treated by other effective methods using an approved product intended for this purpose.

ANGOUMOIS GRAIN MOTHS, CHOCOLATE MOTHS, INDIANMEAL MOTHS, MEDITERRANEAN FLOUR MOTHS: Apply as a space treatment at a rate of 5 - 10 sec per 1,000 cu ft. Close all doors and windows. Direct into all parts of room, especially around stored product containers, pallets and darkened areas. Keep area closed for 15 min following treatment. Open and ventilate before reoccupying. Treatment may be repeated every 3 days if necessary. Infested stored products should be fumigated or treated by other effective methods using an approved product intended for this purpose.

FLEAS IN BUILDINGS: Hold product 36" above floor and direct toward floor and lower walls at a rate of 10 sec per 100 sq ft. making sure that all floor, sofa and chair surfaces are contacted. Keep area closed for 15 min. Open and ventilate before reoccupying. Treatment may be repeated daily if necessary.

TICKS: Holding container 36" above the floor, direct PT 565 PLUS XLO toward pet beds and resting quarters. Treat adjacent cracks and crevices such as behind baseboards and edges of carpet and floor covers. Treat higher cracks such as upper door cases and window framing, behind pictures and other areas where females may crawl to lay eggs. Treatment may be repeated daily if necessary.

OUTDOOR TREATMENTS

GROUND APPLICATION (Flies, Gnats and Mosquitoes): Treat open areas near buildings and in campgrounds. Best results are obtained when wind speed is 5 MPH or less. Apply at a rate of 1.5 - 2 sec per 1,000 sq ft. Apply in wide swaths across area to be treated. Allow treatment to penetrate dense foliage. Treatment may be repeated daily if necessary.

MUD DAUBER AND WASP NESTS: Apply mist directly on insects and their nests from approximately 18". Contacted insects will fly away from fog. Applications should be made in late evening when insects are at rest. Disperse into hiding and breeding places contacting as many insects as possible.

STORAGE & DISPOSAL

Do not contaminate water, food or feed by storage or disposal.

STORAGE: Store in a cool area away from heat or open flame.

PESTICIDE DISPOSAL: Waste resulting from use of this product may be disposed of on site or at an approved waste disposal facility.

CONTAINER DISPOSAL: Do not puncture or incinerate! Empty container by using the product according to the label directions. Offer empty container for recycling, if available, or place in trash if allowed by state and local regulations. If container is partly full, contact your local solid waste agency or call 1-800-CLEANUP for disposal instructions.

Contains no CFCs or other ozone depleting substances. Federal regulations prohibit CFC propellants in aerosols.

A Prescription Treatment® brand Insecticide from:
Whitmire Micro-Gen Research Laboratories, Inc
3568 Tree Court Industrial Blvd.
St. Louis MO 63122-6682
© 2005 Whitmire Micro-Gen Research Laboratories, Inc

NOTE: This specimen label is for informational purposes only. All uses may not be approved in all states. See labeling which accompanied product for Directions for Use or call 800-777-8570 for more information. **For automatic specimen label updates, register at www.wmmg.com.**

070711-94

Pest Control Program

DIRECTIONS FOR USE

It is a violation of Federal law to use this product in a manner inconsistent with its labeling.

READ THIS LABEL and follow all use directions and use precautions.

IMPORTANT: Do not expose children, pets or other nontarget animals to rodenticides. To help prevent accidents:
1. Store product not in use in a location out of reach of children and pets.
2. Apply bait in locations out of reach of children, pets, domestic animals and nontarget wildlife, or in tamper-resistant bait stations. These stations must be resistant to destruction by dogs and children under six years of age, and must be used in a manner that prevents such children from reaching into bait compartments and obtaining bait. If bait can be shaken from stations when they are lifted, units must be secured or otherwise immobilized. Even stronger bait stations are needed in areas open to hooded livestock, raccoons, bears, other potentially destructive animals, or in areas prone to vandalism.
3. Dispose of product container, unused, spoiled, and unconsumed bait as specified on this label.

USE RESTRICTIONS: This product may be used to control Norway rats, roof rats and house mice. Do not place bait in areas where there is a possibility of contaminating food or surfaces that come in direct contact with food. Do not broadcast bait.

Urban areas: This product may be used in and around homes, food processing facilities, industrial, commercial and public buildings; in transport vehicles (ships, trains, aircraft) and in and around related port or terminal buildings. This product may also be used in alleys.

Non-urban areas: This product may only be used inside of homes and agricultural buildings.

SELECTION OF TREATMENT AREAS: Determine areas where rats or mice will most likely find and consume bait. Generally, these are along walls, by gnawed openings, in or beside burrows, in corners and concealed places, between floors or walls, or in locations where rodents or their signs have been seen. Remove as much alternative food as possible.

APPLICATION DIRECTIONS:

Mice: Apply one block per placement. 2 blocks may be needed at points of very high mouse activity. Space placements 8 to 12 feet apart. Maintain a constant supply of fresh bait for 15 days or until fresh signs of mouse activity cease.

Rats: Apply 3 to 16 blocks per placement. Space placements 15 to 30 feet apart. Maintain a constant supply of fresh bait for 10 days or until fresh signs of rat activity cease.

For burrows, insert one or two blocks deep into each active burrow. Check treated burrows each morning and evening to make sure that bait has not been pushed out of burrows.

For **sewer** applications, run a wire through the holes in the bait blocks and attach the wire to a stationary structure such as the bottom step of a manhole ladder or to a sewer grate, allowing just enough wire for the blocks to rest on manhole benching. If benching is not present, suspend bait blocks a few inches above the high water line or place blocks on a board supported by opposing steps of the ladder. Securing blocks in this manner will minimize chances for removal by rats or water. Place at least 12 bait blocks per manhole.

Follow-up: Replace contaminated or spoiled bait immediately. Collect and dispose of all visible dead animals and leftover bait properly. To discourage reinfestation, limit sources of rodent food, water and harborage as much as possible. If reinfestation occurs, repeat treatment. For a continuous infestation, set up permanent bait stations and replenish bait as needed.

(02105)

MINI BLOCKS

KILLS NORWAY RATS, ROOF RATS, HOUSE MICE AND WARFARIN-RESISTANT NORWAY RATS

NORWAY RATS & HOUSE MICE CAN CONSUME A LETHAL DOSE IN ONE NIGHT'S FEEDING WITH FIRST DEAD RODENTS APPEARING FOUR OR FIVE DAYS AFTER FEEDING BEGINS

Active Ingredient: bromadiolone 0.005%
Inert Ingredients 99.995%
Total 100.000%

KEEP OUT OF REACH OF CHILDREN
CAUTION: See side panel for additional precautionary statements.

LIPHATECH

Liphatech, Inc.
3600 W. Elm Street
Milwaukee, WI 53209
(414) 351-1476

PRECAUTIONARY STATEMENTS
Hazard to Humans and Domestic Animals

CAUTION: Keep away from humans, domestic animals and pets. May be harmful if swallowed or absorbed through the skin because this material may reduce the clotting ability of blood and cause bleeding. Do not get in eyes, on skin or on clothing. Wash arms, hands and face with soap and water after applying and before eating and smoking. If bait is handled, wear gloves.

FIRST AID

Have this label with you when obtaining treatment advice.

If swallowed: Call a poison control center or doctor immediately for treatment advice.

If in eyes: Hold eye open and rinse slowly and gently with water for 15-20 minutes. Remove contact lenses, if present, after the first 5 minutes, then continue rinsing eye. Call a poison control center or doctor for treatment advice.

If on skin or clothing: Take off contaminated clothing. Rinse skin with plenty of cool water for 15-20 minutes. Call a poison control center or doctor for treatment advice.

If inhaled: Move person to fresh air. If person is not breathing, call 911 or an ambulance, then give artificial respiration, preferably mouth-to-mouth, if possible. Call a poison control center or doctor for treatment advice.

Note to Physician: If ingested, administer Vitamin K_1 intramuscularly or orally as indicated in bishydroxycoumarin overdoses. Repeat as necessary based on monitoring of prothrombin times.

ENVIRONMENTAL HAZARDS: This product is toxic to fish and wildlife. Predatory and scavenging mammals and birds might be poisoned if they feed upon animals that have eaten the bait. Do not apply directly to water, or to areas where surface water is present or to intertidal areas below the mean high water mark.

STORAGE AND DISPOSAL

Do not contaminate water, food or feed by storage or disposal.

Storage: Store in original container in a cool, dry place inaccessible to children and pets.

Pesticide Disposal: Wastes resulting from the use of this product may be disposed of on site or at an approved waste disposal facility.

Container Disposal: Dispose of empty container in a sanitary landfill, or by incineration, or if allowed by state and local authorities, by burning. If burned, stay out of smoke.

WARRANTY: Seller makes no warranty, expressed or implied, concerning the use of this product other than indicated on the label. Buyer assumes all risk of use and/or handling of this material when such use and/or handling is contrary to label instructions.

EPA Reg. No. 7173-202
EPA Est. No. 7173-WI-1

DATE 11/07

CONTRAC®
All-Weather Blox

KILLS RATS AND MICE

KILLS WARFARIN RESISTANT NORWAY RATS

Norway rats and house mice may consume a lethal dose in one night's feeding with first dead rodents appearing four or five days after feeding begins.

ACTIVE INGREDIENT:
Bromadiolone (CAS #28772-56-7): 0.005%
INERT INGREDIENTS*: 99.995%
                                                100.000%

*Contains Denatonium Benzoate

KEEP OUT OF REACH OF CHILDREN
CAUTION

FIRST AID
HAVE LABEL WITH YOU WHEN OBTAINING TREATMENT ADVICE

IF SWALLOWED:
- Call a poison control center, doctor, or 1-877-854-2494 immediately for treatment advice.
- Have person sip a glass of water if able to swallow.
- Do not induce vomiting unless told to do so by the poison control center or doctor.

IF ON SKIN:
- Wash with plenty of soap and water.

NOTE TO PHYSICIAN OR VETERINARIAN
If swallowed, this material may reduce the clotting ability of the blood and cause bleeding. If ingested, administer Vitamin K₁ intramuscularly or orally as indicated in bishydroxycoumarin overdoses. Repeat as necessary based on monitoring of prothrombin times.

NET WEIGHT: 4 lbs. (1.8 Kg)

Manufactured by:

Bell Laboratories, Inc.
Madison, WI 53704

EPA REG. NO. 12455-79   EPA EST. NO. 12455-WI-1

DIRECTIONS FOR USE
It is a violation of Federal law to use this product in a manner inconsistent with its labeling.
READ THIS LABEL: Read this entire label and follow all use directions and use precautions.

IMPORTANT: Do not expose children, pets, or nontarget animals to rodenticides. To help to prevent accidents:
1. Store unused product out of reach of children and pets.
2. Apply bait in locations out of reach of children, pets, domestic animals and nontarget wildlife, or in tamper-resistant bait stations. These stations must be resistant to destruction by dogs and by children under six years of age, and must be used in a manner that prevents such children from reaching into bait compartments and obtaining bait. If bait can be shaken from bait stations when they are lifted, units must be secured or otherwise immobilized. Stronger bait stations are needed in areas open to hoofed livestock, raccoons, bears, or other potentially destructive animals, or in areas prone to vandalism.
3. Dispose of product container and unused, spoiled, or unconsumed bait as specified on this label.

USE RESTRICTIONS: For control of Norway rats, roof rats and house mice. Do not place bait in areas where there is a possibility of contaminating food or surfaces that come in direct contact with food. When used in USDA-inspected facilities, this product must be applied in tamper-resistant bait stations. Do not broadcast bait.

Urban Areas: This product may be used in and around the periphery of homes, industrial, commercial, and public buildings. May also be used in transport vehicles (ships, trains, aircraft) and in and around related port or terminal buildings. May also be used in alleys. Do not use in sewers.

Non-Urban Areas: This product may be used inside of homes and agricultural buildings.

SELECTION OF TREATMENT AREAS: Determine areas where rats or mice will most likely find and consume the bait. Generally, these areas are along walls, by gnawed openings, in or beside burrows, in corners and concealed places, between floors and walls, or in locations where rodents or their signs have been seen. Protect bait from rain and snow. Remove as much alternative food as possible.

APPLICATION DIRECTIONS:
Each bait block in this container weighs approximately one ounce.

DIRECTIONS FOR USE (Continued from other panel)
APPLICATION DIRECTIONS (Continued from other panel)
RATS: Place 3 to 16 bait blocks (usually at intervals of 15 to 30 feet) per placement. Maintain an uninterrupted supply of fresh bait for at least 10 days or until signs of rat activity cease.

MICE: Place 1 block per placement. Space placements at 8 to 12 foot intervals. Two blocks may be needed at points of very high mouse activity. Maintain an uninterrupted supply of fresh bait for at least 15 days or until signs of mouse activity cease.

FOLLOW-UP: Replace bait contaminated or spoiled bait immediately. Collect and dispose of all dead, exposed animals and leftover bait. To prevent reinfestation, limit sources of rodent food, water, and harborage as much as possible. If reinfestation does occur, repeat treatment. Where a continuous source of infestation is present, establish permanent bait stations and replenish as needed.

PRECAUTIONARY STATEMENTS
HAZARDS TO HUMANS AND DOMESTIC ANIMALS
CAUTION: Harmful if swallowed. Wash thoroughly with soap and water after handling.

ENVIRONMENTAL HAZARDS
This product is toxic to fish, birds and other wildlife. Do not apply this product directly to water or to areas where surface water is present or to intertidal areas below the mean high water mark.

STORAGE AND DISPOSAL
Do not contaminate water, food or feed by storage or disposal.
Storage: Store only in original container in a cool, dry place inaccessible to children and pets. Keep containers closed and away from other chemicals.
Pesticide Disposal: Wastes resulting from the use of this product may be placed in trash or delivered to an approved waste disposal facility.
Pesticide Container: Do not reuse empty container. Dispose of empty container by placing in trash, at an approved waste disposal facility or by incineration or, if allowed by state and local authorities, by burning. If burned stay out of smoke. Call your Local Waste Agency for any questions on proper disposal.

WARRANTY: Seller makes no warranty, expressed or implied, concerning the use of this product other than indicated on the label. Buyer assumes all risk of use and/or handling of this material when such use and/or handling is contrary to label instructions.

101504

MATERIAL SAFETY DATA SHEET

SECTION 1 — PRODUCT & COMPANY IDENTIFICATION

Maki® Mini Blocks
EPA Reg. No. 7173-202
Maki® Paraffin Blocks
EPA Reg. No. 7173-189

Other Designation:	Anticoagulant rodenticide with Bromadiolone
Manufacturer:	Liphatech, Inc.
	3600 W. Elm Street, Milwaukee, WI 53209
Emergency Phone:	414-351-1476 Monday-Friday, 8:00 am-4:30 pm CST
After Hours:	Call CHEMTREC at 1-800-424-9300

SECTION 2 — INGREDIENT INFORMATION

Hazardous Ingredient:	CAS Number:	OSHA PEL:	ACGIH TLV:	ACGIH STEL:
Bromadiolone	28772-56-7	Not assigned	Not assigned	Not assigned

SECTION 3 — HAZARDS IDENTIFICATION

Emergency Overview: May be harmful if swallowed or absorbed through the skin, because this material may reduce the clotting ability of the blood and cause bleeding.
Primary Entry Routes: Oral (swallowing), dermal (absorption through skin)
Acute Effects (Signs and Symptoms of Overexposure):
- **Eyes:** May cause temporary eye irritation.
- **Skin:** May be harmful if absorbed through skin. Symptoms of toxicity include lethargy, loss of appetite, reduced blood clotting ability and bleeding.
- **Inhalation:** Due to this product's solid form, inhalation is unlikely.
- **Ingestion:** May be harmful if swallowed. Symptoms of toxicity include lethargy, loss of appetite, reduced clotting ability of blood, and bleeding.

Chronic Effects: Prolonged and/or repeated exposure to small amounts of product can produce cumulative toxicity. Symptoms of toxicity include lethargy, loss of appetite, reduced clotting ability of blood, and bleeding.
Medical Conditions Aggravated by Exposure: Bleeding disorders
Target Organs: Blood
Carcinogenicity: Contains no known or suspected carcinogens.
HMIS: Health – 2, Flammability – 0, Reactivity – 0

SECTION 4 — FIRST AID MEASURES

Eyes: Flush with water. Get medical attention if irritation persists.
Skin: Wash with soap and water. Get medical attention if irritation persists.
Inhalation: If inhaled, remove person to fresh air and Get medical attention.
Ingestion: Call a physician or poison control center immediately. Have the product label available for medical personnel to read.

Induce vomiting under the direction of medical personnel. Drink 1 or 2 glasses of water and induce vomiting by touching the back of throat with finger. If syrup of ipecac is available, give 1 tablespoon (15 ml) followed by 1 or 2 glasses of water. If vomiting does not occur within 20 minutes, repeat this dosage once. Do not induce vomiting or give anything by mouth to an unconscious person.

Note to Physician: This rodenticide contains an anticoagulant ingredient. If ingested, administer vitamin K_1 intramuscularly or orally, as indicated in bishydroxycoumarin overdoses. Repeat as necessary based on monitoring of prothrombin times.

For information on this pesticide product (including health concerns, medical emergencies, or pesticide incidents) call the National Pesticide Information Center at 1-800-858-7378.

SECTION 5 — FIRE FIGHTING MEASURES

Flash Point:	None
Autoignition Temp.:	Not determined
Explosive Limits:	LEL: Not applicable
	UEL: Not applicable
Extinguishing Media:	Use media suitable for the surrounding fire
Unusual Fire or Explosion Hazards:	None known
Fire Fighting Instructions:	Firefighters should wear self-contained breathing apparatus (full facepiece) and full protective clothing. Contain runoff to prevent pollution.

NFPA: 0 / 2 / 0

SECTION 6 — ACCIDENTAL RELEASE MEASURES

Large Spill/Leak Procedures: Isolate and contain spill. Limit access to the spill area to necessary personnel. Do not allow spilled material to enter sewers, streams or other waters. Scoop up spilled material and place in a closed, labeled container for use or disposal.
Small Spills: Scoop up material for use according to label instructions.

SECTION 7 — STORAGE AND HANDLING

Storage Requirements: Store in original container in a cool, dry area out of reach of children, pets and domestic animals. Do not contaminate water, food or feed. Keep container tightly closed. Do not remove or destroy the product label.
Handling Precautions: Read the entire product label before using this rodenticide. Carefully follow all cautions, directions and use restrictions on the label. Avoid contact with eyes, skin or clothing.

SECTION 8 — EXPOSURE CONTROLS/PERSONAL PROTECTION

Ventilation: Special ventilation is not required for the normal handling and use of this product when following the label instructions.
Protective Clothing/Equipment: Wear gloves when handling bait.
Respirator: None required when used according to label instructions.
Contaminated Equipment: Damaged or unwanted bait stations and bait holders should be wrapped in paper and discarded in trash.
Comments: Never eat, drink or smoke in work areas. Practice good personal hygiene after using this product. Wash arms, hands and face with soap and water after handling this product, and before eating and smoking. Launder contaminated clothing separate from street clothes.

SECTION 9 — PHYSICAL & CHEMICAL PROPERTIES

Physical State:	Solid blocks	Water Solubility:	Negligible
Color:	Tan (Mini Blocks)	% Volatile (Volume):	Not applicable
	Green (Paraffin Blocks)	Specific Gravity:	1.12 g/cc
Odor:	Raw grain odor	Vapor Pressure:	Not applicable
Melting Point:	Not available	Vapor Density:	Not applicable
Freezing Point:	Not applicable	pH:	Not applicable

SECTION 10 — STABILITY AND REACTIVITY

Stability: Stable
Conditions to Avoid: None
Hazardous Polymerization: Will not occur
Chemical Incompatibilities: Alkaline materials
Hazardous Products of Decomposition: Oxides of carbon

SECTION 11 — TOXICOLOGICAL INFORMATION

Eye Effects/Eye Irritation:	Mild, transient irritant
Acute Oral Effects:	LD_{50} (oral-rat): >5000 mg/kg
Acute Inhalation Effects:	No data available
Acute Dermal Effects:	LD_{50} (dermal-rabbit): >2000 mg/kg
Skin Irritation:	Non-irritating
Skin Sensitization:	Not a skin sensitizer

SECTION 12 — ECOLOGICAL INFORMATION

This product is toxic to fish and wildlife. Do not apply this product directly to water, where surface water is present or to intertidal areas below the mean high water mark. Carefully follow label cautions and instructions to reduce hazards to children, pets and non-target wildlife.

SECTION 13 — DISPOSAL CONSIDERATIONS

Disposal: Wastes resulting from the use of this product according to the label instructions must be disposed of as specified on the product label.
RCRA Waste Status: This product is not regulated as a hazardous waste under RCRA. State and local regulation may affect the disposal of this product. Consult your state or local environmental agency for disposal of waste generated other than by use according to label instructions.

SECTION 14 — TRANSPORT INFORMATION

Transportation Data (49 CFR): This product is not regulated as a hazardous material for all modes of transportation within the U.S.
Hazard Class: Not applicable   **ID No.:** Not applicable

SECTION 15 — REGULATORY INFORMATION

TSCA: All components of this product are listed on the TSCA inventory.
SARA Section 313: Contains no reportable components.
OSHA Hazard Classification: Chronic health hazard.
Proposition 65: Contains no components subject to warning requirement.

SECTION 16 — OTHER INFORMATION

Prepared by: T. Schmit   **Date:** 10/1/2005
Information presented on this Material Safety Data Sheet is believed to be accurate at the time of publication. No warranty, expressed or implied, is made with regard to this information. This information may not be adequate for every application, and the user must determine the suitability of this information due to the manner/conditions of use, storage or local regulation.

CONTRAC® All-Weather Blox MSDS

MANUFACTURER'S ADDRESS: BELL LABORATORIES, INC. 3699 Kinsman Blvd, Madison, WI 53704	PREPARED BY: PSM/CAR	TELEPHONE NO: (608) 241-0202	EMERGENCY PHONE NOS: Medical (877) 854-2494
PRODUCT NAME: CONTRAC® All-Weather Blox			Transportation (Spills) (800) 424-9300 CHEMTREC
USE: Anticoagulant Rodenticide	BAIT FORM: Formulated Dry Bait		EPA REGISTRATION NO: 12455-79

SECTION I. HAZARDOUS INGREDIENTS

INGREDIENT NAME	% BY WEIGHT	CURRENT TLV
Bromadiolone [3-[3-(4'-Bromo-[1,1'-biphenyl]-4-yl)-3-hydroxy-1-phenylpropyl]-4-hydroxy-2H-1-benzopyran-2-one] CAS No. 28772-56-7	0.005 %	N/A

This product contains no components subject to the reporting requirements of Section 313 of the Superfund Amendment and Reauthorization Act (SARA) of 1986

SECTION II. PHYSICAL DATA

APPEARANCE: Polygonal Block	COLOR: Blue	ODOR: Sweet, grain-like	SPECIFIC GRAVITY: 0.629 gm/cc
VAPOR DENSITY: N/A	MELTING POINT: N/A	WATER REACTIVITY: N/A	EVAPORATION RATE: N/A
VAPOR PRESSURE: N/A	BOILING POINT: N/A	SOLUBILITY: Not soluble in water	BULK DENSITY: N/A

SECTION III. FIRE AND EXPLOSION DATA

FLASH POINT (Method Used): N/A	FLAMMABLE LIMIT: Upper Limit: N/A Lower Limit: N/A	AUTOIGNITION TEMP: N/A

EXTINGUISHING MEDIA:
Extinguish with water, foam or inert gas.

SPECIAL FIREFIGHTING PROCEDURES:
Firefighters should be equipped with protective clothing and self-contained breathing apparatus.

UNUSUAL FIRE OR EXPLOSION HAZARDS:
None

SECTION IV. REACTIVITY HAZARD DATA

STABILITY: Stable	CONDITIONS TO AVOID: None
POLYMERIZATION: Will not occur	CONDITIONS TO AVOID: None
INCOMPATIBILITY (MATERIALS TO AVOID): Strongly alkaline materials	HAZARDOUS DECOMPOSITION PRODUCTS: Oxides of carbon

SECTION V. TOXICITY DATA

LD50, ORAL (INGESTION): >5000 mg/kg (rats)	LD50, DERMAL (SKIN CONTACT): > 5001 mg/kg (rats)	LC50, INHALATION: N/A
EYE IRRITATION: None (rabbits)	SKIN IRRITATION: None (rabbits)	DERMAL SENSITIZATION: Not considered a Sensitizer

CONTRAC® All-Weather Blox MSDS

SECTION VI. HEALTH HAZARDS

PRIMARY ROUTE OF ENTRY:
Ingestion

SIGNS & SYMPTOMS OF EXPOSURE:
Nausea, vomiting, loss of appetite, extreme thirst, lethargy, diarrhea, bleeding.

EMERGENCY FIRST AID PROCEDURES:
Eyes: Flush with cool water for at least 15 minutes. If irritation develops, obtain medical assistance.
Skin: Wash with soap and water.
Ingestion: Call physician or emergency phone number immediately. Do not give anything by mouth or induce vomiting unless instructed by physician.
Inhalation: None.

NOTE TO PHYSICIAN: If ingested, administer Vitamin K_1 intramuscularly or orally as indicated by bishydroxycoumarin overdoses. Repeat as necessary as based upon monitoring of prothrombin times.

SECTION VII. CONTROL AND PROTECTIVE MEASURES

RESPIRATOR TYPE:
Not required

EYE PROTECTION:
Not required

GLOVES (Recommended):
Rubber Gloves

VENTILATION:
Not required

OTHER PROTECTIVE MEASURES:
Not required

NATIONAL FIRE PROTECTION ASSOCIATION (NFPA) RATINGS:
HEALTH: 1 (Caution)   **FIRE:** 0 (Will not burn)   **REACTIVITY:** 0 (Stable)   **SPECIFIC HAZARD:** None

HAZARDOUS MATERIALS IDENTIFICATION SYSTEM (HMIS) RATINGS:
HEALTH: 2 (Moderate)   **FLAMMABILITY:** 0 (Minimal)   **REACTIVITY:** 0 (Minimal)   **PROTECTIVE EQUIPMENT:** B

SECTION VIII. SPILL OR LEAK PROCEDURES

STEPS TO BE TAKEN IN THE EVENT MATERIAL IS RELEASED OR SPILLED:
Sweep up spilled material, place in properly labeled container for disposal or re-use.

WASTE DISPOSAL METHOD:
Wastes resulting from use may be disposed of on-site or at an approved waste disposal facility. Dispose of all wastes in accordance with all Federal, state and local regulations.

SECTION IX. SPECIAL PRECAUTIONS AND STORAGE DATA

STORAGE TEMPERATURE:
Room temperature

AVERAGE SHELF LIFE:
Bait is stable for a minimum of 1 year when stored at room temperature.

SPECIAL SENSITIVITY (HEAT, LIGHT, MOISTURE):
Avoid exposure to light and extreme humidity

PRECAUTIONS TO BE TAKEN IN HANDLING AND STORAGE:
Store in a cool, dry place inaccessible to children, pets and wildlife. Keep container tightly closed when not in use. Avoid contamination of lakes, streams and ponds by use, storage or disposal. Wash thoroughly with soap and water after handling.

SECTION X. SHIPPING DATA

DOT SHIPPING NAME:
None required

DOT HAZARD CLASSIFICATION:
Non-hazardous

DOT LABELS REQUIRED:
None required

FREIGHT CLASSIFICATION:
LTL Class 60

WARRANTY: The information provided in this Material Safety Data Sheet has been obtained from sources believed to be reliable. Bell Laboratories, Inc. provides no warranties, either expressed or implied, and assumes no responsibility for the accuracy or completeness of the data contained herein. This information is offered for your consideration and investigation. The user is responsible to ensure that they have all current data relevant to their particular use.

MATERIAL SAFETY DATA SHEET

SECTION 1 — PRODUCT & COMPANY IDENTIFICATION

Maki® Paraffinized Pellets
EPA Reg. No. 7173-187

Maki® Pellets Place Packs
EPA Reg. No. 7173-188

Other Designation:	Anticoagulant rodenticide with Bromadiolone
Manufacturer:	Liphatech, Inc.
	3600 W. Elm Street, Milwaukee, WI 53209
Emergency Phone:	414-351-1476 Monday-Friday, 8:00 am-4:30 pm CST
After Hours:	Call CHEMTREC at 1-800-424-9300

SECTION 2 — INGREDIENT INFORMATION

Hazardous Ingredient:	CAS Number:	OSHA PEL:	ACGIH TLV:	ACGIH STEL:
Bromadiolone	28772-56-7	Not assigned	Not assigned	Not assigned

SECTION 3 — HAZARDS IDENTIFICATION

Emergency Overview: May be harmful if swallowed or absorbed through the skin, because this material may reduce the clotting ability of the blood and cause bleeding.

Primary Entry Routes: Oral (swallowing), dermal (absorption through skin)

Acute Effects (Signs and Symptoms of Overexposure):
- **Eyes:** May cause temporary eye irritation.
- **Skin:** May be harmful if absorbed through skin. Symptoms of toxicity include lethargy, loss of appetite, reduced blood clotting ability and bleeding.
- **Inhalation:** Due to this product's solid form, inhalation is unlikely.
- **Ingestion:** May be harmful if swallowed. Symptoms of toxicity include lethargy, loss of appetite, reduced clotting ability of blood, and bleeding.

Chronic Effects: Prolonged and/or repeated exposure to small amounts of product can produce cumulative toxicity. Symptoms of toxicity include lethargy, loss of appetite, reduced clotting ability of blood, and bleeding.

Medical Conditions Aggravated by Exposure: Bleeding disorders

Target Organs: Blood

Carcinogenicity: Contains no known or suspected carcinogens.

HMIS: Health – 2, Flammability – 0, Reactivity – 0

SECTION 4 — FIRST AID MEASURES

Eyes: Flush with water. Get medical attention if irritation persists.

Skin: Wash with soap and water. Get medical attention if irritation persists.

Inhalation: If inhaled, remove person to fresh air and Get medical attention.

Ingestion: Call a physician or poison control center immediately. Have the product label available for medical personnel to read.

Induce vomiting under the direction of medical personnel. Drink 1 or 2 glasses of water and induce vomiting by touching the back of throat with finger. If syrup of ipecac is available, give 1 tablespoon (15 ml) followed by 1 or 2 glasses of water. If vomiting does not occur within 20 minutes, repeat this dosage once. Do not induce vomiting or give anything by mouth to an unconscious person.

Note to Physician: This rodenticide contains an anticoagulant ingredient. If ingested, administer vitamin K_1 intramuscularly or orally, as indicated in bishydroxycoumarin overdoses. Repeat as necessary based on monitoring of prothrombin times.

For information on this pesticide product (including health concerns, medical emergencies, or pesticide incidents) call the National Pesticide Information Center at 1-800-858-7378.

SECTION 5 — FIRE FIGHTING MEASURES

NFPA: 2 / 0 / 0

Flash Point:	None
Autoignition Temp.:	Not determined
Explosive Limits:	LEL: Not applicable
	UEL: Not applicable
Extinguishing Media:	Use media suitable for the surrounding fire
Unusual Fire or Explosion Hazards:	None known
Fire Fighting Instructions:	Firefighters should wear self-contained breathing apparatus (full facepiece) and full protective clothing. Contain runoff to prevent pollution.

SECTION 6 — ACCIDENTAL RELEASE MEASURES

Large Spill/Leak Procedures: Isolate and contain spill. Limit access to the spill area to necessary personnel. Do not allow spilled material to enter sewers, streams or other waters. Scoop up spilled material and place in a closed, labeled container for use or disposal.

Small Spills: Scoop up material for use according to label instructions.

SECTION 7 — STORAGE AND HANDLING

Storage Requirements: Store in original container in a cool, dry area out of reach of children, pets and domestic animals. Do not contaminate water, food or feed. Keep container tightly closed. Do not remove or destroy the product label.

Handling Precautions: Read the entire product label before using this rodenticide. Carefully follow all cautions, directions and use restrictions on the label. Avoid contact with eyes, skin or clothing.

SECTION 8 — EXPOSURE CONTROLS/PERSONAL PROTECTION

Ventilation: Special ventilation is not required for the normal handling and use of this product when following the label instructions.

Protective Clothing/Equipment: None required for normal handling.

Respirator: None required when used according to label instructions.

Contaminated Equipment: Damaged or unwanted bait stations and bait holders should be wrapped in paper and discarded in trash.

Comments: Never eat, drink or smoke in work areas. Practice good personal hygiene after using this product. Wash arms, hands and face with soap and water after handling this product, and before eating and smoking. Launder contaminated clothing separate from street clothes.

SECTION 9 — PHYSICAL & CHEMICAL PROPERTIES

Physical State:	Solid pellets	Water Solubility:	Negligible
Color:	Green	% Volatile (Volume):	Not applicable
Odor:	Raw grain odor	Specific Gravity:	1.27 g/cc
Melting Point:	Not available	Vapor Pressure:	Not applicable
Boiling Point:	Not applicable	Vapor Density:	Not applicable
Freezing Point:	Not applicable	pH:	Not applicable

SECTION 10 — STABILITY AND REACTIVITY

Stability: Stable

Conditions to Avoid: None

Hazardous Polymerization: Will not occur

Chemical Incompatibilities: Alkaline materials

Hazardous Products of Decomposition: Oxides of carbon

SECTION 11 — TOXICOLOGICAL INFORMATION

Eye Effects/Eye Irritation:	Mild, transient irritant
Acute Oral Effects:	LD_{50} (oral-rat): >5000 mg/kg
Acute Inhalation Effects:	No data available
Acute Dermal Effects:	LD_{50} (dermal-rabbit): >2000 mg/kg
Skin Irritation:	Non-irritating
Skin Sensitization:	Not a skin sensitizer

SECTION 12 — ECOLOGICAL INFORMATION

This product is toxic to fish and wildlife. Do not apply this product directly to water, where surface water is present or to intertidal areas below the mean high water mark. Carefully follow label cautions and instructions to reduce hazards to children, pets and non-target wildlife.

SECTION 13 — DISPOSAL CONSIDERATIONS

Disposal: Wastes resulting from the use of this product according to the label instructions must be disposed of as specified on the product label.

RCRA Waste Status: This product is not regulated as a hazardous waste under RCRA. State and local regulation may affect the disposal of this product. Consult your state or local environmental agency for disposal of waste generated other than by use according to label instructions.

SECTION 14 — TRANSPORT INFORMATION

Transportation Data (49 CFR): This product is not regulated as a hazardous material for all modes of transportation within the U.S.

Hazard Class: Not applicable   **ID No.:** Not applicable

SECTION 15 — REGULATORY INFORMATION

TSCA: All components of this product are listed on the TSCA inventory.

SARA Section 313: Contains no reportable components.

OSHA Hazard Classification: Chronic health hazard.

Proposition 65: Contains no components subject to warning requirement.

SECTION 16 — OTHER INFORMATION

Prepared by: T. Schmit   **Date:** 12/1/2005

Information presented on this Material Safety Data Sheet is believed to be accurate at the time of publication. No warranty, expressed or implied, is made with regard to this information. This information may not be adequate for every application, and the user must determine the suitability of this information due to the manner/conditions of use, storage or local regulation.

MATERIAL SAFETY DATA SHEET

SECTION 1 — PRODUCT & COMPANY IDENTIFICATION

Maki® Mini Blocks
EPA Reg. No. 7173-202

Maki® Paraffin Blocks
EPA Reg. No. 7173-189

Other Designation:	Anticoagulant rodenticide with Bromadiolone
Manufacturer:	Liphatech, Inc.
	3600 W. Elm Street, Milwaukee, WI 53209
Emergency Phone:	414-351-1476 Monday-Friday, 8:00 am-4:30 pm CST
After Hours:	Call CHEMTREC at 1-800-424-9300

SECTION 2 — INGREDIENT INFORMATION

Hazardous Ingredient:	CAS Number:	OSHA PEL:	ACGIH TLV:	ACGIH STEL:
Bromadiolone	28772-56-7	Not assigned	Not assigned	Not assigned

SECTION 3 — HAZARDS IDENTIFICATION

Emergency Overview: May be harmful if swallowed or absorbed through the skin, because this material may reduce the clotting ability of the blood and cause bleeding.

Primary Entry Routes: Oral (swallowing), dermal (absorption through skin)

Acute Effects (Signs and Symptoms of Overexposure):
- **Eyes:** May cause temporary eye irritation.
- **Skin:** May be harmful if absorbed through skin. Symptoms of toxicity include lethargy, loss of appetite, reduced blood clotting ability and bleeding.
- **Inhalation:** Due to this product's solid form, inhalation is unlikely.
- **Ingestion:** May be harmful if swallowed. Symptoms of toxicity include lethargy, loss of appetite, reduced clotting ability of blood, and bleeding.

Chronic Effects: Prolonged and/or repeated exposure to small amounts of product can produce cumulative toxicity. Symptoms of toxicity include lethargy, loss of appetite, reduced clotting ability of blood, and bleeding.

Medical Conditions Aggravated by Exposure: Bleeding disorders

Target Organs: Blood

Carcinogenicity: Contains no known or suspected carcinogens.

HMIS: Health – 2, Flammability – 0, Reactivity – 0

SECTION 4 — FIRST AID MEASURES

Eyes: Flush with water. Get medical attention if irritation persists.

Skin: Wash with soap and water. Get medical attention if irritation persists.

Inhalation: If inhaled, remove person to fresh air and Get medical attention.

Ingestion: Call a physician or poison control center immediately. Have the product label available for medical personnel to read.

Induce vomiting under the direction of medical personnel. Drink 1 or 2 glasses of water and induce vomiting by touching the back of throat with finger. If syrup of ipecac is available, give 1 tablespoon (15 ml) followed by 1 or 2 glasses of water. If vomiting does not occur within 20 minutes, repeat this dosage once. Do not induce vomiting or give anything by mouth to an unconscious person.

Note to Physician: This rodenticide contains an anticoagulant ingredient. If ingested, administer vitamin K_1 intramuscularly or orally, as indicated in bishydroxycoumarin overdoses. Repeat as necessary based on monitoring of prothrombin times.

For information on this pesticide product (including health concerns, medical emergencies, or pesticide incidents) call the National Pesticide Information Center at 1-800-858-7378.

SECTION 5 — FIRE FIGHTING MEASURES

NFPA: 0 / 2 / 0

Flash Point:	None
Autoignition Temp.:	Not determined
Explosive Limits:	LEL: Not applicable
	UEL: Not applicable
Extinguishing Media:	Use media suitable for the surrounding fire
Unusual Fire or Explosion Hazards:	None known
Fire Fighting Instructions:	Firefighters should wear self-contained breathing apparatus (full facepiece) and full protective clothing. Contain runoff to prevent pollution.

SECTION 6 — ACCIDENTAL RELEASE MEASURES

Large Spill/Leak Procedures: Isolate and contain spill. Limit access to the spill area to necessary personnel. Do not allow spilled material to enter sewers, streams or other waters. Scoop up spilled material and place in a closed, labeled container for use or disposal.

Small Spills: Scoop up material for use according to label instructions.

SECTION 7 — STORAGE AND HANDLING

Storage Requirements: Store in original container in a cool, dry area out of reach of children, pets and domestic animals. Do not contaminate water, food or feed. Keep container tightly closed. Do not remove or destroy the product label.

Handling Precautions: Read the entire product label before using this rodenticide. Carefully follow all cautions, directions and use restrictions on the label. Avoid contact with eyes, skin or clothing.

SECTION 8 — EXPOSURE CONTROLS/PERSONAL PROTECTION

Ventilation: Special ventilation is not required for the normal handling and use of this product when following the label instructions.

Protective Clothing/Equipment: Wear gloves when handling bait.

Respirator: None required when used according to label instructions.

Contaminated Equipment: Damaged or unwanted bait stations and bait holders should be wrapped in paper and discarded in trash.

Comments: Never eat, drink or smoke in work areas. Practice good personal hygiene after using this product. Wash arms, hands and face with soap and water after handling this product, and before eating and smoking. Launder contaminated clothing separate from street clothes.

SECTION 9 — PHYSICAL & CHEMICAL PROPERTIES

Physical State:	Solid blocks	Water Solubility:	Negligible
Color:	Tan (Mini Blocks)	% Volatile (Volume):	Not applicable
	Green (Paraffin Blocks)	Specific Gravity:	1.12 g/cc
Odor:	Raw grain odor	Vapor Pressure:	Not applicable
Melting Point:	Not available	Vapor Density:	Not applicable
Freezing Point:	Not applicable	pH:	Not applicable

SECTION 10 — STABILITY AND REACTIVITY

Stability: Stable
Conditions to Avoid: None
Hazardous Polymerization: Will not occur
Chemical Incompatibilities: Alkaline materials
Hazardous Products of Decomposition: Oxides of carbon

SECTION 11 — TOXICOLOGICAL INFORMATION

Eye Effects/Eye Irritation:	Mild, transient irritant
Acute Oral Effects:	LD_{50} (oral-rat): >5000 mg/kg
Acute Inhalation Effects:	No data available
Acute Dermal Effects:	LD_{50} (dermal-rabbit): >2000 mg/kg
Skin Irritation:	Non-irritating
Skin Sensitization:	Not a skin sensitizer

SECTION 12 — ECOLOGICAL INFORMATION

This product is toxic to fish and wildlife. Do not apply this product directly to water, where surface water is present or to intertidal areas below the mean high water mark. Carefully follow label cautions and instructions to reduce hazards to children, pets and non-target wildlife.

SECTION 13 — DISPOSAL CONSIDERATIONS

Disposal: Wastes resulting from the use of this product according to the label instructions must be disposed of as specified on the product label.

RCRA Waste Status: This product is not regulated as a hazardous waste under RCRA. State and local regulation may affect the disposal of this product. Consult your state or local environmental agency for disposal of waste generated other than by use according to label instructions.

SECTION 14 — TRANSPORT INFORMATION

Transportation Data (49 CFR): This product is not regulated as a hazardous material for all modes of transportation within the U.S.

Hazard Class: Not applicable   **ID No.:** Not applicable

SECTION 15 — REGULATORY INFORMATION

TSCA: All components of this product are listed on the TSCA inventory.
SARA Section 313: Contains no reportable components.
OSHA Hazard Classification: Chronic health hazard.
Proposition 65: Contains no components subject to warning requirement.

SECTION 16 — OTHER INFORMATION

Prepared by: T. Schmit   **Date:** 10/1/2005

Information presented on this Material Safety Data Sheet is believed to be accurate at the time of publication. No warranty, expressed or implied, is made with regard to this information. This information may not be adequate for every application, and the user must determine the suitability of this information due to the manner/conditions of use, storage or local regulation.

MATERIAL SAFETY DATA SHEET

Prescription Treatment® brand 565 PLUS XLO® Formula 2

EMERGENCY PHONE NUMBERS:
MEDICAL: 800-225-3320 (Prosar)　　**TRANSPORTATION:** 800-424-9300 (Chemtrec)

SECTION 1. PRODUCT AND COMPANY IDENTIFICATION

Product Name: Prescription Treatment® brand 565 PLUS XLO® Formula 2
EPA Reg. No.: 499-290
Product Code: 02-0505 (12 x 20 oz)
EPA Signal Word: CAUTION
Distributed by: Whitmire Micro-Gen Research Laboratories, Inc.
3568 Tree Court Industrial Blvd.
St. Louis MO 63122-6682

SECTION 2. COMPOSITION/INFORMATION ON INGREDIENTS

COMPOSITION INFORMATION

ACTIVE INGREDIENTS (2.5%)	%	CAS NO.
Pyrethrins	0.5%	8003-34-7
Piperonyl Butoxide, Technical	1.0%	51-03-6
n-Octyl Bicycloheptene Dicarboximide	1.0%	113-48-4
OTHER INGREDIENTS* (97.5%)	**%**	**CAS NO.**
Dimethyl Ether	proprietary	115-10-6
2-propanol	proprietary	67-63-0

* All ingredients may not be listed. Ingredients not listed do not meet the reporting requirements of the OSHA Hazard Communication Standard (HCS) as specified in 29 CFR 1910.1200

EXPOSURE INFORMATION

Material	OSHA PEL STEL	OSHA PEL TWA	ACGIH TLV STEL	ACGIH TLV TWA
Pyrethrins	NE	NE	10 mg/m²	5 mg/m²
Piperonyl Butoxide, Tech.	NE	NE	NE	NE
n-Octyl Bicycloheptene dicarboximide	NE	NE	NE	NE
Dimethyl Ether	NE	NE	NE	1,000 ppm
2-propanol	500 ppm	400 ppm	500 ppm	400 ppm

SECTION 3. HAZARDS IDENTIFICATION

SIGNS/SYMPTOMS OF EXPOSURE
Prolonged exposure may cause skin irritation, and drying of nose and throat. May cause eye irritation.

ROUTES OF ENTRY
Primary: Skin　　**Secondary:** Ingestion　　**Tertiary:** Inhalation

HAZARDOUS DECOMPOSITION PRODUCTS
Thermal decomposition in open flame may result in carbon dioxide and carbon monoxide.

UNUSUAL FIRE, EXPLOSION AND REACTIVITY HAZARDS
Extremely Flammable: Contents under pressure. Exposure to temperatures above 130°F may cause bursting.

SECTION 4. FIRST AID MEASURES

Have the product container or label with you when calling a poison control center or doctor or going for treatment. Describe any symptoms and follow the advice given.

Ingestion: Immediately call a poison control center or doctor. Do not induce vomiting unless told to do so by a poison control center or doctor. Do not give ANY liquid to the person. Do not give anything by mouth to an unconscious person.
Skin Contact: Take off contaminated clothing. Rinse skin immediately with plenty of water for 15 - 20 min. Call a poison control center or doctor for treatment advice.
Eye Contact: Hold eyes open and rinse slowly and gently with water for 15 - 20 min. Remove contact lenses, if present, after the first 5 min, then continue rinsing eyes. Call a poison control center or doctor for treatment advice.
Inhalation: Move person to fresh air. If person is not breathing, call 911 or an ambulance, then give artificial respiration, preferably by mouth-to-mouth, if possible. Call a poison control center or doctor for further treatment advice.
Medical Conditions Generally Aggravated by Exposure: None known
Emergency Telephone Number of Prosar: 800-225-3320 (for medical emergencies)

SECTION 5. FIRE FIGHTING MEASURES

FIRE AND EXPLOSION
Flash Point (TOC): ≤ 20°F
Flame Extension: > 18"
Explosive Limits of propellant in Air (% by volume):
　Lower (LEL) = 3.4%　　Upper (UEL) = 18.0%
NFPA 30B Flammability: Level I Aerosol

UNUSUAL FIRE, EXPLOSION AND REACTIVITY HAZARDS
Extremely Flammable: Contents under pressure. Exposure to temperatures above 130°F may cause bursting.

IN CASE OF FIRE
Extinguisher Media: CO_2, Dry Chemical, Foam
Special Fire Fighting Procedures: None required

SECTION 6. ACCIDENTAL RELEASE MEASURES

IN CASE OF SPILL OR LEAK
If container begins to leak (through puncture, etc.), allow to discharge completely in well ventilated area, then dispose of as directed below.

This product contains the Comprehensive Environmental Response, Compensation, and Liability Act (CERCLA) listed chemical *Pyrethrins* which has a reportable quantity (RQ) of 1 lb. A release of more than 200 lb of this product (approx. 160 containers) is reportable to the National Response Center (800-424-8802).

Emergency Telephone Number of Chemtrec: (800) 424-9300 (for transportation spills)

PROTECTIVE EQUIPMENT FOR CLEANUP PERSONNEL
Eyes: Use proper protection – safety glasses are recommended.
Skin: Wear chemical-resistant gloves. (Good practice requires that gross amounts of any chemical be removed from the skin as soon as practical, especially before eating or smoking.)
Inhalation: Use of a respirator may be appropriate when working with spills in enclosed or confined spaces, or when prolonged exposure to product vapor or spray mist may occur. When using a respirator, wear a NIOSH approved respirator with an organic vapor (OV) cartridge or canister with any R, P or HE prefilter.

WASTE DISPOSAL METHOD
Take full or leaking containers to a local disposal company for biological treatment or incineration. Review all local, state, and federal regulations concerning health and pollution to determine approved disposal procedures. Do not contaminate water, food or feed by storage or disposal.

SECTION 7. HANDLING AND STORAGE

HAZARDS TO HUMANS AND DOMESTIC ANIMALS
Keep out of reach of children. CAUTION – Harmful if swallowed or absorbed through skin. Avoid contact with eyes, skin or clothing. Wash thoroughly with soap and water after handling.

PHYSICAL OR CHEMICAL HAZARDS
Extremely flammable. Contents under pressure. Keep away from heat, sparks and heated surfaces. Do not puncture or incinerate container. Exposure to temperatures above 130°F may cause bursting.

STORAGE
Store in a cool area away from heat or open flame and inaccessible to children. Do not contaminate water, food or feed.

SECTION 8. EXPOSURE CONTROL / PERSONAL PROTECTION

PERSONAL PROTECTIVE EQUIPMENT (PPE)
Respiratory Protection: None required for typical use exposure. See Section 6 for atypical exposure.
Protective Gloves: None required. Chemical-resistant gloves are recommended if prolonged or repeated skin contact is likely. (Good practice requires that gross amounts of any chemical be removed from the skin as soon as practical, especially before eating or smoking.)
Eye Protection: None required. Use proper protection - safety glasses are recommended.
Other Protective Equipment: None required

MATERIAL SAFETY DATA SHEET

Prescription Treatment® brand 565 PLUS XLO® Formula 2

VENTILATION
Local Exhaust: None required
Mechanical: None required
Special: None required
Other: Provide adequate ventilation. In hospitals and nursing homes, rooms should be ventilated for 2 hours after spraying. Do not return patients to room until after ventilation.

SECTION 9. PHYSICAL AND CHEMICAL PROPERTIES

Appearance: Sprays as a strong, dry mist
Odor: Characteristic Pyrethrin and IPA odor
Solubility in Water: Slight
Vapor Pressure: 55 psig @ 70°F
Viscosity: NE
Vapor Density: NE
Boiling Point: NE
Freezing/Melting Point: NE
pH (Conc.): 6.5 - 7.5 (1% in H_2O)
Density (Conc.): 0.85 g/ml @ 20°C

SECTION 10. STABILITY AND REACTIVITY

REACTIVITY
Stability: Indefinite when used according to label directions.
Conditions to Avoid: Do not spray into open flame or onto very hot surfaces.
Incompatibility (Material to Avoid): None known
Hazardous Polymerization: Will not occur

HAZARDOUS DECOMPOSITION PRODUCTS
Thermal decomposition in open flame may result in carbon dioxide and carbon monoxide.

SECTION 11. TOXICOLOGICAL INFORMATION

ROUTES OF ENTRY
Primary: Skin **Secondary:** Ingestion **Tertiary:** Inhalation

ACUTE
Eyes: May cause eye irritation. Produced slight, reversible conjunctival irritation in rabbits.
Skin: Prolonged exposure may cause skin irritation. Produced an average primary skin irritation score of 0.08 in rabbits. Not a dermal sensitizer to guinea pigs.
Ingestion: Acute oral toxicity (rats) LD_{50} = 4,730 mg/kg
Inhalation: Not likely due to the product being pressurized and producing particles large enough not to be respirable. Acute inhalation toxicity (rats) LC_{50} > 7.4 mg/L. A few cases of extrinsic asthma from pyrethrin mixtures have been reported. Prolonged exposure may cause drying of nose and throat.

CHRONIC / CARCINOGENICITY
Neither this product nor any of its ingredients are classified as carcinogens by the National Toxicity Program (NTP), the International Agency for Research on Cancer (IARC) or the Occupational Safety and Health Administration (OSHA).

SECTION 12. ECOLOGICAL INFORMATION

This product is toxic to fish. Do not apply directly to water, or to areas where surface water is present or to intertidal areas below the mean high water mark. Do not contaminate water by cleaning of equipment or disposal of wastes.

SECTION 13. DISPOSAL CONSIDERATION

Do not contaminate water, food or feed by disposal. Dispose of container and wastes in accordance with all federal, state and local regulations.
Container Disposal: Do not puncture or incinerate! Empty container by using the product according to the label directions. Offer empty container for recycling, if available, or place in trash if allowed by state and local regulations. If container is partly full, contact your local solid waste agency or call 1-800-CLEANUP for disposal instructions.
Waste Disposal: Wastes resulting from the use of this product may be disposed of on site or at an approved waste disposal facility.

SECTION 14. TRANSPORT INFORMATION

SHIPMENT BY GROUND WITHIN U.S. (DOT CLASSIFICATION)
Proper Shipping Description: Consumer Commodity, ORM-D
Package Marking: Consumer Commodity, ORM-D
Package Labeling: No label required
Certified Packaging: Not required for Consumer Commodity exception

SHIPMENT BY WATER (IMDG CLASSIFICATION)
Proper Shipping Description: Aerosols, 2.1 UN 1950
Package Marking: Dangerous Goods in Limited Quantities of Class 2.1
Package Labeling: No label required
Certified Packaging: Not required for Limited Quantity exception

SHIPMENT BY AIR (ICAO/IATA CLASSIFICATION)
Proper Shipping Description: Aerosols, flammable, 2.1, UN 1950
Proper Shipping Marking: Aerosols, 2.1, UN 1950, LTD QTY
Package Labeling: Class 2, Flammable Gas
Certified Packaging: Not required for Limited Quantity exception
Other Requirements or Instructions: Complete air bill with packaging instruction "Y203, ERG code 10L."

SECTION 15. REGULATORY INFORMATION

CERCLA
This product contains the CERCLA listed chemical *Pyrethrins* which has a reportable quantity (RQ) of 1 lb.

SARA TITLE III SECTION 311/312 HAZARD CLASS
This product has been reviewed according to the EPA "Hazard Categories" promulgated under Section 311 and 312 of the Superfund Amendment and Reauthorization Act of 1986 (SARA Title III) and is considered, under applicable definitions, to meet the following categories:
Acute Health Hazard
Fire Hazard
Release of Pressure (puncture) Hazard

SARA TITLE III SECTION 313 CHEMICALS
This product contains the following substances subject to the reporting requirements of Section 313 of Title III of the Superfund Amendments and Reauthorization Act of 1986 and 40 CFR Part 372:
Piperonyl Butoxide, Technical   1.0% by weight

TSCA
All components of this product are listed or excluded from listing on the US Toxic Substance Control Act (TSCA) Chemical Substance Inventory.

SECTION 16. OTHER INFORMATION

NFPA HEALTH RATING INFORMATION
HEALTH - 1   FLAMMABILITY - 2   REACTIVITY - 1

HMIS HAZARD RATING INFORMATION
HEALTH - 1   FLAMMABILITY - 2   REACTIVITY - 1

KEY: 4 = Severe
3 = Serious
2 = Moderate
1 = Slight
0 = Minimal

The information and recommendations contained herein are based upon data believed to be correct. However, no guarantee or warranty of any kind, expressed or implied, is made with respect to the information contained herein. **For automatic MSDS updates, register at www.wmmg.com.**

Questions concerning the safe handling of the product should be referred to the Whitmire Micro-Gen Customer Service Department at 800-777-8570.

NA - Not Applicable
NE - Not Established
PEL - Permissible Exposure Limit
TLV - Threshold Limit Value
STEL - Short Term Exposure Limit (15 min.)
TWA - Time Weighted Average (8 hrs.)

WHITMIRE MICRO-GEN
RESEARCH LABORATORIES, INC.

Effective Date: 9/9/05
Review Date: 9/9/05
Supersedes: 6/8/04
Text ID: 060413-6
Code #: 165-010
Part Number: 19-0131-6

Smith's Candy Company Inspection Form
Plant 1: 1997 Sinclair Lane

Date: _____   Inspector: _____

RODENT/PEST INSPECTION FORM

_____
_____ (plant # or plant address)

Ketch-alls: 1____, 2____, 3____, 4____, 5____, 6____, 7____, 8____, 9____, 10____, 11____, 12____, 13____,
14____, 15____, 16____, 17____, 18____, 19____, 20____, 21____, 22____, 23____, 24____, 25____, 26____,
27____, 28____, 29____, 30____, 31____, 32____, 33____, 34____, 35____, 36____, 37____, 38____, 39____,

Bail Stations: 1____, 2____, 3____, 4____, 5____, 6____, 7____, 8____, 9____, 10____, 11____, 12____, 13____,
14____, 15____, 16____

Insectecutors: A____, B____, C____

Comments: _____
_____
_____

Contract Signature Check: 1____, 2____, 3____, 4____, 5____, 6____, 7____, 8____, 9____, 10____, 11____,
12____, 13____, 14____, 15____, 16____, 17____, 18____, 19____, 20____, 21____, 22____, 23____, 24____,
25____, 26____, 27____, 28____, 29____, 30____, 31____, 32____, 33____, 34____, 35____, 36____, 37____,
38____, 39____

Legend: Ketch-alls •   Bait Stations x   Insectecutors A,B,C

COLORADO DEPARTMENT OF AGRICULTURE
700 Kipling St., Suite 4000, Denver, CO 80215-5894

CERTIFIED OPERATOR LICENSE

Expiration Date: 08/04/2004
I.D. Number: 03851
Name: DAVID M. ANAYA

is licensed, and therefore certified to purchase and use restricted use pesticides in these categories:

CATEGORIES

305 •

SIGNATURE

COLORADO DEPARTMENT OF AGRICULTURE
700 Kipling St., Suite 4000, Denver, CO 80215-5894

QUALIFIED SUPERVISOR LICENSE

Expiration Date: 08/04/2004
I.D. Number: 03851
Name: DAVID M. ANAYA

is licensed, and therefore certified to purchase and use restricted use pesticides in these categories:

CATEGORIES

301 • 302 • 303 • 304 • • • • • • • • • • • • • • • •

SIGNATURE

Pest Control Program

PEST CONTROL SERVICE REPORT

Service Date _____ Day _____ Time In _____ Time Out _____

RODENT STATION INSPECTION/SERVICE:

Station Number	Activity Noted:	Comments:
_____	_____	_____
_____	_____	_____
_____	_____	_____
_____	_____	_____
_____	_____	_____
_____	_____	_____
_____	_____	_____
_____	_____	_____
_____	_____	_____
_____	_____	_____
_____	_____	_____
_____	_____	_____

SERVICE PERFORMED:

Insecticides Used:	Concen.:	Quantity:	Location:	Target Pest:
_____	_____ %	_____	_____	_____
_____	_____ %	_____	_____	_____
_____	_____ %	_____	_____	_____
_____	_____ %	_____	_____	_____
_____	_____ %	_____	_____	_____
_____	_____ %	_____	_____	_____
_____	_____ %	_____	_____	_____
_____	_____ %	_____	_____	_____
_____	_____ %	_____	_____	_____

Rodenticides Used:	Concen.:	Quantity:	Comments:
_____	_____ %	_____	_____
_____	_____ %	_____	_____
_____	_____ %	_____	_____
_____	_____ %	_____	_____
_____	_____ %	_____	_____
_____	_____ %	_____	_____
_____	_____ %	_____	_____
_____	_____ %	_____	_____

Signed _____   Signed _____

9 Allergen Program

PROGRAM TYPE: REQUIRED

THEORY

One of the fastest growing areas of concern within the food industry is that of allergen identification and control. This stems from separate yet equal driving forces squeezing each manufacturer, processor, distributor, and importer both financially and legally. On one hand, there are numerous reports relating to the increasing number of people who are allergic or sensitive to compounds commonly found in foods; on the other hand, the federal government increasingly forces manufacturers to recall or withdraw or suggests that they recall or withdraw mislabeled or harmful foodstuffs for allergenic reasons. To prevent significant harmful financial disaster due to these types of ingredients, it is imperative that a manufacturer understands its incoming raw materials, the source of the raw materials, and the pathway that the ingredients travel from source to end product. Before a company establishes an allergen detection and prevention program, it is important to understand the substances to be targeted, human sensitivity, and the current regulatory guidelines.

The substances to be targeted by the allergen detection and prevention program can be separated into two groups: those that the federal government recognizes and an additional set of ingredients that cause reactions but are not on the federal list. Currently, there are eight major foods or food groups that account for 90% of food allergies. These are milk, eggs, fish, crustacean shellfish, tree nuts, peanuts, wheat, and soybeans. Ingestion of these ingredients causes a portion of the population to have an adverse reaction—generally to a protein contained therein. These reactions may be as simple as coughing, wheezing, runny nose, rash, diarrhea, itchy eyes, or upset stomach or may be as severe as a bleeding colon, anaphylactic shock, and even death. In addition to the "big eight" there are several other ingredients or chemicals that should be of concern to the food manufacturer, such as seeds, monosodium glutamate (MSG), sulfites, aspartame/phenylalanine, and colors. Generally, these are lesser known, but not necessarily less harmful, than the big eight and the reactions to them are usually not allergenic but are due to food intolerance. Unfortunately for the food manufacturer, the ingredients that contain these substances come in many forms, including:

Milk and milk derivatives include but are not limited to whey, whey protein concentrate, nonfat dry milk, skim milk, buttermilk, butter, butter oil, casein sodium caseinate, cheese, curds, custard, cream, yogurt, milk chocolate, most other chocolate, cocoa, lactalbumin, and lactoglobulin, etc.

Egg and egg derivatives include but are not limited to egg whites, egg yolks, whole eggs, egg lecithin, egg albumin, meringue, and mayonnaise.

Fish and fish derivatives include any fish with fins, all fish parts, fish oil (e.g., codfish, turbo, halibut, salmon, and anchovies).

Shellfish and shellfish derivatives include but are not limited to shellfish (e.g., crab, lobster shrimp, oysters, prawns, crayfish, abalone, mussels, clams, scallops, and squid).

Tree nuts and tree nut derivatives include but are not limited to almonds, walnuts, pecans, Brazil nuts, hazelnuts/filberts, pine nuts, macadamia nuts, cashews, chestnuts, pistachios, nut oil, nut butters, and nut extracts (does not include coconut or coconut oil).

Peanuts and peanut derivatives include but are not limited to peanuts, peanut meal, peanut butter, hydrolyzed peanut protein, and peanut oil (unrefined or cold pressed).

Wheat and wheat derivatives include but are not limited to wheat germ, wheat germ oil, wheat starch, modified wheat starches, semolina, durum flour, wheat flour, hydrolyzed wheat protein, farina, pasta, macaroni products, cracker meal, vinegar made from wheat, wheat syrup, graham flour, and cereal extracts.

Soy and soy derivatives include but are not limited to soy protein, soy protein concentrate, soy protein isolate, soy lecithin, hydrolyzed soy protein, soy sauce, soybean oil, miso, tofu, textured vegetable protein, soy nuts, and soy flour.

Seed and seed derivatives include but are not limited to sesame seeds, sesame seed oil, sunflower seeds, sunflower oil, mustard, and celery seeds.

Monosodium glutamate (MSG) includes but is not limited to MSG, soy sauce, and textured vegetable protein.

Yellow #5 includes FD&C yellow #5, tartrazine, and ingredients containing yellow #5.

Yellow #6 includes FD&C yellow #6, sunset yellow, and ingredients containing yellow #6.

Mollusks include but are not limited to clams, mussels, oysters, cockles, scallops, octopus, squid, whelk, periwinkle, cuttlefish, abalone, and snail.

Gluten nonwheat and derivatives include but are not limited to rye, barley, and oats.

Sulfites and derivatives include but are not limited to sodium metasulfite, sodium bisulfite, sulfur dioxide, dried fruits, and wine.

To complicate this even further for manufacturers, each of these components can be a subcomponent of other ingredients and not declared by the original manufacturer. These include the broad classifications of flavors, natural flavors, spices, and processing aids. With the ever increasing level of processing of base ingredients into subcomponents, it is even more important for manufacturers to understand what is in their ingredients and where they came from.

Human sensitivity to these ingredients varies from person to person, published report to published report, advocacy group to advocacy group, and governmental body to governmental body. These sensitivity ranges are from 1 part per million (ppm) up to numbers in excess of 100 ppm. Although these numbers are fascinating and important to manufacturers and governmental agencies, they are extremely important to those customers who do have a reaction. Therefore, it must be extremely important to all manufacturers to determine the allergen-containing components of their ingredients and end products.

Current regulatory position is contained under the Food Allergen Labeling and Consumer Protection of 2004 (FALCPA), which addresses the labeling of the big eight. It states that the word *contains* followed by the name of the food source from which the major food allergen is derived is printed immediately after or is adjacent to the list of ingredients. It also states that a flavoring, coloring, or incidental additive that is, bears, or contains a major food allergen will be subject to the food labeling law. The FDA does not currently have any threshold values for allergen content but is in the regulatory hearing stage of determining a threshold value for gluten. Early indications are that the agency is leaning toward 20 ppm. Unfortunately, because allergens even at the lowest level may pose a health risk to a small portion of the population, it is imperative to determine and correctly label them.

To compose an effective strategy for the identification and handling of allergen-containing ingredients, a three-pronged approach must be taken. First, the allergens contained within the ingredients, packaging, and processing aids used in the facility must be identified. This is done by communicating with each supplier and getting an ingredient statement. The ingredients contained therein should be evaluated to determine if it is possible that subingredients that are not being disclosed are present. If so, the supplier must be contacted to obtain a full listing. If the supplier will not disclose the complete ingredient list, they need to be asked for an allergen-free statement with every lot (see HACCP program, risk analysis). Next, places within the distribution system, the storage system, and the processing system where any cross-contamination does or may occur must be evaluated. This includes vertical and horizontal warehousing, mixed product loads, common air systems, common processing lines, common storage tanks and valves, employee movement, and forklift movement. Specifically, places where ingredients and packaging are transported or change format should be sought. Some common areas of cross-contamination are pallets of product containing an allergen placed above pallets of product that do not contain allergens or producing an allergen-containing product on a line next to a non-allergen-containing product.

Second, all packaging labels should be reviewed to ensure their compliance with FALCPA. It is necessary to verify that wherever substances are present in the ingredients, even at low amounts, the label has a "contains..." statement. Although the tendency for some manufacturers is to put a "contains..." or a "may contain..." statement on all their labels, this opens the company to greater review by regulatory agencies. In short, it is important to determine accurately what is or may be in the product and then accurately label it.

Third, a testing program that accurately verifies the allergen declaration on the label should be developed. Currently, several companies manufacture allergen testing units. These include Neogen, Tepnel Biosystems, Biadiagnostics R-Biopharm AG, Charm Sciences, and bioMarieux SA and the testing is based on enzyme-linked immunosorbent assay technology. Each test is usually specific for a single protein, such as peanut, dairy, tree nut, etc., and most have been validated for most matrices. Depending on the test unit and method, these tests can be conducted rapidly (as quickly as 5 minutes) or for as long as 1 hour or more. Most of this extended time is due to sample preparation but provides an accurate quantitative measure of the allergenic protein. Typically, the accuracy of these tests exceeds 20 ppm, with some reportedly accurate down to 1 ppm.

Integration of this type of testing into the formal company allergen program involves validation of sanitation methods, validation of core ingredients, and verification of finished goods. Validation of the typical food plant cleanup from an allergen standpoint requires the sampling of rinse water, from wet cleanups, or line swabbing, from dry cleanups, and then conducting the appropriate allergen test. In general, only one type of allergen test needs to be completed. For example, if the plant manufactures a product containing peanuts and then cleans in preparation to manufacture a product that does not contain peanuts, it would be appropriate to utilize a peanut allergen kit to determine if the line has been cleaned satisfactorily. On the other hand, if a company were bottling soymilk and then cleaned and sanitized in preparation to produce regular milk, it would be appropriate to conduct a test on the rinse water for soy proteins before running milk. After conducting any of these validation tests, it is important to record the results as part of the preoperational checklist or as part of the processing changeover records.

When all of the preceding steps are completed, the final administrative step in development of the allergen program is to write the allergen control program document. This should include a general statement of the program's compliance with the Food Allergen Labeling and Consumer Protection Act of 2004, a section on the strategy used to identify allergens within the food stream, and a section on how unwanted allergens will be prevented from entering the food stream. There should also be a reference to testing methods, training, program documentation, and who is responsible for development and administration of the program.

As a final step for program implementation, due to the legal, financial, and health implications of allergens, it is important to make sure that every employee on staff understands what allergens are, where they are found, what cross-contamination is, how allergens affect people's health, and finally, what the consequences will be if stray allergens enter the food supply. General or specific meetings should be conducted with the staff and to share the importance of the knowledge of allergens. Upon completion of this training, each employee should sign an initial training verification for the allergen testing program that states that the employee understands the program and will be vigilant in maintaining a separation of allergens and will report allergen contamination to his or her immediate supervisor.

APPLICATION

A risk assessment worksheet for allergen food hazards is provided on the CD (Book 1_HACCP Program\Section 4_Hazard Analysis:Allergen Risk Analysis) and at the end of this chapter. This can be completed during the hazard analysis (see HACCP program development) for each finished product. The name of the product,

Allergen Program

date completed, and product item number at the top of the form should be filled in. Then, for each ingredient in the finished product, the item number and ingredient description should be filled in and the allergens contained in the ingredient should be marked. Most ingredients are pretty straightforward, but they should be evaluated for each of the allergens. For those ingredients that are complex or that the supplier does not make clear which, if any, are contained therein, the supplier should be sent an allergen information request form. An easy template to send to suppliers for the acquisition of the allergen information can be found on the CD (Book 2_Quality Control Manual\Section 5_Allergen Program:Allergen Information Request Form) and at the end of this chapter. Once the analysis is complete, the results should be put into the summary report found on the CD (Book 2_Quality Control Manual:Section_5 Allergen Program:Allergen Inclusion Chart) and at the end of this chapter. This form provides the manufacturer a basic chart for recording the allergens contained within each finished good; when it is complete, it provides a rapid reference for the presence of allergens within the company.

Next, it is necessary to establish the testing regimen for postsanitation inspection and for changeover cleaning regarding allergens and to determine where these results will be recorded. On the CD (Book 2_Quality Control Manual\Section 7_Inspection Program:Daily Plant Sanitation Inspection Form) and at the end of this chapter is a template for a preoperational sanitation form. The allergen testing results can be written under the "other areas" section. This form dictates that production does not occur until the production supervisor (operations) and the quality manager (quality control) agree that the plant is sanitary and meets all allergenic requirements. Another device for the recording of allergen tests can be found on the CD (Book 2_ Quality Control Manual\Section 5_Allergen Program:Allergen Testing Form) and at the end of this chapter. This may be used either in the laboratory or on the production floor for preoperational startup checks or changeover checks.

Another important part of the allergen program is the training of employees. The key to this is to set up training meetings for all employees in which the serious nature of allergens and the company's allergen program are discussed. At these meetings, the leader should explain what allergens are, in what they are contained, and the serious nature of the threat they pose to the general population. On the CD (Book 2_ Quality Control Manual\Section 5_Allergen Program:Initial Training Verification) and at the end of this chapter is a template for the manufacturer's employees to sign stating that they have been trained and understand the program.

The final part of creating the allergen program is to write the program document. The form on the CD (Book 2_Quality Control Manual\Section 5_Allergen:General Overview) and at the end of this chapter can be customized to reflect the details of the allergen program as designed by the company. The quality control manager should sign this document and place it in the quality manual under the allergen program tab.

SUPPLEMENTAL MATERIALS

Allergen Program

HACCP Principle 1

RISK ASSESSMENT WORKSHEET FOR **ALLERGEN** FOOD HAZARDS

Product Name: _____   Date: _____

Product Item Number: _____

Finished Product

	How Stored			Potential Introduction of Hazard Y/N	If yes, is the hazard detected or removed by the customer
	Frozen	Refrigerated	Room Temp		
Company Warehouse					
Distributor					
Retailer					
Customer					

Raw Material Analysis

Item #	Ingredient	Milk	Eggs	Fish	Shellfish	Tree Nuts	Peanuts
1							
2							
3							
4							
5							
6							
7							
8							
9							
10							
11							
12							
13							
14							
15							

Item #	Ingredient	Wheat	Soybeans	Seeds	MSG	Sulfites	Colors
1							
2							
3							
4							
5							
6							
7							
8							
9							
10							
11							
12							
13							
14							
15							

ALLERGEN INFORMATION REQUEST FORM
(ONE FORM PER ITEM NUMBER)

Date:	Manufacturer's Name:
Item #:	Ingredient Name:
Manufacturer's Item #:	Manufacturer's Ingredient Name:

Does the ingredient contain any of the following major food allergens or ingredients associated with food sensitivities, **even if in minor or trace amounts, such as incidental additives or processing aids?** (Please indicate presence even if component is found in ingredient listing.) If this ingredient is produced on shared equipment with other allergens or sensitizers, please indicate.

(Column 1) Allergenic Food Source or Its Derivatives	Does this ingredient contain any foods listed in Column 1?	Provide the specific source food (eg., soy lecithin, peanut extract, fish gelatin)	Does this Ingredient contain any protein from Column 1 Foods? Indicate percentage in this ingredient.		Is this ingredient produced on shared equipment with other products containing protein from foods from Col. 1?
	Yes/No	Common or Usual Name	Yes/No	%	Yes/No
Peanuts					
Dairy / Milk					
Egg					
Fish					
Shellfish					
Soy					
Wheat					
Tree Nuts & Tree Nut Oils					
Seeds & Seed Oils					
Gluten Sources other than wheat					
Sulfites (at any amount)			NA		

Name: _____ Title _____ Date _____

(Person completing form)

Signature_____ Phone: _____

Please send completed form to

_____
_____
_____
_____

Allergen Program

Guidelines for Completion of the Form

Survey your ingredient suppliers using the same type of form as the attached. If the ingredient contains any amount (even less than 1 ppm) of one of the following allergenic food components, report that source on the attached form as being contained in the ingredient. Determine from your supplier's information or through analysis (using any recognized standard method, for example Elisa allergen detection kits), if the component contains <u>any</u> protein from the allergenic source, and list the protein source on the form. Example: Soy Sauce (Soybeans, Wheat and Water): List protein from the soy and from the wheat.

<u>Peanuts and Peanut Derivatives</u>
Includes but is not limited to Peanuts, Peanut Meal, Peanut Butter, Hydrolyzed Peanut Protein, Peanut Oil. Indicate if oil is highly refined, deodorized, and bleached, if known.

<u>Dairy/Milk and Milk Derivatives</u>
Includes but is not limited to Whey, Whey protein concentrate, Nonfat dry Milk, Skim Milk, Milk, Buttermilk, Butter, Butter Oil, Casein, Sodium Caseinate, etc.

<u>Egg and Egg Derivatives</u>
Includes but is not limited to Egg Whites, Egg Yolks, Whole Eggs, Egg Lecithin, Egg Albumin

<u>Fish and Fish Derivatives</u>
Includes but is not limited to any fish with fins such as Codfish, Turbo, Halibut, Anchovies (Common in Worcestershire Sauce), etc., and Fish Oil. Listing should name actual fish name, if known, not just a generic "fish extract".

<u>Shell Fish and Shell Fish Derivatives (includes Mollusks)</u>
Includes but is not limited to Crab, Lobster, Shrimp, Oysters, Prawns, Crayfish, Abalone, Mussels, Clams, Scallops, Squid, Etc. List actual variety; do not use generic names.

<u>Soy and Soy Derivatives</u>
Includes but is not limited to Soy Protein, Soy Protein Concentrate, Soy Lecithin, Hydrolyzed Soy Protein, Soy Sauce (Soybeans, Water), and Any Soybean Oil. Indicate if oil is highly refined, deodorized and bleached, if known.

<u>Wheat and Wheat Derivatives</u>
Includes but is not limited to Wheat Germ, Wheat Germ Oil, Wheat Starch, Modified Wheat Starches, Semolina, Durum Flour, Wheat Flour, Hydrolyzed Wheat Protein, Farina, Pasta, Macaroni Products, Cracker Meal, Vinegar made with Wheat, Wheat Syrup, etc.

<u>Tree Nuts and Tree Nut Derivatives</u>
Includes but is not limited to Almonds, Walnuts, Pecans, Brazil Nuts, Hazelnuts/Filberts, Pine Nuts, Pistachios, etc, Nut Oil (Indicate if highly refined, bleached, and deodorized, if known), Nut Butters, Nut Extracts. Does not include Coconut or Coconut oil.

<u>Seed, and Seed Derivatives:</u>
Includes but is not limited to, Sesame Seeds, Sesame Seed Oil, Sunflower Seeds, Sunflower Oil, Mustard, and Celery. Indicate if seed oil is highly refined, bleached, deodorized, if known.

<u>Gluten Sources other than Wheat</u>
Includes all products (including HVPs, modified starches, buckwheat, brans, and bran extracts) obtained from rye, barley, triticale, spelt, kamut, oats or their hydrolyzed strains. Corn (maize), rice, sorghum, is not included.

<u>Sulfites (at any amount)</u>
Includes but is not limited to Sulfur Dioxide, Sodium Bisulfite, Sodium Sulfite, etc. Please note if source of sulfites is naturally occurring rather than added as a preservative.

Allergen Inclusion Chart

Item Number	Item Description	Soy	Wheat	Peanuts	Almonds	Milk	Eggs	Non-wheat Gluten	Mustard / Celery Seeds	Other Tree Nuts	Fish / Shellfish	MSG	Sulfites	Colors

DAILY PLANT SANITATION INSPECTION FORM

Date: _____   **Time:** _____

Production Room:
_____
_____
_____
_____
_____

Warehouse:
_____
_____
_____
_____
_____

Restrooms:
_____
_____
_____

Dock:
_____
_____
_____

Other Areas:
_____
_____
_____

AUTHORIZATION TO RUN:

Production Supervisor: _____

Quality Manager: _____

Allergen Testing Form

Sample #	Date	Item #	Description	Vendor	Lot Code	Dairy	Peanut	Almond	Gluten	Other	Notes	Tech

Additional Comments: _____

ALLERGEN TESTING PROGRAM

INITIAL TRAINING VERIFICATION

I have read and understand the _____ allergen testing program and will be vigilant in maintaining a separation of allergens as required. I understand that milk, dairy, soy, eggs, tree nuts, fish, shellfish, seeds, and wheat/gluten are the main allergens, but that other compounds such as sulfites, MSG, and colors are also of concern. Furthermore, I will report any actual or possible allergen contamination to my immediate supervisor.

    I understand that allowing or purposefully contaminating a product with an undeclared allergen will result in disciplinary action.

_____   _____
Employee   Date

_____   _____
Quality Control Manager   Date

ALLERGEN TESTING PROGRAM

GENERAL OVERVIEW

SECTION A: GENERAL

In accordance with the Food Allergen Labeling and Consumer Protection Act of 2004 (FALCPA), _____'s finished good shall contain no undeclared food allergens. The purpose of this program is to ensure that all materials purchased and processed for, or by, _____ are free from undeclared food allergens. These allergens include Milk, Peanuts, Eggs, Soy, Fish, Shellfish, Seeds, Tree Nuts, Wheat and Cereal Gluten. _____'s allergen program is characterized by two parts, exclusion and verification.

SECTION B: EXCLUSION

B1: Raw Materials

All raw materials shall undergo a thorough allergen examination. This shall be conducted by the Quality Control Department **prior to an ingredient from ANY NEW SUPPLIER is purchased** and shall be done in accordance with _____'s supplier certification and HACCP programs.

It shall be _____'s policy to exclude all known allergens from our products unless they are known constituents of an ingredient and the allergen is declared on the package.

B2: Manufacturing

_____ shall require that all products are manufactured in a manner that excludes undeclared allergens and allergens introduced due to cross-contamination. These procedures include all or any combination of:

Using specified ingredients from approved suppliers
Manufacturing using allergen excluding scheduling
Warehousing using methods that prevent cross contamination

SECTION C: VERIFICATION

C1: Raw Materials

When a raw material has the potential, either through the nature of the ingredient or the method in which it is processed, to contain an allergen, _____'s quality control shall require that the supplier test the ingredient on a lot by lot basis and provide documentation of it's cleanliness via a Certificate of Analysis COA. COA's shall either be faxed or emailed to _____'s Quality Control **PRIOR** to usage by the supplier.

Raw materials and packaging shall be reviewed upon receipt as part of the HACCP program.

Allergen Program

C2: Manufacturing

Based in the ingredient and allergen matrix for each finished product, allergen testing may be performed to verify the lack of an allergen introduction during manufacturing.

C3: Testing

As a method to determine the efficacy of the allergen program _____'s quality control department shall conduct finished product audits. These will involve the drawing of random samples and testing them for specific allergens. These may include dairy, gluten, peanut, tree nut or other allergens.

SECTION D: TESTING METHODS

Allergen testing shall be conducted using either rapid tests or Elisa tests. Association of Official Analytical Chemist International approved Tepnel, Biadiagnostic or Neogen tests will be used. _____ reserves the right to utilize and outside laboratory for this testing.

SECTION E: TRAINING

All employees shall undergo periodic training to facilitate a complete understanding of _____'s allergen program, what allergens are, their physical ramifications, cross-contamination issues and legal obligations. This training shall be documented and placed in the employees file.

SECTION F: DOCUMENTATION

All allergen testing shall be documented on the daily production reports or in an allergen testing log. A general list of allergens contained within raw materials shall be maintained on a semi-annual basis or as needed.

SECTION H: RESPONSIBILITY

It is the responsibility of the Quality Control department to oversee the allergen program. In conjunction with purchasing and operations, quality control shares the responsibility of enforcing and monitoring this policy.

_____   _____
Quality Control Manager   Date

10 Weight Control Program

PROGRAM TYPE: REQUIRED

THEORY

Every product sold at the retail level in the United States contains a fill designation. This is in the form of weight, ounces or pounds, or as volume fluid ounces. The Code of Federal Regulations 21CFR101 states that the principal display panel of the label must be in U.S. pounds (lbs) and ounces (oz) and metric scale kilograms (kg) and grams (g) for a dry product and fluid ounces (fl. oz) and metric scale milliliters (ml) for a liquid product. The net weight must be listed as the minimum net weight, not the average net weight. Therefore, it is important that if the company is selling a product to a customer, it is able to prove that it meets this federal guideline because at various times the product may be purchased by the Department of Weights and Measures, which conducts studies to determine if it truly does meet the standard. To ensure that all products produced meet federal fill guidelines, a weight/fill control program needs to be established, documented, and monitored.

There are several different methods to document an accurate filling system, including online weight control, filler control documentation, and utilization of a check weigh unit. Online weight control involves the random sampling of filled retail products directly off the manufacturing line as they are being produced. It involves the development of a weight control recording chart and then designating an employee to pull samples for measuring. This person may be an operation or quality control employee as long as he or she is trained to pull the samples and record the data correctly.

When developing the weight control chart, it is helpful to design it in such a manner that it has a place to tabulate the numerical weights as well as a place to chart the mean, the standard deviation, or both. Providing a venue for charting yields a visual representation of the weights measured. Employees can utilize both sections as guides as they adjust the filling machines to keep them in tolerance.

Typically, a random sample of five units is pulled, weighed, charted, and placed back on the line. Since there are hundreds of books available to the food manufacturer that deal with the topic of statistical weight control, suffice it to say for this book that, as a general rule, half the samples should be above the target, or label weight, and half should be below. The sample weights should fall between a set upper control limit (UCL) and a set lower control limit (LCL)—typically ±5% of the target weight, with no one sample below the maximum allowable variance (MAV). For smaller companies that do not want to put out a large capital outlay on an automatic filler or a check weight unit, the manual recording of weights and measures might be the most economical way to go. It is more labor intensive but does provide the documents needed to satisfy intense regulatory scrutiny.

The next step up from manual weight control is the use of filler control documentation or using a check weigh machine. Filler control documentation utilizes a computer-controlled filling machine. These can be programmed to record the weight of the product for each package. At the end of the run, this information may be downloaded to an external computer for analysis or printed directly from the filler. A less costly automatic approach is to install an in-line check weigh unit. These are manually programmed to weigh each package and remove all those packages that are outside the upper and lower control limits. At the end of the run, the check weigh unit may be downloaded into another computer for data analysis or the data may be sent directly to a printer for manual analysis. Either way, this is a good method for collecting the data required to verify that the packages meet label weight requirements.

After determining how the data will be collected, it is essential to determine who will be responsible for evaluating it to verify that there were no production problems on the line. This responsibility normally ends up as part of operation's duties, but may be a quality control function as well. If weight discrepancies are found, it is important to put the product on hold for further evaluation using proper hold and release procedures.

When all the components of the weight control program are complete, the next step is to write the formal weight control program document. CD (Book 2_Quality Control Manual\Section 6_Weight Control Program:General Overview) and at the end of the chapter may be used to document the company's weight control program. This usually takes the form of a general statement referring to the regulations requiring accurate measurements and then a section on how the measures will be taken. This is followed by sections on how the weights will be documented, how they will be verified, and who will be responsible for this verification. After the program document has been completed and approved, it should be signed and placed in the quality manual along with a copy of the record keeping sheet.

APPLICATION

The CD (Book 2_Quality Control Manual\Section 6_Weight Control Program:General Overview) and the form at the end of the chapter provide a template for a weight control program general overview. This should be customized to fit the processing situation; specifically, in section B, the method of gathering the data and how often the weights will be taken should be added. In section C, the method of how the weight records will be documented and where the finished weight control data will be stored should be added. In section D, a line should be added that specifies who is responsible for reviewing the collected data.

The CD (Book 2_Quality Control Manual\Section 6_Weight Control Program: Weight Control Chart) and the end of this chapter provide a handy recording chart for use if the method of data collection is manual. Before use, the target weight, upper control limit (UCL), and lower control limit (LCL) should be determined for the product being manufactured and should be written on the chart next to UCL, target, and LCL on the right side of the chart. On the top of the form, the date and the product spaces should be filled in. Then, at the intervals determined by the company, the time of sampling, initials of who is gathering the samples, and the sample number are filled in. The samples are gathered and the weights recorded in spaces one through five. The weights are added together and divided by 5; this is the mean average and it is recorded in the space marked "mean." Then, a dot is placed on the mean chart in relation to where the weight would be to the target on the vertical line directly under where the weights were recorded above. If the standard deviation is being charted, it should be calculated and charted on the bottom chart.

EXAMPLES

A demonstration of how a manual weight control chart is filled out can be found on the CD (Book 2_Quality Control Manual\Section 6_Weight Control Program:Weight Control Chart Example A) and at the end of this chapter. Note the specifications filled in on the right side and the weight breakdown on the left side of the chart.

SUPPLEMENTAL MATERIALS

WEIGHT CONTROL PROGRAM

GENERAL OVERVIEW

Section A: General

In accordance with the 21cfr101, _____ shall ensure that all products produced will conform to federal label weight rules. The purpose of this program is to ensure that all produced are randomly tested for weight using statistical process control.

Section B: Weights

On a scheduled basis finished product shall be weighed using a statistical platform. At a minimum, a random sample will be weighed every 15 minutes of running time if utilizing a manual weight control scheme.

Section C: Documentation

These weights will be documented on an appropriate documentation sheet.

Section D: Verification

Completed documentation sheets shall be reviewed to make sure that the products meet all legal statutes.

Section E: Responsibility

It is the responsibility of the Quality Control department to oversee the weight control program. In conjunction with operations, quality control shares the responsibility of enforcing and monitoring this policy.

Approved by: _____ Date: _____

WEIGHT CONTROL CHART

DATE: _____

Product															
Time															
Operator															
Sample #															
1															
2															
3															
4															
5															
sum															
mean															

MEAN CHART ($\bar{X}$)

UCL
Target
LCL

STANDARD DEVIATION CHART (σ)

EXAMPLES

WEIGHT CONTROL CHART

DATE: 15-Jul-07

Product	Tuna 6oz	Tuna 6oz	Tuna 6oz	Tuna 6oz
Time	7:41	8:06	8:23	8:48
Operator	TR	TR	TR	TR
Sample #	1	2	3	4
1	163.23	170.05	177.66	165.22
2	168.31	163.46	162.34	170.19
3	171.49	176.59	165.89	177.23
4	170.01	173.99	174.44	161.99
5	176.46	162.11	173.23	162.34
sum	849.5	846.2	853.56	836.97
mean	169.90	169.24	170.71	167.39

MEAN CHART (X)

UCL 178.60
Target 170.10
LCL 161.60

181.44
178.60
175.77
172.94
170.10
167.27
164.44
161.60
158.77

STANDARD DEVIATION CHART (σ)

6.500
6.000
5.500
5.000
4.500
4.000
3.500
3.000
2.500
2.000
1.500
1.000

11 Inspection Program

PROGRAM TYPE: REQUIRED

THEORY

Clutter; excessive down time; doors, walls and ceilings in disrepair; poor grounds keeping; utensils lying around; and poorly maintained equipment are symptoms of an underlying failure to design the equipment and facility in a manner that builds in food safety or to maintain the equipment and facility properly, yielding a gradual decline in product protection. Typically, small to medium-sized companies focus on parts of the business that directly generate income. It often seems as if there are competing priorities between making a product that makes money and spending money to maintain the facility and equipment that already work and can make a product that makes money. This forms an endless loop, and although this seems like a good strategy at the time, it can lead to a gradual production of conditions that fail to protect the product from physical, chemical, or microbial control. To prevent this from happening, a more balanced approach needs to be applied.

As a basic component of the company's good manufacturing practices (GMPs) program in support of 21CFR110.10, 110.20 part b (1–6), 110.35 parts a–e, 110.40 parts a–g, and 110.80 part b (15–17), which state that "the plant management shall take all reasonable measures and precautions to ensure...," a formal inspection program needs to be established. This program should involve the development of an inspection team. Typically, this is formed around management from various job functions of the company, including but not limited to plant manager, maintenance manager, quality control manager, owner, warehouse manager, and other staff members who have the authority and responsibility to see that the assigned tasks are completed. Together, the team of employees develops daily and monthly inspection strategies that entail development of inspection documentation and assignment of inspection personnel.

The monthly inspection program is designed to take a hard, close look at all aspects of the operation to determine if they contribute or may contribute to an unsanitary condition that might lead to poor food safety. The approach for this inspection is to gather the designated employees as a team and do a complete inspection of all equipment, floors, walls, ceilings, racks, forklifts, hand-washing stations, restrooms, employee locker rooms, doors, curtains, and all other interior surfaces and components of the business. Then, it is necessary to proceed outside and walk around the outside of the facility inspecting all walls, doors, docks, roads, and other exterior facets of the facility. During this inspection, the team members should be evaluating what they are looking at in four basic categories: maintenance, sanitation, cleanliness, and personnel GMPs.

Maintenance deficiencies are those that pertain to the structure or equipment; they fall into several areas that are specifically designated for evaluation by the federal government as included in the GMPs and should be reviewed before each inspection. Specific questions that address the items the inspection team should look for include:

Does the facility have adequate sanitary facilities and are they in working order?
Does the facility have adequate water to perform correct sanitation procedures?
Is the floor adequately sloped for drainage and cleaning?
Is the floor in good repair?
Does the building have good wall-to-floor and wall-to-ceiling joints?
Are all cracks in the floor, walls, and ceilings sealed?
Is the facility compartmentalized through the adequate use of doors, air curtains, or plastic curtains?
Are there adequate hand-washing stations and are they in correct working order?
Are machines well maintained?
Are outside vents screened?
Do utensils have a designated and sanitary storage location?
Is there proper signage for employee wash stations?
Is there backflow prevention?
Are all plants at least 2 feet from the exterior wall of the building?
Is there an 18-inch barrier between all interior walls, racks, and miscellaneous equipment?

Sanitation issues that the team should look for involve those items that are meant to prevent microbial contamination. These would include:

Is there any buildup of product on any surface of the equipment?
Are cleaning chemicals stored in a separate area away from food?
Are non-food-contact surfaces clean?
Can the equipment be properly cleaned and sanitized?
Are freezers and refrigerators clean?
Are restrooms and locker rooms clean?
Are all surfaces and overheads clean?

Cleanliness issues for the team to review are those related to general day-to-day work habits and would include:

Is garbage or refuse disposal capacity adequate?
Are any products, packaging, or chemicals on the ground rather than on pallets?
Are things organized?
Is there an excess amount of clutter?

Members of the monthly inspection team need to look very closely at the facility and write down everything they see that needs to be repaired or cleaned.

Inspection Program

Upon conclusion of the inspection, the team should sit down and review the list, which should be prioritized based on what poses the greatest risk to the safety of the food. Once the list is prioritized, an employee should be assigned to each item. He or she will be responsible for seeing that the item is corrected and signing off that the repair has been completed.

The daily inspection revolves around the continual verification that the employee GMPs are being followed. Although it is every employee's duty to be vigilant in enforcing GMPs, the task of documenting compliance usually ends up with the quality control staff or a production supervisor. This daily inspection must be documented and submitted with the daily paperwork for evaluation by management. CD (Book 2_Quality Control Manual Section 7_Inspection Program:Daily Plant Inspection Form) and at the end of the chapter may be used for this documentation.

When all the components of the internal inspection program are complete, the next step is to write the formal internal inspection program document. This usually takes the form of a general statement referring to the regulations regarding compliance to the GMPs, followed by statements regarding the daily and monthly inspection procedures and how they will be documented. Finally, it references who is responsible for conducting and auditing the internal inspection program.

APPLICATION

The first step in developing an effective inspection program is to designate the people who will be assigned to the team. The form on the CD (Book 2_Quality Control Manual\Section 7_Inspection Program:Inspection Team Roster) and at the end of this chapter can be used to designate the employees on the team and to identify the departments they represent. This ensures that all departments are represented so that the inspectors approach the task with varying viewpoints. The team leader should also be designated on this form.

Next, the designated inspection team convenes and begins the inspection. A form on the CD (Book 2_Quality Control Manual\Section 7_Inspection Program:Monthly Inspection Form) and at the end of this chapter can be used by the inspection team to record the findings. On the top of the form, the date of the inspection and the names of the inspectors are filled in. Then, for every deficiency, the type—(M) maintenance, (S) sanitation, (C) cleanliness, and (G) GMPs—is filled in, followed by a short description of the problem that was found. This list is continued until a complete inspection of the inside and outside of the facility and the equipment has been completed.

Upon completion of the inspection, the team should sit down with the list and prioritize it with an eye toward those that are most important to complete to prevent contamination of the food, as well as how easy they are to complete. When this is done, a person should be listed as responsible for fixing the deficiency in the third column. It is his or her responsibility to determine how it will be corrected, obtain the resources, and complete the task. When finished, he or she needs to fill out the last three columns on the inspection form: action taken, completed by, and date completed.

Periodically during the month, the inspection team leader should review the inspection form and verify that items on the list are being completed. It is this person's responsibility to work toward completion. This is not to say that each and every deficiency must be completed each month. There must be recognition that some things are easier to correct and that some things are more costly; some things require the plant to be out of operation to be repaired. The key is to keep working at a gradual improvement in maintenance with an eye toward protecting the holding, manufacturing, and distribution of the food.

A general template that outlines the company's internal inspection program can be found on the CD (Book 2_Quality Control Manual\Section 7_Inspection Program:General Overview) and at the end of this chapter. It should be modified to reflect what actually occurs.

SUPPLEMENTAL MATERIALS

INTERNAL INSPECTION PROGRAM

INSPECTION TEAM ROSTER

The following employees have been designated as members of the _____ _____ internal inspection team. The team will meet on a monthly basis to make a list of maintenance, sanitation, cleanliness, and employee GMP deficiencies. Furthermore, responsibility for correcting each of the items on the list will be assigned by the team. It is the responsibility of the team leader to confirm that they have been completed in a timely manner.

Employee Name	Department Represented
_____	_____
_____	_____
_____	_____
_____	_____
_____	_____
_____	_____
_____	_____
_____	_____

The team leader is: _____

Inspection Program

MONTHLY INSPECTION FORM

Inspection Date: _____   Inspectors: _____

Type*	Deficiency	Assigned To	Action Taken	Completed By	Date Completed

* M = maintenance, S = sanitation, C = cleanliness, G = GMP's

INTERNAL INSPECTION PROGRAM

GENERAL OVERVIEW

Section A: General

In accordance with 21CFR110.20 part b (1–6), 110.35 parts a–e, 110.40 parts a–g, and 110.80 part b (15–17) good manufacturing practices, procedures shall established that adequately inspect the production area, building and grounds, and equipment to ensure that food produced therein is manufactured in a clean, wholesome manner. Food safety and sanitation inspection shall be performed via a two-pronged inspection program, daily inspection, and monthly inspection.

Section B: Daily Inspection

As part of the daily production run, employee GMPs shall be evaluated. Areas to be examined will be clothing, personal items, jewelry, gum and other food, personal sanitation, hairnets, gloves, garments, and area cleanliness.

Documentation
The daily inspection shall be documented on the line check form or another designated form. This will be turned in with the daily production/quality documentation.

Responsibility
It is the responsibility of the production supervisor or quality control technician to inspect the employees and their area for compliance to the good manufacturing practices program.

Section C: Monthly Inspection

At the beginning of each month, the entire facility shall be inspected by operations and quality management. This inspection will categorize deficiencies found into four classifications:

Maintenance issues relate to the structural integrity of the manufacturing facility itself. Examples of maintenance issues are doors not sealing; racks, walls, and equipment-damaged by forklifts; integrity issues with the roof and walls; lights blown out; and other damage.

Sanitation issues involve poor cleaning practices. They differ from cleanliness and GMP issues because they build up over time during normal production and warehousing. Examples of sanitation issues are warehouse and freezer floors with product buildup and dirty forklifts.

Cleanliness issues relate to product buildup that has occurred that day or recently. They are generally issues that are not buildups over time. Examples of cleanliness issues are overfilled trash cans, products on the floor, and spilled ingredients.

Inspection Program

GMP issues involve good manufacturing practices. Examples of GMP issues are people not wearing hairnets, beard nets, or correct garments; hot water in restrooms not satisfactory; no towels or soap in the restrooms; and trash clutter around workstations.

Documentation

Monthly inspections shall be documented on the monthly inspection form.

Responsibility

It is the responsibility of the plant manager and quality control manager to inspect the facility. During this inspection, responsibility for correcting the deficiency shall be assigned. When completed, it will be noted on the form. After all of the deficiencies are corrected, the form is filed with the daily paperwork.

Approved by: _____ Date: _____

DAILY PLANT SANITATION INSPECTION FORM

Date: _____   Time: _____

Production Room:

_____
_____
_____
_____
_____

Warehouse:

_____
_____
_____
_____
_____

Restrooms:

_____
_____
_____

Dock:

_____
_____
_____

Other Areas:

_____
_____
_____

Authorization to run:

Production Supervisor: _____

Quality Manager: _____

12 Sanitation Program

PROGRAM TYPE: REQUIRED

THEORY

One of the key components of the microbiological control program is that of the sanitation program. Along with the certificate of analysis program and the product testing program, the sanitation program provides a chemical or mechanical pathway or some combination of both to eliminate the substrate needed for microbiological organisms to grow and thrive. In some form, this is a program that each and every company must have to eliminate risk of a contamination. Its importance is outlined in 21CFR110.35b.2d(1–3):

> (1) Food-contact surfaces used for manufacturing or holding low-moisture food shall be in a dry, sanitary condition at the time of use. When the surfaces are wet-cleaned, they shall, when necessary, be sanitized and thoroughly dried before subsequent use. (2) In wet processing, when cleaning is necessary to protect against the introduction of microorganisms into food, all food-contact surfaces shall be cleaned and sanitized before use and after any interruption during which the food-contact surfaces may have become contaminated. Where equipment and utensils are used in a continuous production operation, the utensils and food-contact surfaces of the equipment shall be cleaned and sanitized as necessary. (3) Non-food-contact surfaces of equipment used in the operation of food plants should be cleaned as frequently as necessary to protect against contamination of food.

Although it is easy for a company to take the stance that it can just put things away at the end of the day and make the equipment and processing area look clean, to meet the preceding criteria, a significantly more structured approach must be taken. This includes determining the general method of cleanup, establishing a master cleaning schedule, designing and documenting the specific cleaning methods to be used, and establishing a sanitation inspection program.

To determine the general method of cleanup within the processing area, the manufacturer or processor must decide between a dry and a wet cleanup or develop a combination thereof. This decision is made by evaluating the types of products manufactured, ingredients used, and the type and complexity of the processing or manufacturing equipment. Dry cleanups are typically used for manufacturing facilities where the raw materials carry a very low microbial load or that use a single-ingredient process. Products with low microbial loads would include a flour drying, grinding, and blending operation that heat-treats the ingredients, mills them, and then blends and packages them dry. The process uses the heat as a microbial kill step prior to packaging. The processing line and facility can be properly cleaned by sweeping, vacuuming, and wiping all surfaces without the use of water. Single ingredient processes such as coffee or cocoa processing that use beans that are dried, roasted, and then ground are also best suited for a dry cleanup. In the case of cocoa, the use of water as a cleaning medium causes multiple other processing problems. Another example of a situation where a dry cleanup would be appropriate is a repack operation where the raw materials come into the facility after undergoing a bacteriological kill step. These would be repacked aseptically and a dry cleanup could be employed, unless the ingredients contained organic material that adhered to the food contact surfaces.

The strategy for conducting a dry cleanup in the manufacturing plant is to use a top-down approach and then carefully clean the residual accumulated on the bottom surface. This means that the cleaning team should begin by wiping, vacuuming, blowing, or sweeping off the highest points to be cleaned first. The particles are knocked down onto a lower surface, then the next lower layers are cleaned, and so on until all of the layers to the floor are cleaned. Finally, the floor is carefully swept so as not to kick any dust into the air that might land back on the upper surfaces. Care should be taken to ensure that the material from a lower level is not spread back to an upper layer.

A wet cleanup is typically used for products that have a high bacterial load and high moisture or fat content, are a blend of numerous other ingredients, or adhere to the food contact surfaces during processing, leaving a buildup. Some products that lend themselves to this type of cleanup are dairy products, fruit and vegetable processing, fresh lettuce and spinach processing, jam and jelly manufacturing, sauce manufacturing, candy chopping, cookie dough manufacturing, meat processing, premade meals, and ice cream manufacturing. Each of these products contains moisture or fat levels that adhere to surfaces throughout the facility, thus making it necessary to use mechanical or chemical cleaning methods. These products also contain a microbial load that must be eradicated on a daily or periodic basis to prevent cross-contamination to the finished product or subsequent products.

The strategy for a wet cleanup is to compartmentalize the operation and then utilize clean-in-place (CIP) and clean-out-of-place (COP) systems. Clean-in-place systems are closed loops within the processing line that involve pumping water and/or cleaning solutions through tanks, lines, valves, and pumps. Equipment that is not within the loop is removed and placed in a COP tank and cleaned manually. The rest of the processing facility should be COP cleaned by a water and chemical wash-down with mechanical abrasion. All cleaning follows the rinse, clean, rinse, and then sanitize protocol.

Clean-in-place systems require four components: proper chemicals, adequate time, adequate temperature, and online system documentation. The chemicals used in CIP systems are separated into two groups: cleaners and sanitizers. Cleaners are those substances used to digest residual organic material and typically require specific application temperatures to improve their efficacy and specific contact times to ensure their effectiveness. Most common cleaners are separated into two groups: alkaline cleaners and acid cleaners. Alkaline cleaners are, by definition, cleaners that have pH values greater than 7.0, although they are usually in the 9- to 14-pH range. They may be anionic, cationic, or neutral. Neutral cleaners are lower foaming and are more suitable for sprayers and agitators; cationic and anionic cleaners take more rinsing than neutral cleaners. Alkaline cleaners are normally derivatives of sodium or potassium hydroxide and are excellent for most processing applications. Acid cleaners are those that have a pH below 7.0 but are normally in the 1- to 3-pH range. Both acid and alkaline cleaners can be prepared as foaming or not foaming. Foaming cleaners adhere in the foamed state to the surface to be cleaned, so their application is normally on exterior surfaces. Nonfoaming cleaners lend themselves to CIP cleaning systems.

Sanitizers are chemicals that kill or inhibit the growth of microbial flora on food contact surfaces and are used after the caustic cleaner is completely rinsed off. They come in two forms—leave on and rinse off—and have ammonia, iodine, or chlorine as a basis. Currently, most food processors utilize leave-on sanitizers on all surfaces due to the glove-like protection they provide. These can be left on the surfaces for several hours and still maintain their high effectiveness.

Wet COP cleanups not involved in either the CIP or COP tank systems are completed through a four-step operation: warm-water rinse, chemical wash, hot-water rinse, and sanitizer. First all remaining product is removed using a warm-water rinse. This may be done using a high-pressure in-line wall-mounted system or with a portable unit such as a Hotsey. Often, to remove the heaviest buildup of food particles, mechanical methods must be involved. This includes the use of brushes of various sizes and shapes, scrapers, and "elbow grease." It is imperative that all of the surfaces have the particulates removed to increase the efficacy of the chemical cleaner.

Next, a chemical cleaner is applied as a means to remove all remaining organic surface material. The choice of chemical depends largely on the type of organic material to be removed. After application, the chemical is left standing on the surfaces to digest the particulates. Depending on the type of chemical, temperature of the chemical applied, and type of surface, this standing time can range between 10 and 30 minutes.

Some facilities may choose to utilize a combination of a dry cleanup and a wet cleanup. An example of this type of manufacturing facility is a bakery that is separated into a mix room and a baking room. In the mix room, there are raw materials that contain moisture and possibly a high microbial load. These ingredients are mixed and deposited onto baking trays that are subsequently passed into an oven located in a baking room. This situation lends itself to using a wet cleanup in the mix room and a dry cleanup in the baking room.

Obviously, there are many options and combinations regarding what method to use for cleanup. One resource the manufacturer has at its fingertips to bring clarity and expedience to the development of the sanitation program is to contact an outside chemical and sanitation system provider (CSSP) such as EcoLab (www.ecolab.com). Companies such as this will come to the manufacturing facility and offer expertise on many levels of sanitation system setup, procedure documentation, and chemical safety system implementation. Depending on the expertise of internal personnel and the state of the equipment and documentation in place, the CSSP will evaluate the current cleaning methods and help the company decide the best method of cleaning the entire facility; what chemicals, if any, should be used; how the necessary chemicals should be stored and handled; a staffing level estimate; training for sanitation personnel; and manuals to support the entire system. Utilizing the CSSP offers the company a rapid shortcut to the development of a complete and thorough sanitation program specifically designed to minimize the risk of microbial contamination from static product buildup within the facility.

Once the type of cleaning method is determined, the next step in developing an effective sanitation program is the development of the master sanitation schedule (MSS). This is a document that lists all of the equipment and areas inside and outside the facility, including walls, floors, overheads, forklifts and pallet jacks, processing equipment, restrooms, office areas, outside areas, warehousing areas, and miscellaneous storage areas. It initially lists all of the main equipment and functional areas; subsequently each is broken into its subparts and then broken down until all of the parts, pieces, and areas are listed. In most plants and for most first-time MSS developers, this process can become overwhelming, which can lead to an incomplete or unwieldy document. To overcome this, the developer must communicate carefully with all departments and utilize patience. Working closely with each department lets them understand what the goal is and gives the developer a clear and complete picture of all the components to be added to the final document along with the challenges faced on a daily basis to accomplish the job. As this document is prepared, care should be taken to ensure that it is organized and all inclusive to ensure that every area of the company is cleaned.

A simple way to document the MSS is through the use of either Microsoft Excel or third-party software such as KLEANZ by Nexcor Technologies. Excel can be used to document the entire list of components to be cleaned on the first tab of the workbook; then, on subsequent tabs, the items can be listed based on how often they are to be cleaned. Each tab can be printed as needed and the document can be altered easily as the business grows and changes. The KLEANZ software is more complex, but offers the developer a sanitation administration program that automates scheduling, creates sanitation documents, tracks training, and manages chemical inventories. Software programs can be costly and yet convenient for the small to medium size food processing and manufacturing company; however, the use of them does not remove the responsibility of understanding all of the components within the plant that need to be cleaned and does not eliminate the time needed to develop and organize the MSS.

Begin the development of the master sanitation schedule by listing the major parts or processes of the facility. Next, under each major component, list the subcomponents of the area or machine. Continue to evaluate each subcomponent until the smallest parts are listed. An example would be:

    Processing room
        Conveyor 1 belt
            Belt–food contact surface
            Motor
            Wheels
        Urschel RA-A chopper
            Blades–food contact surface
            Motor
            Wheels

Sanitation Program

Screener
    Trays
    Screens
    Anticlogging rings
    Motor
    Wheels
Conveyor 2 bucket
    Buckets–food contact surface
    Chain track
    Motor
    Base
Ishida scale
    Buckets
    Chutes
    Support pieces
    Electronics
    Stand
    Walkway
    Stairs
Hasson twin tube form filler
    Chute
    Electronics
    Track
    Stand
    Wires and hoses
    Seal mechanisms
Conveyor 3 belt
    Belt–food contact surface
    Motor
    Wheels
Metal detector
    Belt
    Stand
    Wheels
    Electronics
Tape machine
Walls
Floors
    Drains
Ceiling
    Ceiling
    Fixtures
    Fire sprinkler pipes
    Air vents
COP tank
Tables

Because it is not possible to know or understand how all of the equipment within the plant is taken apart for cleaning, it is imperative that the MSS developer work within each department to gain an understanding of its goals and procedures.

After listing all of the parts of the facility and equipment, next to each item list how frequently it should be cleaned to ensure that no product or other material buildup is allowed to form. For example, all of the food contact surfaces need to be cleaned on a daily basis, but the walls may need to be cleaned only on a weekly basis and the overheads may only need to be cleaned on a monthly basis. When deciding how frequently each item needs to be cleaned, it is important to discuss the issue with others within the company as well as to make an independent assessment; too often, what is being done currently is not a thorough cleaning method or not occurring frequently enough. It is critical during this assessment that the frequency decided on for each item is conservative due to the overwhelming negative consequences of not cleaning often enough, including regulatory intervention and loss of all business. Once the basic list is developed, it can be sorted based on frequency and then printed for use as needed.

The third step in establishing the complete sanitation program is designing and documenting the specific cleaning methods to be used for each item listed on the master sanitation schedule. This involves developing a "cleaning document" for each item that includes what is to be cleaned, tools needed, chemicals needed, safety precautions, quantity of personnel needed, how the cleaning will be inspected, who will inspect it, and what will be done if it is found to be out of spec. Although this documentation step takes a considerable amount of time, especially for a complex plant, it is critical; failure to complete this portion will lead to inadequate sanitation and inspection.

The final step in developing the sanitation program is the establishment of methods for inspecting how thorough the cleaning process is. These can be as simple as a visual examination or as scientific as a microbiological or bioluminescence analysis, but they must be appropriate to the level of risk taken when the area is insufficiently clean. For example, areas such as the warehouse, office, or closets are appropriately inspected visually because they do not affect the integrity of the finished food stuff. Food contact surfaces, however, should be inspected stringently. This can be by inspecting the time and temperature records for the CIP system, along with a supporting visual inspection of the surfaces, or by using a standard swabbing method. Swabs are available in various styles from several manufacturers and utilize bioluminescence technology.

The underlying science of this tool is that the swab collects organic material residue from the surface and is complexed with the enzyme luciferase, commonly found in the flash of a firefly. When this meets adenosine triphosphate (ATP), the primary source of energy in living organisms, bioluminescence occurs. The immediate readable light emission is directly proportional to the amount of ATP present. The bioactivity is measured either in relative light units (RLUs) or in total bacteria cell count by a scintillation spectrometer or a luminometer. Studies indicate that dirty surfaces will differ in their levels of ATP depending on what type of product is present. However, clean surfaces, which should contain little or no ATP, should be similar in level of light output regardless of the type of food processed. Although the presence of indicator organisms is not necessarily a danger, that presence signals the potential existence of more serious pathogens

This technique does not indicate that the surface is free from microbiological organisms but rather acts as an indicator of how thorough the cleaning process was. It should be used in conjunction with a thorough visual inspection program. Whichever method or combination of methods is used for inspection, the fact that the area was inspected should be recorded on an appropriate document. This can be the monthly inspection form, the nightly cleanup shifts scheduled duties form, the pre-operational inspection form, or the master sanitation form—but it must be recorded.

A key component of the inspection process is determining and documenting all actions to be taken if the inspected area is not sufficiently cleaned. In most cases this will be as simple as recleaning the food contact surface or area and then reinspecting it; however, it may be as complicated as disassembling a piece of equipment to reclean it adequately and then conducting a microswab test. Once the area is recleaned, it must be reinspected and documented. The inspection–documentation procedure provides a closed-loop system of verifying that no shortcuts are taken during sanitation, which leads to minimizing safety hazards from getting introduced into the food supply.

APPLICATION

The first step in developing the sanitation program is to create the master sanitation schedule. A template found on the CD (C:\Final CD\Book 2_Quality Control Manual\Section 8_Sanitation Program:Master Sanitation Schedule-Combined) and at the end of this chapter can be used to build the master sanitation list of all the components to be cleaned. On the first tab, master sanitation schedule, all of the areas to be cleaned should be listed. Next, how frequently they are to be cleaned needs to be identified. Once this is done, the list is sorted by frequency and each set is copied to its respective tabs. These documents can be printed out for the sanitarian to initial and date once the area has been cleaned according to frequency.

After completing the master sanitation schedule, cleaning protocols should be developed for each area to be cleaned. These should include a general statement explaining what the standard is, the blow-by-blow explanation of how to clean the designated area, a list of equipment and chemicals (if applicable), and a statement of who has responsibility for enforcing the standard. A simple form that can be used to develop each procedure can be found on the CD (Book 2_Quality Control Manual\Section 8_Sanitation Program:Sanitation Procedure/Template) and at the end of this chapter. When the blow-by-blow explanation is written, as much detail as necessary, including pictures and diagrams of the parts, should be included.

The next step following the development of the cleaning procedures is to train the employees in how to clean each of the areas. Once this is done, the training verification form found on the CD (Book 2_Quality Control Manual\Section 8_Sanitation Program:Sanitation Procedure Training Verification) and at the end of this chapter should be signed by each employee indicating that he or she has been trained and will follow the written procedure. This document should be placed in the employee's file for later reference.

Finally, the general program overview should be developed. A template for this document is provided on the CD (Book 2_Quality Control Manual\Section 8_Sanitation Program:Sanitation Program Overview) and at the end of this chapter. This should be customized and placed in the Quality Manual along with copies of the master cleaning schedule and the individual sanitation procedures.

SUPPLEMENTAL MATERIALS

MASTER SANITATION SCHEDULE

	ITEM	FREQUENCY	SOP COMPLETE
1			
2			
3			
4			
5			
6			
7			
8			
9			
10			
11			
12			
13			
14			
15			
16			
17			
18			
19			
20			
21			
22			
23			
24			
25			
26			
27			
28			
29			
30			
31			
32			

Sanitation Program

	ITEM	FREQUENCY	SOP COMPLETE
33			
34			
35			
36			
37			
38			
39			
40			
41			
42			
43			
44			
45			
46			
47			
48			
49			
50			
51			
52			
53			
54			
55			
56			
57			
58			
59			
60			
61			
62			
63			
64			
65			
66			

	ITEM	FREQUENCY	SOP COMPLETE
67			
68			
69			
70			
71			
72			
73			
74			
75			
76			
77			
78			
79			
80			
81			
82			
83			
84			
85			
86			
87			
88			
89			
90			
91			
92			
93			
94			
95			
96			
97			
98			
99			
100			

Sanitation Program

MASTER SANITATION SCHEDULE
Daily Production

Week Starting _____ Week Ending _____

ITEM	FREQ DAILY	DATE	INITIALS	DATE	INITIALS	DATE	INITIALS	DATE	INITIALS	DATE	INITIALS	DATE	INITIALS	DATE	INITIALS	DATE	INITIALS	DATE	INITIALS	DATE	INITIALS
1	x																					
2	x																					
3	x																					
4	x																					
5	x																					
6	x																					
7	x																					
8	x																					
9	x																					
10	x																					
11	x																					
13	x																					
14	x																					
15	x																					
16	x																					
17	x																					
18	x																					
19	x																					
20	x																					
21	x																					
22	x																					

MASTER SANITATION SCHEDULE
Daily Non-Production

Week Starting _____   Week Ending _____

ITEM	FREQ. DAILY	DATE	INITIALS	DATE	INITIALS	DATE	INITIALS	DATE	INITIALS	DATE	INITIALS	DATE	INITIALS	DATE	INITIALS
1	x														
2	x														
3	x														
4	x														
5	x														
6	x														
7	x														
8	x														
9	x														
10	x														
11	x														
12	x														
13	x														
14	x														
15	x														
16	x														
17	x														
18	x														
19	x														
20	x														
21	x														
22	x														

Sanitation Program

MASTER SANITATION SCHEDULE
Weekly

Week _____

ITEM	FREQ. WEEKLY	DATE	INITIALS	DATE	INITIALS	DATE	INITIALS	DATE	INITIALS	DATE	INITIALS
1	x										
3	x										
4	x										
5	x										
6	x										
7	x										
8	x										
9	x										
10	x										
11	x										
12	x										
13	x										
14	x										
15	x										
16	x										
17	x										
18	x										
19	x										
20	x										
21	x										
22	x										

MASTER SANITATION SCHEDULE
Monthly

Month _____

ITEM	FREQ. MONTHLY	DATE	INITIALS	DATE	INITIALS	DATE	INITIALS	DATE	INITIALS	DATE	INITIALS	DATE	INITIALS
1	x												
2	x												
3	x												
5	x												
6	x												
7	x												
8	x												
9	x												
10	x												
11	x												
12	x												
13	x												
14	x												
15	x												
16	x												
17	x												
18	x												
19	x												
20	x												
21	x												
22	x												

MASTER SANITATION SCHEDULE
Yearly Quaterly Semi-Yearly

Yearly													
ITEM	FREQ. QUARTERLY	FREQ. TWICE YEARLY	FREQ. YEARLY	ASSIGNED TO	DATE	INITIALS	DATE	INITIALS	DATE	INITIALS	DATE	INITIALS	DATE	INITIALS
1													
3													
4													
5													
6													
7													
8													
9													
10													
11													
12													
13													
14													
15													
16													
17													
18													
19													
20													
21													
22													

SANITATION PROCEDURE

Area to be cleaned: _____ MSS frequency: _____

Section A: General

As a supporting document to the master sanitation schedule, a procedure will be developed for each area to be cleaned. This standard outlines the cleaning methods for _____. It will be followed according to the frequency listed above.

Section B: Procedure

_____
_____
_____
_____
_____
_____
_____
_____
_____
_____
_____
_____
_____
_____

Section C: Equipment Needed

The following equipment is needed to complete the above cleaning procedure:

_____   _____
_____   _____
_____   _____

Section D: Responsibility

The responsibility for training employees to follow this procedure and complete it at the designated time is assigned to the _____ position.

Approved by: _____ Date: _____

VERIFICATION DOCUMENTATION
SANITATION PROCEDURE TRAINING

I, _____, have read and understand the procedure for cleaning the _____. I have been trained by _____ and will comply with the procedures as trained. If at any time I feel that a deviation from the written procedure needs to be made or I have questions regarding the procedure, I will notify my immediate supervisor.

Date: _____

Name: _____

Instructor: _____

SANITATION PROGRAM OVERVIEW

Section A: General

In accordance with 21CFR110, American Institute of Baking, and Good Manufacturing Practices, _____ will establish and follow specific guidelines for the cleaning and sanitizing of all areas with in the manufacturing facility. This includes the process area, refrigerators and freezers, warehousing areas, restrooms and exterior areas. All sanitation personnel will be trained in the methods and procedures needed to properly clean and sanitize. They will also be trained in the location and utilization of Material Safety Data Sheet's.

Section B: Cleaning Schedule

All parts of the manufacturing facility will be listed on a Master Sanitation Schedule. Equipment and areas are categorized by the frequency with which they need to be cleaned

Section C: Cleaning Methods

Specific cleaning methods are developed for each specific piece of equipment and area. These methods focus on adherence to current food safety practices and emphasize removal of organic material, rinsing and then sanitizing.

Section D: Postcleaning Inspection

Upon completion of each cleaning procedure a post cleaning inspection shall be done. This will be documented and signed by the sanitation supervisor.

Section E: Record Storage

All completed post sanitation inspection sheets and the completed master cleaning schedule are to be turned into the QC Department for proper retention for a period of seven years. These will be stored by year in the record storage area.

Section F: Accountability

Operations shall be responsible for the cleaning of the equipment and facility. Quality control shall be responsible for auditing sanitation.

Approved by: _____ Date: _____

Sanitation Program

MASTER SANITATION SCHEDULE
Daily Production

Week Starting _____ Week Ending _____

ITEM	FREQ. DAILY	DATE	INITIALS	DATE	INITIALS	DATE	INITIALS	DATE	INITIALS	DATE	INITIALS	DATE	INITIALS	DATE	INITIALS
1	x														
2	x														
3	x														
4	x														
5	x														
6	x														
7	x														
8	x														
9	x														
10	x														
11	x														
13	x														
14	x														
15	x														
16	x														
17	x														
18	x														
19	x														
20	x														
21	x														
22	x														

MASTER SANITATION SCHEDULE
Daily Non-Production

Week Starting _____ Week Ending _____

ITEM	FREQ. DAILY	DATE	INITIALS	DATE	INITIALS	DATE	INITIALS	DATE	INITIALS	DATE	INITIALS	DATE	INITIALS	DATE	INITIALS
1	x														
2	x														
3	x														
4	x														
5	x														
6	x														
7	x														
8	x														
9	x														
10	x														
11	x														
12	x														
13	x														
14	x														
15	x														
16	x														
17	x														
18	x														
19	x														
20	x														
21	x														
22	x														

Sanitation Program

MASTER SANITATION SCHEDULE
Weekly

Week Starting _____   Week Ending _____

ITEM	FREQ WEEKLY	DATE	INITIALS	DATE	INITIALS	DATE	INITIALS	DATE	INITIALS	DATE	INITIALS	DATE	INITIALS	DATE
1	x													
3	x													
4	x													
5	x													
6	x													
7	x													
8	x													
9	x													
10	x													
11	x													
12	x													
13	x													
14	x													
15	x													
16	x													
17	x													
18	x													
19	x													
20	x													
21	x													
22	x													

MASTER SANITATION SCHEDULE
Monthly

ITEM	FREQ. MONTHLY	DATE	INITIALS	DATE	INITIALS	DATE	INITIALS	DATE	INITIALS	DATE	INITIALS	DATE	INITIALS	DATE	INITIALS
1	x														
2	x														
3	x														
4	x														
5	x														
6	x														
7	x														
8	x														
9	x														
10	x														
11	x														
12	x														
13	x														
14	x														
15	x														
16	x														
17	x														
18	x														
19	x														
20	x														
21	x														
22	x														

Week Starting _____   Week Ending _____

13 Metal Detection Program

PROGRAM TYPE: REQUIRED

THEORY

Pursuant to 21CFR110.80(b)8 ("Effective measures shall be taken to protect against the inclusion of metal or other extraneous material in food. Compliance with this requirement may be accomplished by using sieves, traps, magnets, electronic metal detectors, or other suitable effective means."), all products produced in a food manufacturing or processing facility will be subjected to some form of metal detection or exclusion. In most plants metal can be introduced through several means, such as from suppliers, in raw materials, from the manufacturing process, from metal on metal, and from in-plant carelessness, maintenance, or employee action. The fact of the matter is that with the vast quantity of metal in the operation, the source ingredients, and human error, sooner or later a piece of metal will show up in the product.

Amid this type of certainty of contamination it is the responsibility of the manufacturer or processor to design and develop systems to eliminate metal from reaching the end user. Currently, there are three basic ways to do this: x-ray technology, electronic technology, and magnetic separators. Among these options, only electronic technology or metal detectors are a proven viable alternative for the small to medium-sized food manufacturer or processor because x-ray technology is very costly and can suffer from matrix interference and physical separation, and magnets can allow small particles to pass. Metal detectors come in various shapes and sizes and from many manufacturers and can detect and remove common industrial ferrous and nonferrous metals. They can sense the presence of aluminum, brass, copper, tin, lead, and stainless steel of varying grades and can be sensitive enough to detect particles down to 1 mm in some matrices. Once the manufacturer decides on purchasing a metal detector, the placement, size, testing method, and rejection method of the metal detector need to be decided.

PLACEMENT

Most metal detectors are designed to fit at varying points within the processing line, depending on the type of product and type of packaging. For instance, in some cases the manufacturer will need to install the metal detector at a point prior to packaging such as between the scale and form-fill bagger. This allows for increased sensitivity but precludes the packaging from being tested; with the amount of recycled packaging material being used, it is possible that the packaging does contain metal. Conversely to in-line placement, the manufacturer can place the metal detector after packaging. This would be after the product has been placed either in primary packaging such as a cookie in a form-fill pouch or ice-cream bar in a horizontally wrapped film, or in secondary packaging such as the carton for a six-pack of individually wrapped cookies. The advantage of this placement is that the product cannot be subsequently contaminated because it is packaged and the package is checked for metal. The drawback to this placement is that it sacrifices some sensitivity due to the size of the aperture needed to accommodate the package.

SIZE

The size of the metal detector needed is determined by the placement in the line, the size of package passing through the aperture, and the sensitivity desired. The latter parameter is fundamental to the purchasing decision. As a rule, the size of the metal particles that the detector can detect is determined by the aperture size, the type of packaging structure, and the matrix of the finished product. For most companies, the metal detector should be sized to provide a maximum sensitivity for the matrix and pack size. The smaller the opening is, the better the situation; however, maintaining flexibility within the packing line is a must. There is no set or legal requirement for the piece size that the metal detector must detect. A typical acceptable sensitivity for an appropriately sized aperture would be 1.0 mm ferrous, 1.5 mm stainless steel, and 2.0 mm nonferrous. Obviously, as the size of the aperture increases, these numbers will increase. Suffice it to say that the more sensitive the detector is, the lower is the risk of any metal passing into the finished product, thus lowering the risk liability.

TESTING

After the metal detector has been sized to fit the application, is installed, and is set up correctly, the next step is to determine a testing regimen. For this task most metal detector manufacturers will supply some form of test wand. These are plastic or rubber sticks, cards, or tubes that contain a piece of metal that has been buried under the surface. The size and shape of the test wands are usually designed according to how the detector will be tested. For instance, for an in-line metal detector used in a milk plant, the wand would be at the end of a long rod that could be carefully inserted between the metal detector collar and tube. For a metal detector that is used on-line in a packaged cookie operation, the test pieces might be just credit-card-sized plastic pieces inserted with the cookie and sent on the belt through the detector. Each set of wands is sometimes color coded and usually marked with the size and type of the metal embedded into the test piece. One test piece is supplied for each specific type of metal tested: ferrous or iron-based metals, nonferrous or non-iron-based metals, and stainless steel. Sometimes, the exact type of stainless steel tested is listed based on its alloy composition, such as 18-8, 304, or 316 stainless steel. Although this is good information, the differences among the three types of stainless steel do make them more or less resistant to chloride pitting; it does not significantly impact the sensitivity of the metal detector or the ability to test its functionality adequately.

The testing regimen applied has two parts: method and frequency. The method used can be to place the wand with the product as it passes through the metal detector or by itself without product. In most cases the preferred method is to pass it with the product through the metal detector. This tends to eliminate false readings and lets the detector look for the metal piece as if it were in actual product. The drawback to this method is that when a test wand is placed in an actual product and passed though the metal detector, it is easy to lose it at the speeds at which products move in today's modern food plants. Therefore, it is important to design record keeping systems to prevent this eventuality. The second alternative—passing the test pieces through the metal detector without any product—removes the chance of losing the test piece, but also removes the effect of the packaging and product on the reading. This is most important when products in larger packaging are being detected due to the layering effect from the product if the test piece is placed in the center of the container. To understand this difference and ensure that there are no false negatives, try passing the test piece through the detector in the middle of the package and then pass just the test piece alone and verify that it is detected both ways. If it is not, then the testing method should be identified as having to be passed through the detector inside the package.

The frequency for testing the metal detector is determined by how much financial risk the company wants to accept if metal is found in its product and it is forced to conduct a recall. As a rule of thumb, the minimum number of tests that should be run is two. These are at the beginning of the shift and at the end of the shift. This technique only ensures that the metal detector was working at the beginning and at the end of the shift—no more. It does not guarantee that the entire run of product was checked for metal, given that things do happen in production environments such as power outages, machines getting jammed and bumped, and other production "gremlins." It is recommended that the metal detector be checked at least prior to startup, every 2 hours of production running, and then at the end of the run or shift. This breaks the problem down to 2-hour blocks, thus limiting the quantity of problem products that must be recalled to the number within those 2-hour blocks.

REJECTION

Once the metal detector is installed and its operation verified, the next step is to determine what will happen when the unit detects metal in the product. Most available options for metal rejection involve an employee notification system and a product alteration option. Employee notification systems include bells, whistles, horns, sirens, and lights. Singly or in combination, they notify the employee that a piece of metal has been found in the product. Care must be taken when installing these devices to make sure that they are loud enough and/or bright enough to get the employee's attention. The more obnoxious the alert system is, the better it is because employees will become complacent over the course of the day and not pay attention to alarms unless they break them from their sight and sound routine.

The other half of the rejection system is that of the physical rejection or alteration of the detected product. This is important in making sure that the product that has the metal in it gets separated and subsequently isolated from the non-metal-containing product. There are several options for product rejection, including physical rejection, where the product is pushed, blown, or dropped off the main line; package deviation, such as the bags not cut on a form-fill machine; or line stoppage, where the entire line is stopped when a piece of metal is detected. Any of these techniques will suffice as long as the product that has the metal in it can be isolated for further examination. One way to isolate the product is to have a catch bin, tray, or trough into which the rejected product can fall.

An important step in the development of this program is the determination of how the rejected product will be handled. For maximum information and process improvement, this product should be divided in half and each half passed back through the metal detector. This should be done separately from good product that is passing through the metal detector. The half that sets the metal detector off should be further broken in half and each half passed through the metal detector. Continue this process until the metal is found. Remember that the metal detector is very sensitive and the operator may need to use a magnifying glass to identify the tramp metal in some product samples. After the metal detector is installed, all employees within the company should be trained to know what the metal detection alarm system sounds and looks like and how they should respond to it. The training should be directed specifically to the position of each employee and the responsibility he or she has in producing the product. For example, a line operator should be trained to listen for the alarm and watch for the light, isolate the product, and notify a supervisor. In contrast, a warehouse person should be trained to know what the alarm system sounds and looks like and to notify someone if it is activated. However this training is conducted, it is important to document it.

The next step in developing the metal detection program is to write the program manual. This includes a metal detector operation document, a metal detector check sheet, a foreign material investigation sheet, an employee training verification document, and a metal detection program document. In conjunction with the manufacturer of the metal detector, an operation document should be developed that explains exactly how the detector is set up, what it tests for, how its operation is verified, how the rejected material will be handled, who is responsible for setting and adjusting the sensitivity, and who is responsible for reviewing the daily documentation. It should be developed in conjunction with the hazard analysis critical control point (HACCP) program. For most companies, the HACCP program will dictate that the metal detector is a critical control point because it controls a physical hazard. There are some within the industry and auditing companies, however, who will argue that the metal detection program is merely a control point (CP) or quality control point (QCP) based on the false concern that the entire plant will have to be shut down if the metal detector goes down. This argument minimizes the importance of the metal detector in the HACCP program and bears no weight because it is clearly possible to run products and put them on hold until they can be run through the metal detector after it is repaired or replaced.

APPLICATION

Food manufacturing and processing companies can easily build a metal detector program by installing the unit, creating the needed documentation, and then training the employees. A metal detector program general overview customizable to fit the exact situation can be found on the CD (Book 2_Quality Control Manual\Section 9_Metal Detection Program:General Overview) and at the end of this chapter. This document ties all of the requirements from several other programs into this one. When this document is customized, the sensitivity of the metal detector and how the detected product is handled need to be updated. This document should be signed and placed in the proper section of the Quality Control Manual.

A generic operation document for the metal detector is located on the CD (Book 2_Quality Control Manual\Section 9_Metal Detection Program:Metal Detection Operation) and at the end of this chapter. It should be customized to reflect the brand and serial number of the detector that is being used, the sensitivity of the unit, and whether the metal detector is a critical control point within the confines of the HACCP program. Next, the unit testing regimen, unit failure, and product rejection sections should be updated to make sure that they reflect exactly how the particular system is set up. Finally, the record keeping section should be updated to reflect how the testing will be recorded and how found metal will be documented. This operation document should be placed behind the metal detector general overview in the quality control manual.

The next step is to develop the metal detector testing recording document. A simple recording sheet for this purpose can be found on the CD (Book 2_Quality Control Manual\Section 9_Metal Detection Program:Metal Detector Check Sheet) and at the end of this chapter. It provides spaces for the documentation of each test piece check, the sensitivity of the unit when tested, any action taken during the testing, and that the test piece was recovered. This can be customized to reflect the correct sensitivity and line and whether it is a critical control point.

The form located on the CD (Book 2_Quality Control Manual\Section 9_Metal Detection Program:Foreign Material Investigation Log) and at the end of this chapter is to be used upon finding any type of foreign material in the product—specifically, for metal rejected by the metal detector. All metal that is isolated from the rejected product should be logged on this document along with any corrective actions taken. This document should be retained by the quality control department for future reference.

The final implementation step for the metal detection program is to train employees. This should be done on a case-by-case basis as discussed earlier; however, all employees should be required to sign a document stating that they have been trained in the operation of the metal detector and that they understand how they should react should they hear the alarm go off. The form on the CD (Book 2_Quality Control Manual\Section 9_Metal Detection Program:Metal Detector Training Verification) and at the end of this chapter can be adapted for this implementation. The signed document should be placed in employees' files.

SUPPLEMENTAL MATERIALS

METAL DETECTION PROGRAM

GENERAL OVERVIEW

Section A: General

A In accordance with 21CFR110.80b8 effective measures shall be taken to protect against the inclusion of metal or other extraneous material contamination in food. Metal contamination normally comes from one of four sources.

- Raw Materials: Typical examples include aluminum foil packaging, box staples, machine parts and wire strapping from material containers.
- Personal Effects: Typical examples include buttons, jewelry, coins, keys, hair clips, thumb tacks, paper clips, cell phones, pagers.
- Maintenance: Typical examples include screwdrivers and similar tools, welding slag and swarf following repairs, copper wire offcusts following electrical repairs, miscellaneous items resulting from inefficient cleanup or carelessness and metal shavings from pipe repair.
- In-Plant Processing: The danger exists every time the product is handled or passes through a process that metal fragments will wear off moving parts. Tramp metal can come from any metal to metal contact.

The metal detection program shall include exclusion, inspection, detection, source elimination and documentation.

Section B: Raw Materials

All suppliers will be required to produce raw materials using a method that excludes foreign and metal objects. Supplier specifications and documentation to this effect will be kept on file. On a case-by-case basis, a supplier may be asked to provide documentation with each shipment stating that it is foreign-object free. All raw materials will be inspected upon receipt for extraneous metal outside the shipping container through a visual inspection.

Section C: Handling

Raw materials shall be visually inspected before processing. This will eliminate many large, easily detected pieces before being processed.

Section D: Metal Detection

A metal detector shall be used to detect and remove metal prior to final packaging. The sensitivity shall be 1.0 mm ferrous, 1.5 mm stainless steel and 2.0 mm non-ferrous. The detector shall be tested prior to startup, every two hours and at the end of production. An alarm and warning light are used to notify the employees that a positive reject has been made. Finished product that registers positive is rejected into a holding bin for further analysis. It shall be operating at all times during production. No product shall be shipped without passing through the metal detector.

Section E: Record Keeping

All metal detector checks will be recorded and submitted with the daily production records.

Section F: Rejected Product

Product that is rejected shall be carefully inspected and the metal isolated. A thorough inspection will be made to determine the source of the tramp metal. Affected areas of the company are notified regarding metal contamination. All rejected metal is logged on a Metal Contamination Log.

Section G: Record Storage

All completed post-sanitation inspection sheets and the completed master cleaning schedule are to be turned into the QC Department for proper record retention for a period of seven years.

Section H: Accountability

Operations shall be responsible for the installation and maintenance of the metal detector. Quality control will be responsible for program development and employee training. All employees are responsible for listening and watching for positive rejects and protecting the product an alarm occurs.

Approved by: _____ Date: _____

METAL DETECTOR OPERATION—CCP (M-1)

Section A: General

_____ uses _____ (model) ___ (serial # _____) metal detectors between the packaging and casing of the finished product. Operation of these machines is detailed as follows.

Section B: Setup and Sensitivity Testing

The metal detector is to be installed between the packaging and casing stage of the production line and turned on. At the beginning of each production run, the test wands will be dropped down the side of the chute with product and then dumped down through the metal detector. If both the light and the alarm go off, the metal detector is activated for the particular metal in the wand. The metal detector by pushing in the reset button. If the light, the alarm, or both fail to go off, the metal detector is not activated for the metal in the wand. **Enter all tests on the metal detection sheet** (tab section: metal detection program; title: metal detection sheet). If the metal detector is not activated for any of the wands, increase the sensitivity until all three wands activate the detector. Repeat the test until all wands set off the alarm and light. **NOTE: The wands must be in the product stream when the metal detector is checked.**

Do not run any product through the metal detector if it is not working properly!

Notify operations or quality control of any changes in the sensitivity of either detector.

Sensitivity specifications:
- 1.0 mm ferrous
- 1.5 mm 316 stainless steel
- 2.0 mm nonferrous

Section C: Periodic Testing

At the start of the shift, every period (first period: start of shift to end of first break; second period: end of first break to end of lunch; third period: end of lunch to end of second break; fourth period: end of second break to end of shift) and at the end of the shift, the metal detectors are to be tested and the results recorded on the metal detection sheet (tab section: metal detection program; title: metal detection sheet).

The testing procedure is as follows:

> At the beginning of the period, test wands shall be placed on the side of the scale chute and dropped down through the metal detector with product flowing through the system. If both the light and the alarm go off, the metal detector is activated for the particular metal in the wand. Reset the metal detector by pushing in the reset button located on the mezzanine. If the light, alarm, or both fail to go off, the metal detector is not activated for the metal in the wand.

Precautions must be taken to ensure that any test samples not rejected do not become "lost" on the production line—that is, do not forget to recover the test wands.

Section D: Action Required if Testing Fails

If a failure is established during routine testing, all production since the previous testing **must be** set aside and put on hold for further testing. Fix the machine immediately, verify its operation, document the test, and continue production. All rejected product should be reinspected by passing the product through the metal detectors. **No product is to be shipped without successfully passing through a properly working metal detector!**

Section E: Action Required for Rejected Product

If a product is found by the metal detector to contain a metal object:

Form-fill product will set off the alarm and light and create a double bag.
Bulk product will stop the line and set off the alarm and light.

Every time the metal detector "detects" metal, the product needs to be segregated. After removing the product from the line, divide it into smaller and smaller pieces and reinspected until the metal can be located.

This procedure must be done in isolation. **Do not reinspect by putting metal-detected product into the normal product flow.**

Section F: Record Keeping

The results of testing and a record of any adjustments and repairs made to the detector should be entered on the metal detection sheet (tab section: metal detection program; title: metal detection sheet). Completed sheets will be retained for 7 years.

Each metal particle found will be shown to the quality control manager and details will be recorded on the metal contamination log (tab section: metal detection program; title: metal contamination log). If the source is known, it should also be recorded. If not, investigations are extremely useful in preventing a reoccurrence and can result in a change in maintenance procedures or even a change in raw material suppliers. Locating and retaining the particles has the added advantage that if a screen or blade, for example, is known to have broken into the product, the individual pieces detected can be collected and the component reassembled to ensure nothing has been missed.

Metal Detection Program

Metal Detection Check Sheet - CCP (M-1)

Line: Main Production

Test Standard

Ferrous:	1.0 mm	Yellow
Stainless:	1.5 mm	Red
Non-ferrous:	2.0 mm	Green

KEY: Checkmark indicates pass; x indicates fail

DATE	TIME	DETECTED (CHECK)			REJECTED (CHECK)			SENSITIVITY	ACTION	Recovered			CHECKER
		f	ss	nf	f	ss	nf			1	1.5	2	

Foreign Material Investigation Log

Date	Time	Foreign Object Found	Source of Foreign Object	Corrective Actions Taken	Initials

VERIFICATION DOCUMENTATION
METAL DETECTION TRAINING

I, _____, have read and understand the section on metal detection and operation at _____. I have been trained in the operation of the metal detector and will properly test and operate it. If I have questions or if it fails to operate correctly, I will immediately notify my supervisor.

Date: _____

Name: _____

Instructor: _____

14 Regulatory Inspection Program

PROGRAM TYPE: REQUIRED

THEORY

The Food and Drug Administration, U.S. Department of Agriculture (USDA) Food Inspection Service, State Department of Agriculture, local health departments, Occupational Safety and Health Administration, Environmental Protection Agency, Department of Labor, weights and measurements agencies, and many other agencies within federal, state, and local governments are a constant yet necessary presence continuously surrounding food manufacturing and processing entities. 21CFR5.35 recognizes the authority of the Food and Drug Administration, and the USDA Food Safety Inspection Service operates under the authority of the Federal Meat Inspection Act, the Poultry Products Inspection Act, and the Egg Products Inspection Act to inspect food manufacturing and processing facilities. Each state operates its own agriculture department and most local entities and states have health departments. With the vast quantity of regulatory bodies, every food company in America is regulated in one or more ways. Which regulatory agency oversees the facility is determined by the type of food product produced and the location of the food plant.

Those companies that are regulated and inspected by the USDA undergo constant inspection and thus generally do not have unannounced inspections. For non-USDA-inspected companies, a regulatory inspector may come at any time and thus the company should be prepared. Although it can be argued that these agencies can force numerous miscellaneous expenses to be incurred and that they can be heavy handed at times, they are a significant force behind the design, development, implementation, and validation of all food safety systems. As a constant presence and with the full weight and authority of the respective government agency behind them, it is incumbent on the manufacturer to have a detailed plan to manage the visitation from regulatory personnel properly. This regulatory inspection plan should contain three distinct parts: the initial introduction, the inspection, and the inspection's conclusion.

As a company plans for a routine inspection, all of the preparations should be directed toward establishing a visitation protocol and training the employees in it. A hierarchy of inspector "guides" should be developed. Normally, the primary contact and tour guide within the company is the quality control manager; however, every company structure is different, so this person may be the plant manager, operations manager, or any other upper level manager. For this important role it helps if the company key contact understands the operations and the quality systems in place so that he or she does not misspeak, causing increased scrutiny on the operations. The people who are chosen should be management representatives who cover all hours of the plant's operation because an inspector may inspect during any scheduled hours of the plant's operation. Each key contact should undergo regulatory inspector management training.

The company's visitation protocol outlines exactly how the inspectors will be treated when they arrive. This involves determining who will greet them, who will sign them in and get their identification, where they will sit, what restroom they will use, and who will request a copy of the Notice of Inspection. Typically, this is done by the front office receptionist due to proximity to the front door, but in the case of is individual is not present, all employees should be versed in greeting and signing the inspector in. All of these details might seem petty and trivial, but it is often at this point that the inspector gets a first impression of the company and may obtain information that may later be used against the company. At this point, it is important to state that inspectors are not the enemy and, in most cases, are not trying to work against the company; they do have a job to perform and even the smallest piece of information can be taken out of context and lead to unwanted consequences. Therefore, total control over the course of the inspection is imperative.

Due to the high importance of this first meeting, there are some things that the greeter should do and not do. Do greet the inspector courteously and request to see identification. Do have the inspector sign in. Do ask for a copy of the Notice of Inspection. Do inform him or her that someone will be there shortly and have him or her sit and wait a short time while the guide is notified. Do not discuss the operation, quality systems, management, customers, personnel, shipments, products, or problems with the inspector. If he or she requests to use the restroom while waiting, direct him or her to a stand-alone restroom if one is available. If not, say that the guide will show the inspector where it is. Often, if there is even the smallest problem occurring within the plant when the inspector arrives, it will be being discussed between employees in the restroom and the inspector will overhear and this will predispose him or her. It is important for employees to watch what they say at all times.

After settling the inspector in the waiting area, the greeter informs the guide that the inspector has arrived. The guide should immediately proceed to the waiting area, introduce himself or herself, and offer a business card to the inspector. The inspector should be moved immediately to a conference room or room that has no open documents or accessible information. If an office is used, any sensitive or confidential information should be put away prior to bringing the inspector to the office. This is the setting for the pre-inspection conference.

During the pre-inspection conference the guide asks the inspector the purpose of the inspection, the product or processes to be covered, and if there was something specific that prompted the visit (e.g., a customer complaint or lab test). Based on this initial conversation, the guide determines who will accompany the inspector on the inspection tour, which and how questions will be answered, and which records will be shown if requested by the inspector. The guide will then take the inspector on a tour of the facility and escort the inspector throughout the plant, including the restrooms. During the inspection the guide should take detailed notes regarding items the inspector pointed out and comments he or she made. If the inspector points out something that can be fixed immediately, the guide should quietly have it done, if possible. The guide should let the inspector do the talking and ask questions, keep the answers short and to the point, and not volunteer information or point things out, whether they are considered good or bad.

During the routine inspection the inspector may request to take photos of the plant, equipment, or product. Do not allow photographs of any kind. If the inspector does request to take photographs, courteously inform him or her that it is company policy not to allow photographs to be taken in the facility.

The inspector may also request to see records of various kinds. Do not show or offer to show production or quality control records unless the inspector has a written request. 21CFR1(j)1.361 states:

> When FDA has a reasonable belief that an article of food is adulterated and presents a threat of serious adverse health consequences or death to humans or animals, any records and other information accessible to FDA under section 414 or 704(a) of the act (21 U.S.C. 350c and 374(a)) must be made readily available for inspection and photocopying or other means of reproduction. Such records and other information must be made available as soon as possible, not to exceed 24 hours from the time of receipt of the official request, from an officer or employee duly designated by the Secretary of Health and Human Services who presents appropriate credentials and a written notice.

To meet the written request, do not provide recipe or formula information, financial data, pricing data, personnel data, research data, or sales data. Inspection personnel do have the authority to inspect interstate shipping records such as bills of lading that show carrier, destination, and product carried. 21CFR1(j)1.362 states, "The establishment and maintenance of records as required by this subpart does not extend to recipes for food as defined in 1.328; financial data, pricing data, personnel data, research data, or sales data (other than shipment data regarding sales)."

The other request that the inspector might make is for samples. According to 21CFR(a)2.10(a)1, the FDA has the authority to take samples. When this occurs, the inspector will draw the sample and provide the manufacturer with a written receipt describing the samples taken. Anytime an inspector takes samples, the manufacturer should ask the inspector the reason for taking the sample and the tests to be performed on it, put the product lot from which the sample was taken on hold, identify any customers that were sent product from the same lot and be prepared to contact them, retain duplicate or split samples for future testing as needed, and obtain analyses on the sample from an independent testing lab. Besides product samples, an inspector is entitled to copies of labels and promotional materials. It is not required that a receipt be given for these materials.

Upon conclusion of the inspection, the inspector is required by law to leave the manufacturer a written report setting forth any conditions or practices that, in the inspector's judgment, do not conform to regulations. It is important that the guide understand all comments written on the report. If anything is not understood, it must be clarified so that all problems and concerns may be fixed. If the guide disagrees with any or part of the report he or she might, after supplying additional information, get the inspector to change or strike the contested points from the record. Otherwise, the guide should request the inspector to note the points of disagreement. When the inspector has completed the report, the guide should escort the inspector to the door and thank him or her for coming. The report should immediately be given to the highest technical manager and any other appropriate upper management.

APPLICATION

In preparation for any regulatory inspection, the company should make sure that it has in place the visitor GMPs, part of the GMP program, and a visitor sign in sheet, part of the biosecurity program, and has a designated waiting space. A customizable regulatory inspection program overview document can be found at on the CD (Book 2_Quality Control Manual\Section 10_Regulatory Inspection Program:Regulatory Inspection Program) and at the end of this chapter. This outlines the basics of how the company should prepare for and interact with regulatory inspectors.

SUPPLEMENTAL MATERIALS

REGULATORY INSPECTION PROGRAM

GENERAL OVERVIEW

Section A: General

In accordance with 21CFR5.35, _____ recognizes the authority of the Food and Drug Administration to inspect the facility with proper notification. To ensure that all inspectors are treated uniformly by _____'s personnel, the following procedures will be followed.

Section B: Initial Introduction

The first person who will usually have contact with the inspector will be the receptionist, who will:

- ask the inspector sign in;
- request proper identification;
- ask for a copy of the Notice of Inspection;
- inform the inspector that someone will be with him or her shortly;
- offer the inspector a chair;
- contact, in order, one of the following to act as a "guide": _____, _____, or _____; and
- inform the guide that an inspector is sitting in the office and is here to inspect.

The receptionist **will not:**

- answer any questions concerning the operation, management, programs, customers, products, shipments in or out, or personnel; or
- engage in conversation with any other employee, phone caller, or visitor concerning the above topics.

If the inspector asks to use the restroom before starting, the receptionist will inform him or her that the guide will show him or her where it is.

Section C: Inspection

The guide will:

- come to the office and introduce himself or herself;
- inquire about the nature of the inspection, the product or processes to be covered, and the reasons for the visit (e.g., a customer complaint);
- advise the inspector of any and all safety precautions and plant rules before touring the plant; and
- have the inspector read and sign the GMP rules.

The guide will decide who, if anybody else, will accompany them on the inspection; determine which and how questions will be answered; and which records will be shown if requested by the inspector. The guide will then take the inspector on a tour of the facility and escort him or her throughout the plant, including the restrooms. During the inspection, the guide will take detailed notes regarding items the inspector points out and comments that are made. If the inspector points out something that can be fixed immediately, the guide should quietly have it done. The following is a summary of how to handle different situations that may arise.

Photographs
Photographs of any kind should not be allowed. If the inspector asks to take a photograph, the courteous reply should be that it is company policy not to allow photographs to be taken in the facility.

Records
The guide should not show or offer to show production or quality control records. The FDA is not granted the authority to inspect, without the owner's permission, any financial records, sales figures other than shipping data, pricing information, personnel files other than data relating to the qualifications of professionals, and research reports. This includes product formulas, raw material specifications, complaint files, and quality control records.

The guide should provide interstate shipping records when asked. This would only apply to bills of lading that show carrier, destination, and product being carried.

Samples
Explicit permission is given to the FDA by statute to take samples of materials. Before the inspector leaves the establishment, a receipt describing the samples taken must be given to the manufacturer. Whenever an FDA inspector takes a sample, the manufacturer is advised to take the following steps:

1. Ask the inspector the reason for taking the sample and the tests to be performed on it.
2. Put on hold the lot from which the sample was taken.
3. Immediately notify customers of any product shipped to them from the sampled lots.
4. Retain duplicate or split samples for future reference.
5. Obtain analyses on the sample from an independent testing laboratory.

Besides product samples, an inspector is entitled to copies of labels and promotional materials. It is not required that a receipt be given for these materials.

Section D: Inspection Conclusion

At the conclusion of an inspection, the FDA official is required by law to leave the manufacturer a written report setting forth any conditions or practices that, in the inspector's judgment, do not conform with regulations. It is important that the guide understand all comments written on the report. If anything is not understood, it must be clarified so that all problems and concerns may be addressed. If the guide disagrees with any or part of the report, after supplying additional information he or she might get the inspector to change or strike the contested points from the record. Otherwise, the guide should request the inspector to note the points of disagreement.

When the inspector has completed the report, the guide should escort the inspector to the door and thank him or her for coming. The report should immediately be given to the highest technical employee and a copy distributed to the president or owner.

Section E: Inspection Follow-up

Following an inspection, the management team will meet with the guide to go over any concerns raised by the inspector. All concerns will be addressed and rectified immediately.

Approved by: _____ Date: _____

15 Lot Coding Program

PROGRAM TYPE: REQUIRED

THEORY

Pursuant to the Bioterrorism Act of 2002 and its subsequent amendments, 21CFR1.345a(3,4), and 21CFR7.59(b), all persons who own food or who hold, manufacture, process, pack, import, receive, or distribute food for purposes other than transportation must maintain records that allow the company to track all product ingredients through finished goods, from the previous sources to the subsequent customers. Specifically, the act requires that the company identify the immediate non-transporting source, whether foreign or domestic, of all foods received and obtain the name of the firm, as well as address, telephone number, fax number, and e-mail address, if available; type of food, including brand name and specific variety; date received; and quantity and type of packaging, as well as identify the immediate transporter's previous sources, including the name, address, telephone number, and if available, fax number and e-mail address. In addition, all persons who manufacture, process, or pack food must include lot or code number or other identifier if the information exists.

Although most of the preceding requirements are covered within the context of the company's receiving and shipping programs, lot tracking program, and hazard analysis critical control point program, to satisfy the last requirement the company must establish and enforce a complete lot coding program. This program involves three parts: establishing a coding scheme, determining how and where to place the code, and deciding where to record the code on the process documents.

Establishing the type of code to be placed on the foodstuff entails making choices between using the date of production as a "best before" date or a random code; Julian date versus Gregorian date; and which other information to add, such as time, plant code, shift code, period code, and line code. There are advantages and drawbacks to each choice. Most companies utilize some form of production date as the master production code; however, for products that are perishable or have a short shelf life, a "best by" or "best before" date is preferred or mandated. The production date is an easy way for the company to track products because all employees can look at the code and determine how old it is and what its remaining shelf life is. A "best by" date can be a positive statement for the consumer and distributor, but may cause customers not to buy the product as it approaches this date and cause distributors to pull the product from the shelf even though it has some effective life left.

A Julian date formatted as YDDD or DDDY is an easy format for the manufacturer to utilize on all documents and in computer systems and is easy to read; however, most customers do not understand this system. Gregorian dates in the format of MM/DD/YY or MM/DD/YYYY are easier for the customer to understand, but this system is more difficult for the manufacturer to read and takes up more places on the coder. Most companies prefer using some form of the Julian date code.

The next decision to be made is that of what other information the company wants to include within the code. This decision is partially dependent on the type of coding apparatus and its limitations, as well as how much space is available on the package to place the code. Some coders are limited in the amount of characters per line, type size, and the amount of lines they can print. The package design also can limit the amount of code that is placed on the package, particularly if the package is smaller. With these considerations in mind, the company can determine if it wants to add additional information, such as plant code, time, shift code, period code, and line code. These are defined as follows:

Plant code is for those companies that have more than one plant, co-packers, or utilize plants that are USDA registered. This code can be chosen randomly—such as a, b, and c—or structured, such as P04-33 to represent Portland (P) plant, fourth (04) plant in the city, and Oregon facility (33), the 33rd state in the Union. USDA plants already have a registration number that may be used as the plant code identifier.

Time is the time of day that the product passes in front of the coder. It can be in military time format, utilizing a 24-hour period per day, or conventional time format, utilizing two 12-hour periods per day. It is in the form of hh:mm.

Shift code identifies the shift on which the product is produced. Typically, shifts are 8 hours with three shifts per day, but they may vary. This code usually takes the form of 1 (for first shift), 2 (for second shift), and 3 (for third shift). In many plants, there are one, two, or three shifts and then some form of cleanup shift.

Period code designates a fractional portion of the shift. In plants where a time stamp is not available due to equipment or package constraints, then the period code is a viable alternative. To establish the period code, most companies break a shift down into roughly 2-hour blocks; a typical breakdown is from start of shift to first break (period 1), first break to lunch (period 2), lunch to second break (period 3), and second break to shift end (period 4).

Line code is used to determine the line on which the product was run. It is important in those cases that have the same product run on two or more lines at the same time. This code is usually in the form of a, b, c or 1, 2, 3.

When making the decision as to which combination of codes to use, the company should keep in mind that the more information the better when considering a product recall.

The final consideration for the company when deciding the appropriate coding scheme is that of the risk associated with the lot size dictated by the code. For instance, the smaller the time period that could be identified from a product to be recalled is, the smaller the lot of product that would have to be contained and returned is. This is important when the affected products might be relegated to a small amount. To understand this concept, it might be helpful to evaluate some lot code examples and the risks associated with them:

Lot code:	Risk exposure in case of recall:
Julian date	Smallest risk is everything produced on the date
Julian date shift	Smallest risk is everything produced on a single shift
Julian date shift time	Smallest risk is everything produced between exact times

Suffice it to say that the smaller the period defined by the lot code is, the easier and more concise a recall can be.

The decision of how and where to place the code is usually dictated by the package design and size and the code complexity and size. If the package surface area is small, the company is limited in the amount of open area for coding after printing the information required by 21CFR101 and the Nutritional Labeling and Education Act. Generally, this leaves space on the top and bottom of the container and on a side panel. When a company is deciding where to place the lot code, a location should be chosen that allows for easy access by the retailer and customer in case of recall and code clarity.

How the lot code is applied can make or break the company's and customer's ability to read the code. In today's manufacturing operations, the use of an ink-jet coding system to apply the lot code is standard, although some manufacturers do use other systems such as embossing or imprinting. Although all systems are prone to fail at some point or other, the ink-jet lot coding system is routinely more accurate, legible, and flexible. These systems are inexpensive and available from numerous manufacturers.

The second consideration regarding how the lot code is applied is that of type size. As a general rule, the type size should be big enough to be legible and contain all of the necessary information, but not so big as to take away from the design of the package. Legibility is the key because it is one of the lynchpins of any effective recall system.

Once a coding scheme and how and where to place the code have been decided, the final part of the lot coding program is to determine where to record the lot information. This decision is usually one of practicality because it must be tied directly to the records of the product being produced on that day. To achieve this level of traceability, the logical choice is to record the lot code on the production monitoring documents that are located on the production line during processing. The advantage of this is that it is accessible to all employees during the run and if the lot code is changed during the run, such as when the period changes, it is easily noted.

APPLICATION

A lot coding program general overview document that can be customized by the company is located on the CD (Book 2_Quality Control Manual\Section 11_Lot Coding Program:General Overview) and at the end of this chapter. It should be altered to reflect in detail the exact coding scheme used by the company. The signed document should be placed in the lot coding section of the quality control manual.

In many cases customers or distributors will require a statement of how to read the company's lot code. It is important that they know this information in case they are called upon to respond to a recall or otherwise have to return or locate a product. The form on the CD (Book 2_Quality Control Manual\Section 11_Lot Coding Program:Lot Code Explanation) and at the end of this chapter provides the company a simple statement sheet of how to read the company's lot code. This document should be adapted to fit the code utilized on the company's product. A copy of this sheet should be placed in the customer's file for future reference.

SUPPLEMENTAL MATERIALS

LOT CODING PROGRAM

GENERAL OVERVIEW

Section A: General

Pursuant to the Bioterrorism Act of 2002 and its amendments, all food processors, manufacturers, and holders of food must establish a lot coding system for the sole purpose of being able to trace the foodstuff throughout the distribution chain. In accordance with federal law, it is the policy of _____ that all products produced will have a verifiable and legible lot code. This will indicate the date of manufacture (basic code) or, if required, the date of expiration.

Section B: Basic Code

The basic date of manufacture code is in the Julian date format. Following the Julian date is a digit indicating the year. Optionally, a period code, time, plant code, and/or shift code will be added. For example, code date 02862 M 15:23 c is translated as the 28th day of 2006, shift 2, Plant M, 3:23 p.m., line c.

Section C: Date of Expiration

For those products requiring an expiration date, it will be in the format: Best Used By: dd/mm/yyyy.

Section D: Printing

All code dates will be clearly printed using nonerasable ink and must be legible. Coding will be done via ink-jet, dot-matrix, laser printer, or hand stamping.

Section E: Record Keeping

The company will record the lot codes for each particular production run on the production documentation. This will be retained for a period of no less than _____.

Section F: Responsibility

It will be the responsibility of operations to ensure that a legible code in the correct format is placed on each package. It is the responsibility of the quality control department to verify that this code is accurate and that it is being recorded.

Approved by: _____ Date: _____

LOT CODING PROGRAM

LOT CODE EXPLANATION

_____ assigns a unique lot code to each product manufactured that is applied via an ink-jet.

For the following products, the lot code is located on the _____ panel:

_____   _____

_____   _____

_____   _____

The basic date of manufacture code is in the Julian date format, DDD.

Following the Julian date is a digit indicating the year.

Optionally, a period code, time, plant code, and/or shift code will be added. For example, code date 02862 M 15:23 c is translated as the 28th day of 2006, shift 2, Plant M, 3:23 p.m., line c.

Approved by: _____ Date: _____

16 Customer Complaint Program

PROGRAM TYPE: REQUIRED

THEORY

It is the goal of every food company to produce a product that is hugely successful financially. To do this, the company designs products that are appetizing and visually appealing to the customer, packaged in a convenient format, and priced at an attractive level. All of these details serve to draw the customer into purchasing the product the first time and, if the company is successful, again and again.

So what happens if the customer does not like the product? For some reason it does not meet expectations—it was too this ... or too that ... the color was off or the package was damaged or too hard to open. One way or the other it was bad. What does the customer do? He or she can chalk it up to a poor buying decision and never purchase it again or, as is human nature, complain. Sometimes this is to family and friends, sometimes it is to the store where it was bought, sometimes it is to the government, and sometimes it is just to the company that made the product. Although complaints are difficult to listen to and accept sometimes, they can provide the company a tool to fuel the product improvement process, assist in marketing, or act as a gauge to determine whether a more serious problem with the product may be present in the marketplace.

21CFR101.5 addresses the correlation between products not meeting customers' expectations and their desire to have a place to complain by requiring that the label contain specific contact information for the customer. Specifically, this section states

> (a) ... the label of a food in packaged form shall specify conspicuously the name and place of business of the manufacturer, packer, or distributor. (b) The requirement for declaration of the name of the manufacturer, packer, or distributor shall be deemed to be satisfied, in the case of a corporation, only by the actual corporate name, which may be preceded or followed by the name of the particular division of the corporation. In the case of an individual, partnership, or association, the name under which the business is conducted shall be used. (c) Where the food is not manufactured by the person whose name appears on the label, the name shall be qualified by a phrase that reveals the connection such person has with such food; such as "Manufactured for _____," "Distributed by _____," or any other wording that expresses the facts. (d) The statement of the place of business shall include the street address, city, state, and ZIP code; however, the street address may be omitted if it is shown in a current city directory or telephone directory. The requirement for inclusion of the ZIP code shall apply only

to consumer commodity labels developed or revised after the effective date of this section. In the case of nonconsumer packages, the ZIP code shall appear either on the label or the labeling including invoice. (e) If a person manufactures, packs, or distributes a food at a place other than his principal place of business, the label may state the principal place of business in lieu of the actual place where such food was manufactured or packed or is to be distributed, unless such statement would be misleading.

Having this information on the label allows the customer to locate a phone number in the local phone book or utilize Internet search services. Another option the company has is to put a consumer affairs or customer service phone number on the package. This optional piece of information can be a very valuable tool for the company because it helps to steer complaining customers directly to the company, which has control of how the information is handled, instead of toward outside entities, where the company has no control. Where complaints are concerned, control is everything. It gives the company an opportunity solve problems before federal, state, or local intervention; even in cases where intervention is required, prior knowledge of the problem provides the company opportunities to organize and coordinate its response. Although many large companies have the financial and labor resources to hire this function out or to have departments that can manage the plethora of customer complaints, the small to medium-sized company needs to develop and manage an efficient and productive customer complaint program. The key components of this program are determining the complaint intake process, complaint path, response, and documentation.

What are complaints and where do they come from? Complaints are any comment regarding the product. They can be positive or negative, subjective or objective, or a combination. Complaints originate with any person outside the company. This may be an end user, store manager, distributor, sales person, or just a family member. All complaints that come to a company should be funneled to an initiator. This is the person who initiates the intake process and completes the customer complaint form. His or her duty is to obtain the facts from the "complainer."

When a complaint is made, the initiator obtains the name, phone number, and other contact information and enters these on the customer complaint form. He or she then inquires what product is involved, what the problem is, and what the lot code is. All of these pieces of information are critical—especially the lot code. Have the customer look vigorously for this code because it provides the numeric basis for the investigation. Lastly, any other information regarding how the product was held and transported by the customer and any other miscellaneous information that might be offered should be documented. If the customer has any of the original packaging or if the complaint involves foreign material, the initiator should request that the customer send this back for examination. In some cases the customer may balk at this request for fear of losing leverage for future legal cases, but if the initiator requests nicely and offers to send a prepaid pouch or envelope in which the customer can send the items back, he or she generally will do so. After obtaining all of the necessary complaint information, the company should take steps to thank the customer for contacting it. This communication to the customer takes a variety of forms and varies from company to company, but it often takes the form of coupons or discounts on future purchases. Regardless of the circumstances relating to the complaint, the initiator should scrupulously pay attention to the following warnings:

- ***Do not admit fault.***
- Do not promise any settlement or performance.
- Do not promise or furnish analytical results on samples received.
- Do not mention insurance coverage.
- Do not mention legal action.

Besides jeopardizing the interests of the company, any admissions or promises during this opening contact can only confuse the issue and the customer. The initiator must defer any question of equitable treatment to management for resolution. Within 24 hours of receiving a complaint, the initiator should give the complaint form to the investigator, who is usually the quality control manager.

The responsibility of the investigator is to determine the seriousness of the complaint, investigate it, and verify that complaint records are maintained. To determine the seriousness of the complainant, the quality manager needs to categorize the seriousness, or risk, that the problem poses to the health of the brand. As a rule of thumb, any complaint that involves a direct or indirect issue of health must be considered extremely serious. These issues would include a customer getting sick from eating the product, having an allergic reaction to eating the product, and/or being harmed from eating or handling the product. When complaints of these types are received, an effort should be made by the quality control manager or other upper management to contact the customer (1) to verify that he or she is all right and (2) to assess the severity of the claim in anticipation of legal recourse. If the complaint portends future legal action, it should be turned over to upper management for further legal advice.

If the complaint is not health threatening, then the quality control manager determines if the complaint falls under the purview of operations, quality control, sales and marketing, product development, or some other department. Based on this determination, the complaint is discussed with all relevant departments for their input. This input is written on the complaint form in the investigation section. There are some complaints that originate with customers for which no follow-up is possible or warranted. These include complaints where the issue is neither a health issue nor one that can be directly fixed by any department. Stated clearly, some customers just want to complain. These complaints get logged and are summarized as part of the customer complaint report.

As the quality control manager investigates the complaint, he or she fills in the pertinent information on the customer complaint form. Any documents or records that pertain to the complaint should be copied and attached to the complaint form for future reference. When a determination is made regarding the cause of the complaint, any corrective action that is taken or will be taken should be filled in as the disposition of the claim itself. Final disposition would include contacting the customer, conducting a product withdrawal, destroying a product, adjusting the process to fix the problem, or just monitoring the situation. For every complaint there must be a disposition. Finally, the quality control manager logs the complaint on the customer complaint log for later tabulation. All customer complaints are to be kept in an organized file or notebook for later reference if needed.

APPLICATION

When the company begins the development of the customer complaint program, it needs to design a customer complaint form for gathering the required information. A basic template for this purpose is provided on the CD (Book 2_Quality Control Manual\Section 12_Customer Complaint Program:Customer Complaint Form) and at the end of this chapter. The initiator fills in the top part of the form as the required information is collected. This form is then sent to the quality control manager for the investigation phase.

During the investigation the quality control manager decides whom to consult and the eventual path of the complaint. The complaint flow diagram found on the CD (Book 2_Quality Control Manual\Section 12_Customer Complaint Program:Complaint Form Diagram) and at the end of this chapter offers a visual diagram to assist in determining this path. Upon completion of the investigation and after completing the customer complaint form, the quality control manager logs the complaint. The form found on the CD (Book 2_Quality Control Manual\Section 12_Customer Complaint Program:Customer Complaint Log) and at the end of this chapter can be used for this purpose. For each complaint received, a complaint number should be assigned. This provides a reference number as well as dictates the complaint filing convention. Fill in the rest of the information on the form and then file the complaint.

Rounding out the development of the customer complaint program is the development of the customer complaint program document. A customizable program document is provided on the CD (Book 2_Quality Control Manual\Section 12_Customer Complaint Program:General Overview) and at the end of this chapter. This should be adapted to represent the company's customer complaint program. After adaptation, it should be signed and placed in the appropriate section of the quality manual along with a blank customer complaint form, a copy of the customer complaint flow diagram, and a blank customer complaint log sheet.

SUPPLEMENTAL MATERIALS

Customer Complaint Form

Complaint Date: _____ Initiator: _____ **Complaint #:** _____

Distribution

    Operations: _____   Quality Control: _____
    Sales: _____   Counsel: _____
    Other: _____   Other: _____

Customer Name and Address:	Product:	Quantity:
	Lot Number:	Invoice #:
	PO Number:	Carrier:

Has a sample been obtained?   [ ] yes [ ] no

Description of problem: Quality [ ] Service [ ] Liability [ ]

                        By: _____   Date: _____

Investigation:

Corrective action:

                        By: _____   Date: _____

Disposition

Complaint logged by: _____   Date: _____

Customer Complaint Flow Diagram

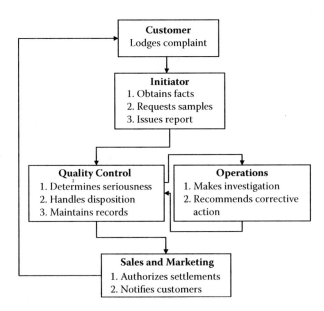

Customer Complaint Log

Complaint #	Complaint Date	Customer	Location	Complaint	Lot #	Initiator	Date Closed

CUSTOMER COMPLAINT PROGRAM

GENERAL OVERVIEW

Section A: General

It is the objective of _____ to provide only perfect products to our customers each and every time. However, in the natural course of business, from time to time, _____ may receive a complaint from a customer regarding quality, price, or service. To provide a timely, adequate, and consistent response, it is imperative that all complaints be channeled through a formal complaint handling system. Each complaint will undergo initiation, investigation, disposition, and documentation phases. This system is diagrammed on _____'s flow diagram for handling customer complaints and involves an initiator, quality control, operations, and sales.

No customer will be unhappy when consuming a product produced by _____.

Section B: Initiator

This person has the initial responsibility of collecting information about a specific complaint. An initiator could be from sales, reception, quality control, or plant manager. It is important that only employees trained to handle customer complaints do so. He or she is to fill out a customer complaint form as completely as possible. Thoroughness of detail is important because this document is the basis for solving problems, improving quality and service, and providing information in case of legal action.

Regardless of the circumstances relating to the complaint, the initiator should pay scrupulous attention to the following warnings:

- Do not admit fault.
- Do not promise any settlement or performance.
- Do not promise to furnish analytical results on samples received.
- Do not mention any insurance coverage.

Besides jeopardizing the interests of the company, any admissions or promises during this opening contact can only confuse the issue and the customer. The initiator must defer any question of equitable treatment to management for resolution. Within 24 hours of receiving a complaint, the initiator should give the complaint form to the investigator, usually the quality control manager.

Section C: Quality Control

The quality control manager will asses the critical nature of the complaint and determines its course of resolution. As needed, the complaint will be sent to operations for investigation. Upon successful gathering of complete information, the quality control manager will determine the disposition of the complaint and will then notify other departments of the results.

In conjunction with operations, the quality control manager will initiate changes in production methods and quality procedures to prevent a recurrence of the problem.

Section D: Sales and Marketing

After notification from the quality control manager, the sales department will contact the customer with the disposition. The customer should always be content with the results. If he or she is not, sales should notify the QC department for further investigation or turn the complaint over to counsel.

Section E: Documentation

For each complaint, a customer complaint form will be filled out by the initiator. When complete, complaints will be tabulated on the customer complaint summary form. A summary report of all complaints will be distributed quarterly to all management.

Approved by: _____ Date: _____

17 Receiving Program

PROGRAM TYPE: REQUIRED

THEORY

Every food manufacturer or processor receives ingredients and packaging from an outside source. Sometimes these suppliers will be as close as next door or across the street; sometimes they will be as far away as across the state, country, or world. Considering today's global marketplace, it is impossible for a manufacturer or processor to control the quality of the products that it receives from another company. To address the potential contamination possibilities and ramifications involved therein, the federal government has established very stringent requirements to be followed when receiving and handling inbound raw materials. Therefore, in compliance with 21CFR110 Current Good Manufacturing Practice in Manufacturing, Packing or Holding of Food section 110.80(a)1, which states, "Raw materials and other ingredients shall be inspected and segregated or otherwise handled as necessary to ascertain that they are clean and suitable for processing into food and shall be stored under conditions that will protect against contamination and minimize deterioration," and 21CFR1.337, which states, "(a) If you are a nontransporter, you must establish and maintain the following records for all food you receive," every food manufacturer should develop and maintain a formalized and documented receiving program.

This program will cover all raw materials and other ingredients, bulk and packaged, and all packaging. Its key components—bill of lading inspection, carrier inspection, and product inspection—occur in order for every carrier, no matter what type, and begin at the arrival time of the carrier. Prior to any receipt inspection, the company must develop the basic record keeping document: the receiving log. This is housed in a receiving book that should be located at the receiving station to provide maximum access for warehouse and quality control employees when ingredients are received.

Ingredients come to manufacturing facilities via conventional carriers, including semitrucks with dry, refrigerated, and frozen compartments; trains with dry and refrigerated cars; and bulk haulers as well as unconventional carriers including contract carriers such as FedEx or UPS and even employee cars. Each carries a potential plethora of logistical, economic, and scientific problems, so each must be prescreened prior to off loading. This takes the form of the bill of lading inspection.

To conduct this inspection the manufacturer should check the bill of lading for the name, address, and phone number of the consignor; the description; item number; unit of measure; quantity of each item; and seal number, if applicable. It should also have the consignee's name, address, phone number, and purchase order number. All of the information on the bill of lading should be verified against current company records to ensure that the company is not accepting ingredients that have not been ordered, are not the correct ingredients, or are not from an approved supplier. Once the company deems that the bill of lading information is correct and verifiable, the basic shipping information should be transferred to the receiving log found in the receiving book. Next, if there is a seal, the number should be entered on the log and compared against the one listed on the bill of lading. If it is the same, the carrier inspection of the program can take place. If the number is not the same, no further action should take place until the shipper has been contacted and the seal number resolved. This seal is placed on the truck as a tool for the receiver to verify that the contents inside have not been tampered with.

After the bill of lading inspection and seal verification are completed, the next step in the receiving program should be the carrier inspection. The company inspector inspects the carrier for overall cleanliness, container integrity, and proper temperature. The inspector should make sure that the container adheres to 21CFR110.80(a)1: "Containers and carriers of raw materials should be inspected on receipt to ensure that their condition has not contributed to the contamination or deterioration of the food." Specifically, holes in the exterior of the container, excess dirt on the floor, and miscellaneous off odors should be checked. If the foodstuff is to be held in a refrigerated or frozen state during transportation, the reefer temperature set point on the outside should be recorded on the log. If the container fails this initial inspection, the carrier should be consulted and the failure documented. If it passes inspection, the company inspector should proceed to the product inspection phase.

Once the bill of lading and carrier are approved for unloading, the company begins removing the ingredients from the container. During this process, the product should be inspected to verify that the ingredient itself, the packaging, or the way it has been transported or handled would not lead to any type of physical or microbiological adulteration. 21CFR110.80(a)1 specifically addresses this: "Raw materials and other ingredients shall be inspected and segregated or otherwise handled as necessary to ascertain that they are clean and suitable for processing into food and shall be stored under conditions that will protect against contamination and minimize deterioration."

At this point of the receiving program, the testing that needs to be done on the incoming ingredient is dependent solely on the type of ingredient and the potential for risk. If the ingredient is on the list designated by the HACCP risk analysis to require a certificate of analysis (COA), it should be checked by lot number to ensure that the supplier has removed or tested for the absence of the hazard. In many cases a small sample of each lot of ingredient will be culled and tested for various analytical parameters prior to acceptance into the factory. These analyses might include physical tests, such as color, granulation, or product integrity; microbiological tests, such as aerobic plate count, yeast, and mold; or a pathogen test. If the testing that has to be completed might take an extended period of time, the product might be accepted on a contingent basis and moved to a "hold" area to facilitate warehouse expediency.

When the product is completely inspected for quality attributes, the ingredients should be evaluated for any potential labeling issues. These include adherence to correct kosher, organic, halal, and ingredient labeling. The last is a potentially mandatory evaluation based on the allergen risk assessment component of the HACCP program.

Receiving Program

These marks and statements should be inspected to verify that they are in compliance and that nothing has been changed by the supplier. Every company should be aware that suppliers may sell similar ingredients to other companies that have different ingredients and markings and might potentially be shipped to them. Therefore, this ingredient verification is critical to preventing unwanted ingredients and ingredient components in the final product, thus ultimately preventing a recall.

Upon completion of sampling and testing, where necessary, warehouse personnel will determine if there are any damaged containers and note them on the bill of lading and on the receiving log. For financial considerations, the carrier should acknowledge the damaged product via signature. If the damaged containers can be salvaged, quality control personnel should determine if they can be scalped and the remainder used or if they should be disposed of in their entirety.

Finally, warehouse personnel should place a tag that contains the item number and the date of receipt on the bottom of each pallet under the wrap. If each ingredient container is to be used separately, each one should get a sticker. The basis for this is to provide warehousing and production a tool to ensure correct ingredient rotation (i.e., first in first out, FIFO).

APPLICATION

The first step to developing a functional receiving program is to build a receiving log. The form on the CD (Book 2_Quality Control Manual\Section 13_Receiving Program:Receiving Log) and at the end of this chapter can be adapted to fit the company's requirements. When completed, multiple copies of it should be placed in tab section 2 of an empty three-ring binder with three tab sections. A copy of the receiving log directions should be placed in section 1. They can be found on the CD (Book 2_Quality Control Manual\Section 13_Receiving Program:Receiving Log Directions) and at the end of this chapter. The kosher mark list should be placed in section 3, if applicable. This book is labeled "Receiving Log" and is placed in the warehouse receiving area for ease of access.

The second step is to create a receiving program document. A template for the company to customize to its unique situation and requirements can be found on the CD (Book 2_Quality Control Manual\Section 13_Receiving Program:General Overview) and at the end of this chapter. When complete, a signed copy of the program document should be placed in the quality control manual in the appropriate tab section.

SUPPLEMENTAL MATERIALS

Receiving Log

Date	Product Name	Item Number	PO Number	Quantity Received	Unit	Carrier	Seal #	Truck Inspected	Rejected?	Quantity Damaged	How Damaged	Reason for Rejection or Other Comments	Ingred. Checked	Kosher Symbol	Initials

RECEIVING LOG DIRECTIONS

Section A: General

To ensure accurate and consistent documentation of the ingredients, packaging, samples, and other receipts, a receiving log will be maintained. For *every* product that is received, including all samples, parts, and miscellany, a new row is to be filled in. The following are instructions regarding what information to put in each column of the receiving log.

Section B: Column Instruction

Date: Put in the date the product was received. Use the format mm/dd/yy.

Product name: Write in the name of the product. If you need the correct name, look at the name on the item list in the front of the receiving book.

Item number: Write in the item number of the product received. It is located on the item list in the front of the receiving book. It is in the format of A###-###.

PO number: Write the purchase order number in the box. This should be located on the bill of lading. Every product ordered has a purchase order number. If you cannot find it on the bill of lading, then ask the person who ordered it to get it for you.

Quantity received: Enter the number of units counted when the product was unloaded. This number should be in terms of the unit in the unit column.

Unit: This is the unit of measure of the product received. Typically, this will be "case," but it could be "each" or "pound."

Carrier: Enter the name of the trucking company that brought the product.

Seal #: Enter the number of the seal. Every full truck load of product should have a seal. You must watch the driver cut the seal off, compare the number with the one on the bill of lading, and if they match, enter it in this column. If they do not match, you are to notify the Quality Control manager immediately.

Truck inspected: Put your initials here after you have inspected the truck according to the receiving program.

Rejected: If the truck is rejected, write "yes," if not, write "no."

Quantity damaged: Write the number of units damaged on the truck.

How damaged: Detail how the product was damaged and its disposition (i.e., left on the truck or brought into the warehouse).

Reason for rejection or other comments: Fill in the reason the truck was rejected and any other comments related to the load or product.

Ingredients checked: Put your initials here after you have verified that the ingredient statement on the box is *exactly* the same as the approved ingredient statement in the ingredient book.

Kosher symbol: Compare the kosher symbol on the case with the symbol required from the kosher list in the receiving book.

Initials: If you are the person who signed the bill of lading *and* entered the information in the receiving book, then put your initials in this box. If more than one person unloads the truck, put both sets of initials in the box.

Section C: Record Storage

All completed receiving log pages are to be turned into the QC department for proper retention for a period of 7 years. These will be stored by year in the record storage area.

VERIFICATION DOCUMENTATION

Receiving Ingredient/Packaging Training

I, _____, have read and understand the section on receiving procedures. I have been trained in how to complete the receiving log, truck inspection, ingredient verification, and kosher verification. If I have questions or if something is out of specification, I will immediately notify my supervisor.

Date: _____

Name: _____

Instructor: _____

RECEIVING PROGRAM

GENERAL OVERVIEW

Section A: General

In accordance with 21CFR110.80a and 21CFR1.345, _____ adheres to the following receiving procedures for the receipt of raw materials.

Section B: Record Keeping/Documentation

All deliveries will be documented completely on the receiving log.

Section C: Inspection

1. The bill of lading (BOL) will be inspected for name, address, and phone number of the consignor and consignee, product number, item number, lot numbers, and quantities.
2. Prior to opening the doors, the seal needs to be cut off and the number recorded on the BOL.
3. All trucks delivering raw materials will be inspected for overall cleanliness. If the truck is damaged, dirty, or infested, the contents will be rejected.
4. If the truck has an unusual odor, the truck driver will be asked what was hauled prior to delivery. If the prior shipment in any way could have contaminated the current contents, the load should be rejected. Some off odors that will affect other products are fish, meat, or chemicals. *Note:* If the truck is carrying or was carrying chemicals, the contents are rejected automatically.
5. If the contents of the truck are frozen or refrigerated, the temperature is to be verified prior to opening the truck doors and recorded on the receiving log.

Section D: Labeling

1. All raw materials received will be marked with the date of receipt on the bottom unit. This date is not to be placed on the stretch wrap.
2. The date of receipt will be ink letters no smaller than a 4 × 6 card.
3. If two or more different ingredients are shipped on the same pallet or if each individual unit is used separately, each must be marked with the date of receipt.

Section E: Quality

For all ingredients, review the ingredient list for special handling requirements.

Section F: Ingredients—CCP (A-1)

For all raw materials and packaging, the ingredient statement will be compared against the ingredients in the ingredient book of standard statements located at the receiving desk. Any differences must be reported to the quality control manager **before** unloading or put on hold by the QC department pending approval.

Section G: Kosher

For all ingredients, review the ingredient list for special kosher requirements. The addendum to agreement with _____ from the _____ certifying agency lists the kosher requirements for each item. This is located in the front of the receiving book. Look under "name of ingredient" and then "kosher symbols required" for the symbol that must be on the container. Look on the container and see if the symbol is there. If it is not exactly what is listed, notify the quality control department before the product is accepted. Enter the symbol and your initials in the receiving book.

Section H: Storage

For all ingredients, review the ingredient list for proper storage and special handling requirements.

Section I: Damaged Goods

When an item has been damaged in transit:

1. Make a note on the receiving log.
2. Make a note on the bill of lading and have the truck driver initial that damaged goods were unloaded.
3. Notify QC so that the ingredient can be inspected for its food safety.

Approved by: _____ Date: _____

18 Shipping Program

PROGRAM TYPE: REQUIRED

THEORY

One of the most exciting things that small to medium-sized companies get to do is to ship products to customers. Whether these customers are in the same city, the next state, or across the country, all customers expect the product they receive to be in perfect condition when it arrives. To meet this expectation, the company must develop systems surrounding its shipping function. Therefore, as the company is required to perform certain tasks regarding inbound freight, so it is tasked with requirements regarding outbound freight and its carriers. 21CFR1.352 states, "If you are a transporter, you must establish and maintain the following records for each food you transport in the United States." Likewise, 21CFR110.93 states, "Storage and transportation of finished food shall be under conditions that will protect food against physical, chemical, and microbial contamination as well as against deterioration of the food and the container." Both of these legal citations require the manufacturer to establish a program to ensure the continued protection against adulteration postproduction. To fulfill these requirements, most companies establish a shipping program that follows the company's shipping procedures and contains three parts: documentation, inspection, and record keeping.

When a truck arrives to pick up products to transport to another company or location, the company should create a bill of lading or packing list containing the following legally required pieces of information:

name, address and telephone of transporter;
company name, address, and phone number;
date of shipment;
description of freight; and
number of packages.

In addition to this information, the bill of lading should list the lot number for each item shipped in conjunction with the quantity of each lot and be signed by the carrier upon pickup.

Next, the shipping log should be obtained and the carrier name and identification number filled in. The carrier is then inspected to verify that it meets the quality requirements as set forth previously. The inspector, whether from quality control or operations, should specifically look for filth on the floor of the container and integrity of the walls, ceiling, floor, and doors of the shipping container. If the product to be shipped is frozen or refrigerated, the refrigeration unit should be inspected and verified that it is working. The inspector should also smell the interior of the shipping container to make sure there are no off odors that might become infused in the product. When the inspection is complete, the log should be completed.

The last part of the company shipping program involves record keeping. As the shipping logs are being completed, they should be stored with the shipping records for future reference if needed. Before retaining, they need to be reviewed to ensure completeness.

APPLICATION

Similar to the company's receiving program, the first part of developing the shipping program is to create a shipping log document. A customizable form for keeping track of any shipments that are sent can be found on the CD (Book 2_Quality Control Manual\Section 14_Shipping Program:Shipping Log) and at the end of this chapter. This should be altered to reflect the types of shipments made and the types of carriers used. For instance, if the shipments are all liquid bulk, a column would be added specifically to address tanker washout confirmation. Likewise, if the shipments are all UPS or FedEx, then the trailer number column could be removed.

The form on the CD (Book 2_Quality Control Manual\Section 14_Shipping Program:Shipping Log Directions) and at the end of this chapter provides the company complete directions for filling in the shipping log. These should be customized to address the changes made to the shipping log. They should be reviewed with shipping personnel prior to shipping to ensure that all of the legal requirements are being met.

Finally, a general shipping program overview should be developed. The requirements of the shipping program are outlined on the CD (Book 2_Quality Control Manual\Section 14_Shipping Program:General Overview) and at the end of this chapter. When it is approved, it should be signed and placed along with a copy of a blank shipping log and the shipping log directions under the appropriate tab section of the Quality Control Manual.

SUPPLEMENTAL MATERIALS

SHIPPING CONTAINER INSPECTION LOG

Date	Company	Trailer #	Cleanliness	Odors	Damage	Refer Unit on? Yes	Temp	Other Defects	Action Taken

SHIPPING LOG DIRECTIONS

Section A: General

To ensure accurate and consistent documentation of the containers used to ship products the shipping log should be filled out for each truck that is loaded with product. The following are instructions on what information to put in each column of the shipping log.

Section B: Column Instruction

Date: Put in the date the product was shipped. Use the format mm/dd/yy.
Company: Write in the name of the trucking company that is getting loaded.
Trailer number: Write in the number of the trailer. This is usually located on the inside wall of the trailer. If you do not find it, ask the driver.
Cleanliness: Inspect the truck for cleanliness and record the findings.
Odors: Note any objectionable odors.
Damage: Document any damage to the container before loading.
Reefer unit on: Fill in whether the reefer is on or not. If it is, fill in the temperature.
Other defects: Write any other defects found during the container inspection.
Action taken: Fill in any actions taken during the inspection. This includes sweeping, washing, and rejection.

Section C: Record Storage

All completed shipping log pages are to be turned in to the QC department for proper retention for a period of 7 years. These will be stored by year in the record storage area.

SHIPPING PROGRAM

GENERAL OVERVIEW

Section A: General

In accordance with 21CFR110.93 Good Manufacturing Practices (GMPs) and 21CFR1.352, _____ will inspect all outbound carriers to determine whether they are fit to carry food products. This inspection will be documented.

Section B: Documentation

All inspections will be documented on the shipping log.

Section C: Inspection

Each trailer that is staged to pick up products will be inspected for the following before any product is loaded:

1. Overall cleanliness: Each truck shall be inspected for overall cleanliness. If the truck is dirty or infested, it will be documented and rejected. Care should be taken to look at the floor and walls.
2. Odors: If the truck has an unusual odor, it will be rejected for cleaning and/or airing out. This will be documented.
3. Damage: If the integrity of the truck has been compromised (e.g., holes in the floors, walls, or ceilings; refrigeration unit not working; doors not closing properly), it is to be rejected. This will be documented.

Minor defects should be corrected before loading. This includes the removal of any debris, sweeping and/or vacuuming the floor, and removing protruding nails from the floor, walls, or ceiling. Objectionable odors may be removed if possible.

Section D: Rejection Notification

If a truck is rejected, the quality control department is to be notified immediately.

Section E: Record Storage

All completed shipping log pages are to be turned in to the QC department for proper retention for a period of 7 years. These will be stored by year in the record storage area.

Approved by: _____ Date: _____

19 Specification Program

PROGRAM TYPE: OPTIONAL

THEORY

Once a company has created a product that it believes the consumer will purchase and like, it is incumbent on it to be able to provide the customer with the same product for each and every experience. To meet this obligation, companies spend a signification amount of time and money on equipment and labor. Although this is important, unless the company has established standards for all facets of production, product quality and consistency will end up being a moving target. To establish these standards, the company needs to evaluate three specific areas and create specifications for each: raw materials, processing, and finished goods. Although they vary in their scope and application, together they provide the needed guidelines and structure for the company to make the same product each and every time.

RAW MATERIAL SPECIFICATIONS

The building blocks of any product are the components, ingredients, and packaging that go into it. As a means for the purchasing department to purchase the same ingredient of the same exact quality from various suppliers correctly, a set of raw material specifications should be developed. This is in conjunction with the product development function of the company and relies on the basic information collected during the information gathering step of the HACCP program. Therefore, for each ingredient and package a specification should be prepared as a guideline for purchasing and receiving the ingredient for use in the facility. This document begins with the company item number, a description of the ingredient, and an ingredient statement where applicable. It also lists the basic package and pallet configuration as well as the storage and shelf life information. The bulk of the specification, however, is a detailed list of physical, chemical, and microbiological requirements for the ingredient or package to meet. All of the requirements must be scientifically measurable, typical characteristics of the item, and, in most cases, unique to the type of ingredient being purchased.

Some of the more common characteristics measured include:

    microbiological: aerobic plate count, yeast and mold, coliform, *Escherichia coli, Salmonella, Listeria,* staph;
    physical: color, granulation, viscosity; and
    chemical: sugar breakdown, fat breakdown, sapponification number.

Finally, the specification lists other parameters, including a nutritional breakdown, shipping information, and any certification requirements such as organic, non-GMO (Genetically Modified Organism), kosher, and halal.

All raw material specification sheets should be reviewed by product development, purchasing, production, and quality control to ensure complete agreement regarding the item's properties, packaging, and storage conditions. It is important that any concerns that arise at the review stage are resolved prior to purchasing the ingredient because these can lead to inconsistent products and eventual waste. After all of the ingredient specifications for a particular finished good are completed, the next step is to develop a processing specification.

Processing Specifications

Processing specifications are the guideposts for the manufacturing environment. They provide operations and quality control an online tool to guide manufacturing in all aspects of machine setup, processing, packaging, palletizing, and storage. They chart the course of the specified ingredients and packaging into the processing environment, document the changes that are made as they are converted into finished products, and then specify the details of the end products. Processing specifications are a combination of inputs leading to changes accompanied by the quality parameters measured when the change takes place. This leads to the next processing change and the next quality parameter and so on until the finished product is stored.

In most cases, processing specs are several pages long and include several additional components, such as drawings and pictures. The company should use any means at its disposal to document every change related to the process so that each time the product is manufactured, it is manufactured in exactly the same way. In those cases where a quality parameter is measured, this measurement should be recorded either on the production documentation or on separate quality control recording sheets. These specifications should be available to the operations team for immediate access prior to and while running the product, and a copy should be held in a processing specification file available to product development and quality control. Often it is prudent to laminate the production copy prior to taking it to the production line so as to keep it from becoming soiled, which can lead to microbial contamination.

Finished Goods Specifications

Once the finished good is in storage, it enters the shipping and sales phase. Finished goods that are sold to distributors for consumers typically will have a sales sheet developed by the sales team for presentations. For those finished goods that are sold to other manufacturers for additional processing or inclusion in other products, a finished good specification should be developed. This specification is item unique and should list the technical parameters that would be important to the end user. These include the item number, possibly the customers and the manufacturers, a description of the product, an ingredient statement, analytical parameters, packaging and palletizing requirements, and shipping, storage, and shelf life information. In many cases customers will dictate specific items they would like to see on the specification sheet by their questions during the initial purchasing cycle. These requests might also include requiring a certificate of analysis to accompany the product on a lot-by-lot basis.

APPLICATION

RAW MATERIAL SPECIFICATION

As the company begins to install a raw material specification program it must first build a specification template to ensure that the basics of all specs are the same. The form on the CD (Book 2_Quality Control Manual\Section 15_Specification Program:Ingredient Specification) and at the end of this chapter can be used as the basis for developing the specifications. At the top of the form the date of specification, vendor information, and the item name and number should be filled in. Next, the ingredient description and statement should be completed. If a continuing food guarantee is required and for each allergen that may be present, check the required frequency.

After that, all of the critical acceptance requirements; the microbiological, physical, and chemical requirements; and the shipping and storage requirements should be added. Finally, the shelf life, pack and pallet configurations, and any required certifications should be filled in. The completed ingredient specification should be submitted to product development, quality control, purchasing, and operations for their approval via signature. Upon completion, the ingredient specification form is filed and used as a guide when needed.

FINISHED GOODS

Similarly to the ingredient specification required for all ingredients, the manufacturer should create a specification for each finished product. A template for collecting all of the information needed for a finished product specification is located on the CD (Book 2_Quality Control Manual\Section 15_Specification Program:Finished Product Specification) and at the end of this chapter. After the information has been gathered, it should be formatted onto a company letterhead and signed prior to sending it to a customer.

SUPPLEMENTAL MATERIALS

INGREDIENT SPECIFICATION

Date:
Description:

Vendor: <Name>
<Address>
<Address>

Item Name:
Item Number:

Contact: <Name>
Phone #:
Fax #:

Customer Item Number:

E-mail:

I Ingredient Description

II Ingredient Statement

III Statements

Continuing food guarantee

☐ Required ☐ By lot ☐ Up front only

Allergens

☐ Milk ☐ Eggs ☐ Tree nuts ☐ Soy ☐ Peanuts ☐ Wheat/gluten ☐ Fish
☐ Crustacean shellfish ☐ Seeds ☐ Colors ☐ Other _____

IV Critical Acceptance Requirements

Characteristics	Standard	Tolerance	Method
Appearance			Visual
Foreign matter			
Taste			Organoleptic
Density			
Color			

V Microbiological/Physical/Chemical

Characteristic	Standard	Tolerance	Method
APC	cfu/g		FDA/BAM
Yeast and mold	cfu/g		FDA/BAM
Coliform	cfu/g		FDA/BAM
E. coli	cfu/g		FDA/BAM
Staphylococcus coagulase Positive-MPN	cfu/g		FDA/BAM
Listeria	Negative		FDA/BAM
Salmonella	Negative		FDA/BAM

VI Microbiological/Physical/Chemical

Characteristic	Standard	Tolerance	Method
Granulation	on US__		
	on US__		
	on US__		
	through US__		

VII Shipping/Storage
Recommended storage conditions—
Recommended shipping conditions—

VIII Shelf Life at Recommended Storage Conditions
Shelf life—

IX Container
Pack size—

X Pallet
Pallet configuration—

XI Certifications
Kosher ☐   Organic ☐   Halal ☐   Non-GMO ☐   Vegan ☐

Approvals

Quality Control: _____   Product Development: _____

Purchasing: _____   Operations: _____

FINISHED PRODUCT SPECIFICATION

Date: Vendor: <Name>
Description: <Address>
 <Address>

Item Name: Contact: <Name>
Item Number: Phone #:
 Fax #:
Customer Item Number: E-mail:

I Product Description

II Ingredient Statement

III Statements

Continuing food guarantee

☐ Required ☐ By lot ☐ Up front only

Allergens

☐ Milk ☐ Eggs ☐ Tree nuts ☐ Soy ☐ Peanuts ☐ Wheat/gluten ☐ Fish
☐ Crustacean shellfish ☐ Seeds ☐ Colors ☐ Other _____

IV Critical Acceptance Requirements

Characteristics	Standard	Tolerance	Method
Appearance			Visual
Foreign matter			
Taste			Organoleptic
Density			
Color			

V Microbiological/Physical/Chemical

Characteristic	Standard	Tolerance	Method
APC	cfu/g		FDA/BAM
Yeast and mold	cfu/g		FDA/BAM
Coliform	cfu/g		FDA/BAM
E. coli	cfu/g		FDA/BAM
Staphylococcus coagulase Positive-MPN	cfu/g		FDA/BAM
Listeria	Negative		FDA/BAM
Salmonella	Negative		FDA/BAM

VI Microbiological/Physical/Chemical

Characteristic	Standard	Tolerance	Method
Granulation	on US__		
	on US__		
	on US__		
	through US__		

VII Shipping/Storage
Recommended storage conditions—
Recommended shipping conditions—

VIII Shelf Life at Recommended Storage Conditions
Shelf life—

IX Container
Pack size—

X Pallet
Pallet configuration—

XI Certifications
Kosher ☐   Organic ☐   Halal ☐   Non-GMO ☐   Vegan ☐

Approvals

Quality Control: _____   Product Development: _____

Purchasing: _____   Operations: _____

CERTIFICATE OF COMPLIANCE

This confirms that, based on the following criteria, all products shipped from the _____ production facility will conform to the following specification.

Product ID	Description	Lot Number	Pack Size

    This product conforms to all elements of the _____ quality systems for food safety and product consistency. This product meets _____'s positive release requirements. All applicable critical control points, product quality consistency, and food safety monitoring criteria have been reviewed and approved. This product is produced, stored, and shipped in compliance with the Good Manufacturing Practices of the U.S. Food and Drug Administration, Title 21 CFR Part 110.

_____

Name and Title

20 Recall Program

PROGRAM TYPE: REQUIRED

THEORY

Recall! No other word sends larger and deeper shudders through a company than this one. Recalls can decimate small and medium-sized companies financially and marketwise and can wreak havoc on all facets of the staff. Recognizing how potentially lethal violative products can be, the federal government has developed regulations that define deficient products and explain its responsibility and a firm's responsibility. Based on these regulations, companies are required to develop formal systems to deal with these types of situations and in tandem forcibly remove from the marketplace products that are substandard. Although this section discusses the regulatory requirements for a company, it does not address the economic or political ramifications with which companies deal regarding recalls.

21CFR1107.40(a) defines *recalls* as

> an effective method of removing or correcting consumer products that are in violation of laws administered by the Food and Drug Administration. Recall is a voluntary action that takes place because manufacturers and distributors carry out their responsibility to protect the public health and well-being from products that present a risk of injury or gross deception or are otherwise defective.

The role of recalls is explained thusly:

> This section and 7.41 through 7.59 recognize the voluntary nature of recall by providing guidance so that responsible firms may effectively discharge their recall responsibilities. These sections also recognize that recall is an alternative to a Food and Drug Administration-initiated court action for removing or correcting violative, distributed products by setting forth specific recall procedures for the Food and Drug Administration to monitor recalls and assess the adequacy of a firm's efforts in recall.

These are voluntary in nature as opposed to a Food and Drug Administration (FDA)-initiated court action and as such they can be initiated by the federal government as well as by the manufacturer. Section (b) states that a "recall may be undertaken voluntarily and at any time by manufacturers and distributors, or at the request of the Food and Drug Administration."

In order to determine what types of problems constitute a violative product, a risk-based system is utilized. Section 7.41 outlines this system of hazard determination:

(a) An evaluation of the health hazard presented by a product being recalled or considered for recall will be conducted by an ad hoc committee of Food and Drug Administration scientists and will take into account, but need not be limited to, the following factors: (1) Whether any disease or injuries have already occurred from the use of the product. (2) Whether any existing conditions could contribute to a clinical situation that could expose humans or animals to a health hazard. Any conclusion shall be supported as completely as possible by scientific documentation and/or statements that the conclusion is the opinion of the individual(s) making the health hazard determination. (3) Assessment of hazard to various segments of the population, e.g., children, surgical patients, pets, livestock, etc., who are expected to be exposed to the product being considered, with particular attention paid to the hazard to those individuals who may be at greatest risk. (4) Assessment of the degree of seriousness of health hazard to which the populations at risk would be exposed. (5) Assessment of the likelihood of occurrence of the hazard. (6) Assessment of the consequences (immediate or long-range) of occurrence of the hazard.

Based on this hazard risk assessment, the type or classification of the recall needed is determined. These are classified as:

class I: a situation in which there is a reasonable probability that the use of, or exposure to, a violative product will cause serious adverse health consequences or death;
class II: a situation in which use of, or exposure to, a violative product may cause temporary or medically reversible adverse health consequences or where the probability of serious adverse health consequences is remote; and
class III: a situation in which use of, or exposure to, a violative product is not likely to cause adverse health consequences.

Each of these classes of recalls requires a firm and focused response from the manufacturer in order to prevent a product seizure. The recommended method of preparing for a recall is to develop an effective recall strategy.

Section 7.42(b) states that the requirements of an effective recall strategy are depth of recall, public warning, and effectiveness checks. Specifically, it states

A recall strategy will address the following elements regarding the conduct of the recall: (1) *Depth of recall*. Depending on the product's degree of hazard and extent of distribution, the recall strategy will specify the level in the distribution chain to which the recall is to extend, as follows: (i) consumer or user level, which may vary with product, including any intermediate wholesale or retail level; or (ii) retail level, including any intermediate wholesale level; or (iii) wholesale level. (2) *Public Warning*. The purpose of a public warning is to alert the public that a product being recalled presents a serious hazard to health. It is reserved for urgent situations where other means for preventing use of the recalled product appear inadequate. The Food and Drug Administration in consultation with the recalling firm will ordinarily issue such publicity. The recalling firm that decides to issue its own public warning is requested to submit its proposed public warning and plan for distribution of the warning for review and comment by the Food and Drug Administration. The recall strategy will specify whether a public warning is needed and whether it will issue as: (i) general public warning through the general news media, either national or local as appropriate, or

(ii) public warning through specialized news media, e.g., professional or trade press, or to specific segments of the population such as physicians, hospitals, etc. (3) *Effectiveness Checks.* The purpose of effectiveness checks is to verify that all consignees at the recall depth specified by the strategy have received notification about the recall and have taken appropriate action. The method for contacting consignees may be accomplished by personal visits, telephone, letters, or a combination thereof .... The recall strategy will specify the method(s) to be used and the level of effectiveness checks that will be conducted.

Effectiveness checks are evaluated for their adequacy by the FDA using the level system and are set forth as

(i) Level A—100 percent of the total number of consignees to be contacted; (ii) Level B—Some percentage of the total number of consignees to be contacted; (iii) Level C—10 percent of the total number of consignees to be contacted; (iv) Level D—2 percent of the total number of consignees to be contacted; or (v) Level E—No effectiveness checks.

Recalls fall into two categories: those initiated by the government and those initiated by the company. If the recall is initiated by the government, upon determining through scientific means that a hazard exists, the federal government notifies the company officially via letter or telegram of the need to proceed with a recall, although it may begin via oral communication. This communiqué will "specify the violation, the health hazard classification of the violative product, the recall strategy, and other appropriate instructions for conducting the recall including a possible request for other information. If the recall is company initiated, the firm will be asked to provide the identity of the product involved, the reason for the removal or correction, the date and circumstances under which the product deficiency or possible deficiency was discovered, an evaluation of the risk associated with the deficiency or possible deficiency, and the total amount of such products produced and/or the time span of the production. The company will also need to provide the total amount of such products estimated to be in distribution, distribution information including the number of direct accounts and, where necessary, the identity of the direct accounts, a copy of the firm's recall communication if any was issued, or a proposed communication if none is issued, a proposed strategy for conducting the recall, and the name and telephone number of the firm official who should be contacted concerning the recall."

Companies that are conducting a recall must promptly notify all of their domestic and foreign direct accounts, including all retail companies, wholesale companies, and distributors to whom the product was sold. This communication should state (7.49):

that the product in question is subject to a recall;
that further distribution or use of any remaining product should cease immediately;
where appropriate, that the direct account should in turn notify its customers who received the product about the recall; and
instructions regarding what to do with the product.

The communication can be delivered by any means possible, including mail, e-mail, and telephone; all communication should be documented. If using "snail mail," the company should clearly mark the envelope, preferably in red ink, "Urgent" and "Recall" and send it registered to establish documentation. When using e-mail for notification, the notification option for verification should be used; when using the telephone, these calls should be followed by official notification via envelope.

The actual communication notification should include (7.49.c) the identity of the product, size, lot number(s), code(s) or serial number(s), and any other pertinent descriptive information to enable accurate and immediate identification of the product. It should explain concisely the reason for the recall and the hazard involved, if any; provide specific instructions on what should be done with respect to the recalled products; and provide a ready means for the recipient of the communication to report to the recalling firm whether it has any of the product. Promotional material or other superfluous information or spam should not be sent.

In tandem with the company notification to its customers regarding product that needs to be recalled, the federal government will publish on its weekly enforcement report the details of the recall. Companies should not worry about this because at this point it only assists the company to notify those customers that are not paying attention to the serious nature of the company's notification. The federal government may not publish this notification immediately or publish a notice at all if it feels that they are market withdrawals or stock corrections.

During and upon conclusion of the recall, the company will be asked to provide progress reports specifically outlining (7.53) the number of consignees notified of the recall and date and method of notification; the number of consignees responding to the recall communication and quantity of products on hand at the time it was received; the number of consignees that did not respond (if needed, the identity of nonresponding consignees may be requested by the FDA); the number of products returned or corrected by each consignee contacted and the quantity of products accounted for; the number and results of effectiveness checks that were made; and the estimated time frames for completion of the recall. How often and how complicated this reporting will be is at the discretion of the government and is based on the type and severity of the risk involved. This reporting will end following the end of the recall, which comes after the government ends it or the company petitions to end it. When doing so, the company must provide a written request outlining (7.55) the effectiveness of the recall, the most current recall status report, and a description of the disposition of the recalled product.

Due to the complexity, potential health and safety risks, and likely negative publicity, it behooves all food manufacturer and processor to take every precaution, no matter the cost, to prevent a recall. Although this may seem obvious to most companies, it is so important that the federal government provides suggestions that should be followed to prevent them. These are found in 21CFR7.59 and state:

> (a) Prepare and maintain a current written contingency plan for use in initiating and effecting a recall in accordance with 7.40 through 7.49, 7.53, and 7.55. (b) Use sufficient coding of regulated products to make possible positive lot identification and to facilitate effective recall of all violative lots. (c) Maintain such product distribution records

as are necessary to facilitate location of products that are being recalled. Such records should be maintained for a period of time that exceeds the shelf life and expected use of the product and is at least the length of time specified in other applicable regulations concerning record retention.

APPLICATION

In light of the seriousness posed by any defective product, it is imperative that each company develop and maintain a formal and structured recall contingency plan. This should take into consideration each of the classifications of recalls, market withdrawals, and stock recoveries. The plan should contain sections on evaluation of the hazard, classification of the recall, communication of the recall, handling the defective material, documentation, effectiveness checks, and record retention. The form on the CD (Book 2_Quality Control Manual\Section 16_Recall Program:Recall Program Overview) and at the end of this chapter provides companies a structured program document that contains all of the required components. This should be customized to reflect the responsible person for every task.

The first section contains a general statement outlining the legal basis for this program. This is followed by a summary of questions regarding how to asses the risk the problem poses to the general public. The answers to these questions determine the classification from section C into which the recall will be placed. The next section outlines how communication regarding the recall will be conducted.

Section E of the program document outlines how the recalled product will be handled. Section F identifies the documentation utilized and section G explains how the effectiveness of the recall will be determined. The final section explains how the recall records will be retained.

A standard customizable letter that the firm can utilize to expedite this communication is located on the CD (Book 2_Quality Control Manual\Section 16_Recall Program:Recall Communication) and at the end of this chapter. It contains multiple fields to be filled in, including product information, specific details of the recall, and the contact information for the recall coordinator.

Another portion of the recall program that each company must prepare is an emergency recall notification list of company personnel that may be called by customers, the media, or regulatory agencies in the event of a recall. The form on the CD (Book 2_Quality Control Manual\Section 16_Recall Program:Emergency Recall Notification List) and at the end of this chapter can be adapted to fit the organizational structure of the company. The completed document should be provided to all distributors, vendors, and regulatory agencies for immediate identification of those individuals at the company responsible for handling a recall.

CONDUCTING A RECALL

Upon identifying that a product produced by the company or for the company contains a defective attribute, whether it is a health hazard or not, the decision of whether to recall it from the marketplace must be made. This decision is risk based; however, any product that is mislabeled, contains or may contain a pathogenic organism, or contains or may contain an undeclared allergen should seriously be considered for a recall or product withdrawal. There are four basic parts to conducting a recall: identification, communication, validation, and disposition. Before a recall is conducted, it must first be determined if the defective product was an ingredient added to a finished good that was contaminated or was an adulteration from an internal process. This decision will dictate the course the recall plan will take: vendor directed or company directed.

If the recall is vendor directed, the vendor will provide some form of communication outlining the affected product. From the communication, the recall proceeds according to the following steps:

1. The lot codes involved must be identified. The supplier or vendor should be asked to provide specific lot codes for the ingredients involved and the dates when the affected lots were sent to the company. The raw material recall sheet located on the CD (Book 2_Quality Control Manual\Section 16_Recall Program:Raw Material Recall Sheet) and at the end of this chapter should be filled in with vendor name, product name, and the lot code. The receiving paperwork (bills of lading) that refers to the affected lots should be copied, stapled together, and placed in a recall file folder.

    Computer tip: Lot codes and date received are entered as raw material receipt information within some manufacturing software.

2. The quantity of the raw material in question is calculated. This should be in the standard unit of measure utilized within the processing specifications. Typically, this would be pounds, ounces, or gallons; however, it could be any unit of measure. This number is written in the starting balance box.

    Computer tip: Some computer systems can report this quantity by lot information as a function of the raw material receipt data.

3. The finished goods in which the ingredient is used should be determined. In a simple production scenario, this might be only one finished product, but in complex systems it potentially could be tens or even hundreds, especially for basic ingredients such as sugar or flour. The finished product code for each configuration of the finished goods made with the ingredient should be written in the product code boxes, taking extreme care to write in all of the finished products because it is easy to overlook products with different packaging and sizes that contain the same base products. For each product identified, the description of the product should be filled in.

    Computer tip: Most manufacturing software reports these data in the form of a "where used" report. If a computer-generated report is used, a copy of this report should be placed in the recall file folder.

4. It must be determined which of the products identified in the previous step were made after the lot in question was received. The production records should be examined and the lot code/date of use for each product that was made with the ingredient recorded in the box on the form on the line of the specific finished good. For those products that were made on multiple dates, new lines should be created for each additional manufacturing date as needed. As the finished goods that were manufactured are filled in, the quantity of the finished good should be written in the finished goods "quantity made" box. Next, the quantity of the raw material used for each product

made should be recorded in the "quantity used" box. This number should be in the same unit of measurement as was the starting balance. When all of the production records have been identified and recorded on the raw material recall sheet, each of the "quantity used" entries should be subtracted from the starting balance.

Computer tip: Manufacturing software can provide this information in report form. Print copies of these reports and place them in the recall file folder.

5. The amount of the ingredient left in stock should be determined and a physical inventory of this lot code conducted within the raw material inventory. This number should be written in the "quantity left in stack" box on the raw material recall sheet and subtracted from the bottom number in the "inventory balance" column; the remaining total is written in the ending balance. This number should be *zero*. If it is not zero, then it is necessary to start from step 2, paying careful attention to the products and quantities involved. The records should be reexamined until the location of all of the raw material has been found or a plausible explanation for any missing ingredient is found. Some commonly overlooked uses are daily production waste, recorded or not recorded; damaged ingredients found at point of receipt that may or may not have been listed on the bill of lading; donations (items given as charity); and internal samples (ingredients taken for product development, sales, quality control, or other needs).

6. The location of each of the finished goods produced with the affected ingredient needs to be identified. For each of the products listed on the raw material recall sheet that were produced, a finished product recall sheet, located on the CD (Book 2_Quality Control Manual\Section 16_Recall Program:Finished Product Recall Sheet) and at the end of this chapter, should be created. Product, item number, lot code, and starting balance for the lot code should be entered in the appropriate spaces. Next, all the shipping records should be reviewed to track how much of and where each of the identified finished goods was shipped. For each shipment, the name of the customer or consignee, the purchase order number, the date of the shipment, and the quantity that was shipped should be filled in. A copy of each of the shipping documents should be placed in the recall file folder.

Computer tip: Computer software can typically provide a report that shows how much of each product was shipped and to whom. Many times it also tracks the lot codes sold for each product. If using computer-generated reports, place them in the recall file folder.

7. It is necessary to determine how much of each finished good lot code remains in stock. A physical inventory of the finished goods by lot should be conducted and that quantity recorded on the "finished product left in stock" box. These quantities should be subtracted from the starting balance as a running tally using the Inventory Balance columns. The ending balance should be zero. If it is not, it is necessary to review all the shipping records. If no other sales can be found, miscellaneous stock deductions, such as samples taken by sales, marketing, product development, and

quality control, need to be examined. Also, damaged products and products on hold should be considered. Note: All products must have a completed recall sheet. This is the only way to account for where all of the initial raw material went.

Computer tip: Manufacturing software will typically run reports that identify how much of and where each individual lot was shipped.

8. *Contact customers.* For each customer that received a product to be recalled, a recall communication letter must be prepared and customized to specify the affected product, lot code, quantity, and reason for recall. A succinct reason why the product is being recalled should be offered. The reason for the recall should not be hidden because the customer might consider the request not worth his or her time and set the letter aside. When it sends this letter, the company should seriously consider including a self-addressed, stamped postcard for the customer to return acknowledging receipt of the letter.

If, in the expediency of time, the company deems that another form of recall communication is to be used, such as e-mail or telephone, it is still recommended that the aforementioned recall communication letter be mailed. 21CFR does allow for other types of communication media, such as personal visits or telephone calls, but these are more difficult to prove to have occurred. At the very least, all communication between customers and the company that pertains to the recall should be documented.

Based on the type and classification of the recall, various other communications might be made at this point. These include print and television media communications that are designed to reach out to the end user. These types of communication will be part of the overall recall plan approved by the government at the beginning of the recall and can be effective to notify the general public regarding potential health hazards.

9. *Determine disposition.* Based on the hazard contained in the product and the risk it poses to the general public, a decision must be made on what to do with the product being recalled. This can range from destruction by the customer, retail or wholesale store, and/or distributors all the way to returning the product for reconditioning. In any case, the customer must be instructed to isolate the product, count it carefully by lot number, and then tag it for disposition. When considering this decision, several questions should be considered, such as whether the hazard can be removed; the costs associated with removing or reconditioning the defective product, including peripheral costs such as return freight, time and materials, and customer goodwill; and the appropriate sales channel for the reconditioned product. When a reasonable disposition is determined, the customer should be informed and the disposition noted on the individual recall communication.

During this process there will be times when different dispositions will be made for different customers. For instance, if a customer has a few cases that need to be destroyed, it would be appropriate to have the customer dispose of them; however, if another customer has several pallets of products, it might be more appropriate to have those sent back for disposal. If the product poses a potential health risk, then it is critical that the company know, understand, and document the disposition of all recalled products. Product disposition is a critical and active task for all members of the management team due to its various food safety, regulatory, and financial implications upon the company and its customers.

10. *Effectiveness checks.* Once all of the customers that received the recalled product have been contacted and a disposition settled on, the next step is to determine the effectiveness of the recall. This is through the use of a recall effectiveness check form and its purpose is to determine the percentage of affected companies that were contacted and the percentage of affected product that was removed from the market. The form on the CD (Book 2_Quality Control Manual\Section 16_Recall Program:Effectiveness of Product Recall Sheet) and at the end of this chapter offers the recalling company a helpful tool to summarize its actions regarding notification and disposition. First, the names of the finished consignees from the finished product recall sheets are transferred to the "name of consignee" column. If the consignee appears on more than one sheet, each entry should be listed separately. Next, the "date notified," "how notified," and "notification verified" columns for each consignee are filled in. Then, the product description, product code date shipped, unit of measure, and the quantity shipped of each lot are written in. Finally, after it is verified that the assigned disposition has been adhered to, the "disposition verified" column should be initialed. The total number of consignees is counted and entered in both boxes at the bottom of the form and then the number notified and the number of consignees that followed the disposition requirements are filled in. Once all of this information is filled in, the percent notified and percent following disposition will be automatically calculated. These numbers should be compared against the required level established during the recall plan development and based on the following federal standards:

Level A—100% of the total number of consignees to be contacted;
Level B—some percentage of the total number of consignees to be contacted;
Level C—10% of the total number of consignees to be contacted;
Level D—2% of the total number of consignees to be contacted; and
Level E—no effectiveness checks.

If the recall is company initiated, the same steps as previously described should be followed, unless the finished product was not contaminated by an ingredient. In this case, only the finished product recall sheet and the effectiveness of product recall sheet should be used to conduct the recall and determine its effectiveness.

SUPPLEMENTAL MATERIALS

RECALL PROGRAM OVERVIEW

Section A: General

In accordance with 21CFR7.59 subpart (c), _____ will have an effective means by which to remove product from the marketplace. This will include provisions for hazard evaluation, recall classification, communication plans, defective product handling, and effectiveness checks. All facets will include a description of the provision, a responsible person, and a time table if required.

Section B: Hazard Evaluation

An evaluation of the hazard will be conducted. This should answer the following questions:

1. Have any deaths occurred or is it likely any deaths will occur?
2. Has disease or injury occurred or will they occur from the use of the product?
3. Is the product misbranded or adulterated by definition?
4. How widespread is the occurrence of the hazard?
5. What code dates are involved?
6. Where were the affected code dates distributed?

Responsible person: quality control manager

Section C: Recall Classification

Each occurrence will be classified as follows:

> Class I (recall) is a situation in which there is a reasonable probability that the use of, or exposure to, a violative product will cause serious adverse health consequences or death. This requires mandatory Food and Drug Administration notification.
> Class II (recall) is a situation in which use of, or exposure to, a violative product may cause temporary or medically reversible adverse health consequences or where the probability of serious adverse health consequences is remote. This is a recall to the consumer level and requires Food and Drug Administration notification.
> Class III (recall) is a situation in which use of, or exposure to, a violative product is not likely to cause adverse heath consequences. This is a recall to the retail level and requires Food and Drug Administration notification.
> Market withdrawal is removal or correction of a distributed product, which involves a minor violation that would not be subject to legal action by the Food and Drug Administration or that involves no violation (e.g., normal stock rotation practices, routine equipment adjustments and repairs). This is a recall to the wholesale level.
> Stock recovery is removal or correction of a product that has not been marketed or that has not left the direct control of the firm (i.e., the product is

located on premises owned by, or under the control of, the firm and no portion of the lot has been released for sale or use).

Responsible person: president/vice president/quality control manager

Section D: Communication Plan

In the event of a recall, market withdrawal, or stock recovery, proper communication and notification will be given to customers, management, and regulatory agencies as follows:

> class I: management, regulatory agencies, distributors, brokers, and customers;
> class II: management, regulatory agencies, distributors, brokers;
> class III: management, distributors, brokers;
> market withdrawal: management, distributors, brokers; and
> stock recovery: management.

Communication with regulatory agencies will be done by authorized personnel. Notification to brokers, distributors, and customers will be verbal, followed by written notification, or in writing. It will include some or all of the following:

> a complete description of the product to be recalled, including the quantity of material sent to each consignee, the dates when it was shipped, and the lot number;
> the fact that the FDA has requested the recall or has been notified and is involved;
> a disclosure of the product's defect, the risks associated with its continued use, and the classification of the recall;
> actions to be taken by the consignee, including cessation of further use and distribution of the product, initiation of the subrecall if necessary, and disposition of unused material in a proper way;
> an offer to assume all costs incurred as a direct result of the recall; and
> a self-addressed, postage-paid card for reporting to the manufacturer the amount of the product consumed and the remaining stock in inventory.

Section E: Defective Product Handling

Isolation of Defective Product

Recalled products that are returned will be segregated in the facility or outside storage and tagged "Hold—Not for Sale," "Rejected," or "Not for Food Use." A hold form is to be filled out upon return, and then a rework/destruction form is to be completed before disposition. Care must be taken to prevent cross-contamination between returned products and new products.

Generally, recalled products should not be left in the hands of customers for them to discard. The risk is too great that the defective product will be diverted to some unauthorized use or into a secondary sales stream. Depending on the circumstances, regulatory agents may want to witness the actual destruction or disposal of the recalled product.

Recall Program

Reconditioning of Defective Product

Insofar as possible, _____ may avail itself of the opportunity to recondition distressed products. This may only be done within FDA guidelines for reconditioning products.

Section F: Documentation

Each recall will be conducted using a product recall sheet.
Responsible person: quality control manager

Section G: Effectiveness Checks

To determine how effective the recall is, an "effectiveness of product" recall sheet will be filled out. A report will be completed that includes:

> number of consignees notified of the recall, and date and method of notification;
> number of consignees responding to the recall and communication and quantity of products on hand at the time recall notice was received;
> number of consignees that did not respond (if needed, the identity of nonresponding consignees may be requested by the FDA);
> number of products returned or corrected by each consignee contacted and the quantity of products accounted for;
> number of results of effectiveness checks that were made; and
> estimated time frames for completion of the recall.

Responsible person: quality control manager

Section H: Record Retention

Product recall sheets, effectiveness of product recall sheets, copies of related bills of lading, and other notes pertaining to the recall will be placed in a manila file folder labeled with the date of recall or mock recall.

April 20, 2008

Attention:
«Address Block»
«Address Block»
«Address Block»

RE: Notification of Product Recall

Dear Name:

Effective April 20, 2008, COMPANY NAME is conducting a voluntary recall of the following product(s) and lot(s). Further distribution of the following product should cease immediately. All subsequent users/customers should be notified that this product is not to be consumed.

This product is being recalled due to _____
_____.

The specific hazard involved is _____.

Recalled products should be handled by «Return? Hold? Destroy?»

Product #	Description	Size	U/M	Lot Number

All products subject to this recall and isolated should be _____
_____.

Please direct all questions to the recall coordinator:

Contact Name
Title: *Title*
Office phone: *Telephone number*
Cellular phone: *Cell number*
Fax: *Fax number*
E-mail: *E-mail address*
Address: *Address*
Address
City, State, Zip

Please respond to this communication by _____.
Thank you for your immediate attention to this matter.

Sincerely,

Name
Title
Company

cc: _____
_____

EMERGENCY RECALL NOTIFICATION LIST

Date: _____

Section A: General

To provide customers, vendors, and regulatory agencies a direct means of contact in case of emergency or recall, the following should be contacted:

Section B: Contacts

President: _____
- Work: _____
- Cell: _____
- Fax: _____
- Home: _____
- E-mail: _____

Vice-president: _____
- Work: _____
- Cell: _____
- Fax: _____
- Home: _____
- E-mail: _____

Plant manager: _____
- Work: _____
- Cell: _____
- Fax: _____
- Home: _____
- E-mail: _____

QC manager: _____
- Work: _____
- Cell: _____
- Fax: _____
- Home: _____
- E-mail: _____

Raw Material Recall Sheet

Vendor: _____

Product: _____

Lot Code: [   ]

Starting Balance [   ]

Product Used In	Product Code	Lot Code / Date of Use	Finished Goods Quantity Made	Raw Material Quantity Used	Inventory Balance
Quantity of Ingredient Left in Stock				[ ]	

Ending Balance ⇒ [   ]

Finished Product Recall Sheet

Product: _____

Item Number: _____

Lot Code: [ ]

Starting Balance: [   ]

Name of Consignee	PO Number	Date of Shipment	Quantity Shipped	Inventory Balance
Finished Product Left In Stock				

Ending Balance ⇒ [   ]

Effectiveness of Product Recall Sheet

Name of Consignee	Date Notified	How Notified	Notification Verified	Product Description	Product Code	Date Shipped	U/M	Quantity Shipped by Lot	Quantity Received by Lot	Disposition Verified

Number of Consignees
Number Notified
Percent Notified

Number of Consignees
Number of Consignees Following Disposition
Percent Following Disposition

21 Supplier Certification Program

PROGRAM TYPE: REQUIRED

THEORY

Every food manufacturer requires suppliers in some form or another. These may be as raw materials, processed materials, or packaging suppliers that provide components that eventually end up in the finished goods. With the use of these suppliers comes an ever increasing risk that they might provide ingredients and packaging that may contain a food safety risk. These risks should be identified prior to purchasing during the risk analysis phase of the HACCP program development. Exclusion of risk coming from suppliers can be managed utilizing a supplier certification program. This is a program that encompasses activities conducted and information garnered via other programs blended with an inspection and legal documentation component.

Supplier certification programs are rooted in 21CFR110.80(a)2 as part of the regulations regarding how raw materials and other ingredients will be kept free from microorganisms or other contaminations. Specifically, it states, "Compliance with this requirement may be verified by any effective means, including purchasing raw materials and other ingredients under a supplier's guarantee or certification." The basic parts of this program are information gathering, auditing, and legal guarantees.

Information gathering is the task that occurs at the beginning of the supplier–company relationship prior to any purchase. It is the process of requesting from the potential supplier documentation that helps the quality control manager to identify the microbiological, chemical, physical, and allergenic risks associated with the raw material. These documents include the ingredient specification sheet, nutritional statement, continuing food guarantee, biosecurity act compliance, and declaration of allergens; they are filed in the ingredient folder as described in the HACCP program section.

Auditing is a task undertaken by the company during the final supplier qualification phase. Its purpose is to determine and understand the quality systems in play at the supplier with regard to food safety; it provides the company a unique understanding of how the supplier assesses, manages, and eliminates risks from entering the final product. The two types of supplier audits are survey and physical audits. A survey supplier audit is used by a company to gain the required information without physically visiting the supplier. It is conducted by mailing, e-mailing, or faxing an ingredient supplier a safety survey form composed of a series of questions. These questions include information about the company and its contacts, product safety and regulatory compliance, pest control, personnel practices, maintenance, record keeping, supplier management, personnel practices, raw materials, equipment conditions, process controls, sanitation, storage and shipping areas, finished product security, and warehousing. It is expected that the supplier will complete the survey and return it to the quality control department prior to the company's first purchase. The advantage of using the survey method of supplier auditing is that it is much less expensive. The disadvantage is that the information obtained is secondhand and may not be correct because the supplier will normally overstate the completeness of its systems, which might possibly put the company at risk.

The second type of auditing—physical auditing—is done through visiting the supplier and conducting a physical audit of the facility and systems that are in place. This audit utilizes a more complete auditing form and when utilized to completion ends up with an audit score that can then be used to qualify or disqualify the supplier. The advantage of this type of audit is that a more thorough, and probably more accurate, representation of the supplier's food safety systems can be measured and documented for eventual use during the hazard analysis. As such it would be ideal to audit all suppliers in this manner. The obvious disadvantage of the physical audit is the overwhelming cost to small and medium-sized companies as they struggle to balance expenditures on all fronts.

For most companies, the type of supplier audits used is usually a blend of both types. For example, startup companies generally conduct no audits or do a cursory survey audit of their suppliers prior to, or sometimes after, making their first purchase. As the company grows, it realizes the potential risks associated with not knowing what the suppliers are sending it and the types of food safety controls the suppliers use, so it begins requiring surveys of all of them. When the company gains an even larger distribution scope, it realizes that its economic risk is greater in the event of a recall, so it reanalyzes its suppliers and begins conducting physical audits on its largest and/or riskiest suppliers. Finally, the company reaches the size where the risk associated with any supplier overshadows the costs associated with conducting physical audits on all of its suppliers. In the end, successful companies rank the risks posed by their suppliers and categorize them into A, B, and C classes. Class A suppliers pose high potential risk from the ingredient and are audited at least yearly. Class B suppliers pose less risk and are audited every 2 to 3 years, and class C suppliers pose minimal risk and are audited only prior to the first purchase and then subsequently only when a significant change occurs such as the sale of the supplier. As a means to recognize and eliminate risk, this strategy works for most companies, but should be evaluated on a case-by-case basis depending on the type and serious nature of the hazard.

The final component of the supplier certification program is that of legal guarantees. Specifically, the two types of legal guarantees that are provided by the supplier are the continuing food guarantee and the insurance review. Continuing food guarantees are referred to in 21CFR 7.12 and 21CFR303(c) 2 as a means for the supplier to guarantee to the company that the item received was not adulterated or misbranded. They may take the form of either a specific shipment guarantee or a continuing guarantee. A specific shipment guarantee is limited to a single shipment. It may be delivered to the receiving company as part of the delivery paperwork or on the back of the packing list or bill of lading or bill of sale. The typical form of this type of guarantee is directed in section 7.13(b)1 as:

(Name of person giving the guaranty or undertaking) hereby guarantees that no article listed herein is adulterated or misbranded within the meaning of the Federal Food, Drug, and Cosmetic Act, or is an article which may not, under the provisions of section 404, 505, or 512 of the act, be introduced into interstate commerce (signature and post-office address of person giving the guaranty or undertaking).

The continuing guarantee, also known as a continuing food guarantee, is a general statement provided at the conception of the supplier–company relationship as part of the document gathering portion of the hazard analysis and it applies to any shipment or other delivery of the ingredient or packaging. Due to its continuing nature, it is considered in full effect at the date of the shipment of the item up to that time in which the item becomes adulterated or misbranded. Its typical form is stated in 7.13(b)2 as:

The article comprising each shipment or other delivery hereafter made by (name of person giving the guaranty or undertaking) to, or in the order of (name and post-office address of person to whom the guaranty or undertaking is given) is herby guaranteed, as of the date of such shipment or delivery, to be, on such date, not adulterated or misbranded within the meaning of the Federal Food, Drug, and Cosmetic Act, and not an article which may not, under the provisions of section 404, 505, or 512 of the act, be introduced into interstate commerce (signature and post-office address of person giving the guaranty of undertaking).

The second legal guarantee is that of the supplier proof of insurance. When the company deals with suppliers, their insurance coverage is important because it indicates their ability to cover the costs associated with a recall caused by an ingredient they supplied. Proof of insurance from the supplier should be on a standard 1986 ISO commercial general liability policy form or equivalent and state the company's limits of general liability/products liability coverage of not less than an amount that covers the theoretical amount that may need to be recovered after the recall. Other specific areas to be considered are a vendor's liability endorsement, the company named as an additional insured, the umbrella limits, and the rating of the policy. In the case of the last, it should be an A. M. BestA-X rating or better.

To validate that the supplier has the appropriate insurance, the company should request a copy for review. At the time of the request, the company should specify the limits necessary and how the additional named should be listed. When approved, these documents should be filed in the supplier master file, typically located in the accounting department.

APPLICATION

Prior to the use of any ingredient, collect the required information. The form on the CD (Book 1_HACCP Program\Planning:Supplier Data Sheet) and at the end of this chapter can be used to organize the required documents. Specific to the supplier certification program, both the continuing food guarantee and evidence of insurance should be requested. The form on the CD (Book 2_Quality Control Manual\Section 17_Supplier Certification Program:Continuing Food Guarantee) and at the end of this chapter can be placed on the company letterhead as a template to provide to the supplier as an example of what a continuing guarantee should look like. In most cases, this is probably not necessary. The form for requesting evidence of insurance on the CD (Book 2_Quality Control Manual\Section 17_Supplier Certification Program:Request for Evidence of Insurance) and at the end of this chapter can be adapted and sent to the supplier to document its insurance coverage. As directed in the HACCP planning chapter, a set of properly labeled folders should be created for the retention of these documents. Care must be taken at this step to make sure that the actual manufacturer is identified even if the ingredient is obtained from a broker because it is the manufacturer's quality control and processes that are to be scrutinized for risks. The more thorough, organized, and clearly tabulated these documents and files are, the easier the supplier certification process will be.

The basic supplier certification program is developed by separating the suppliers into manageable groups, A, B, and C. This designation is based on the degree of risk at which the ingredients or packaging that they supply might affect the company's finished product. "A" suppliers are those that supply products that, by there inherent nature, most likely will contain a physical, chemical, microbiological, or allergenic contaminant. Sometimes these ingredients will be from the largest suppliers and sometimes they will be from the smallest suppliers. Likewise, they might be right next door, but they might just as easily be across town, the country, or the world. The key is to identify the risk potential of the components and to categorize them as such. Examples of some "A" ingredients would be dairy products, fruit products, raw ingredients, ingredients that claim to be allergen free but are processed on the same line as other allergen ingredients, and minimally processed ingredients.

"B" suppliers are those that by sheer volume supply a component that might contain an adulterant but most probably will not. Included in this category would be ingredients that are highly processed on a line that will not contaminate it with undesired allergens. These might include oils and fats, sweeteners, and core ingredients. Category "C" suppliers are those whose ingredients and packaging pose little or no risk. These typically include flavors, extracts, low moisture dry goods like salt and sugar, and very highly processed ingredients.

The form on the CD (Book 2_Quality Control Manual\Section 17_Supplier Certification Program:Supplier Risk Categorization Sheet) and at the end of this chapter can be used to summarize the company's suppliers in respect to ingredients provided and the risk contained therein. For each supplier, the name, location, ingredient or packaging supplied, and the components' item numbers should be listed. Next, each component should be evaluated for risk, whether each is an A, B, or C item should be determined, and this column filled in. After all ingredients and packaging have been analyzed, the list should be sorted by risk supplier and then by risk category. The suppliers are divided into A, B, and C groups using the following rules:

1. If the supplier has any ingredient in the A category, the supplier is an A group risk.
2. If the supplier does not have any A category ingredients but has B category ingredients, then the supplier is a B group risk.
3. If the supplier does not have any B category ingredients but has C category ingredients, then the supplier is a C group risk.

Supplier Certification Program

Once the suppliers are categorized, the decision of whether to audit them using a survey or a physical audit must be made. This decision is often influenced by the economics of the company and not necessarily by the degree of risk that the supplier's ingredients bring to the company. In a perfect world, all suppliers would be audited in person to minimize the risk but this is not practical. In lieu of this approach, some general guidelines may be used:

1. Survey all suppliers as an initial guide to determine the extent of the quality systems in place. This also gives the company a measure of management's commitment to quality and its ability to comply with special quality requests such as certificates of analysis, allergen testing, and microbiological testing.
2. Based on the survey, visit in person those companies that are deemed lacking in systems or provided answers that raised red flags that the ingredients supplied might contain risks.
3. Visit those B and C suppliers during the course of visiting an A supplier. This spreads the cost of auditing across more than one supplier.

The form on the CD (Book 2_Quality Control Manual\Section 17_Supplier Certification Program:Ingredient Supplier_Copacker Food Quality Survey) and at the end of this chapter can be used to obtain initial quality information from all suppliers. The form should be sent to all suppliers and reviewed carefully when it is returned. An actual audit form found on the CD (Book 2_Quality Control Manual\Section 17_Supplier Certification Program:Food Safety Audit Report) and at the end of this chapter can be used during a physical audit of a supplier. This is based on a 1000-point scale weighted on food safety systems.

The final step in developing a supplier certification program is to write the program document. A customizable document can be found on the CD (Book 2_Quality Control Manual\Section 17_Supplier Certification Program:General Overview) and at the end of this chapter. Its sections outline the basic structure of the program, the types of documents that will be collected, and the categorization of suppliers by risk. It also identifies who is responsible for managing this program. When it is completed, this document should be placed in the quality control manual.

SUPPLEMENTAL MATERIALS

Supplier Certification Program   337

Company Name

December 12, 2006

RE: Continuing Food Guarantee

ATTN:

_____, its subsidiaries, and divisions agree to indemnify, hold harmless, and defend _____, its divisions, and/or affiliated companies against any claim made or action brought against Customer by any person alleging that any article shipped or delivered by _____ to Customer or on Customer order was adulterated or misbranded, or that the consumption or use thereof caused illness, injury, death, or damage to property; provided that _____ receives reasonable notice of said claim or action and is afforded the opportunity to defend against the same.

    The article comprising each shipment or other delivery hereafter made by _____ to Customer on Customer order is hereby guaranteed, as of the date of such shipment or delivery:

(a) To be, on such date, not adulterated or misbranded within the meaning of the Federal Food, Drug, and Cosmetic Act, and not an article that may not, under the provisions of Section 404 or 505 of the Act, be introduced into interstate commerce; and

(b) To be, on such date, of merchantable quality.

_____ makes no warranty respecting misbranding where goods are shipped or delivered under labels not subject to _____'s control, nor any warranty or agreement of indemnity respecting unmerchantability, illness, injury, death, or damage to property attributable to Customer act or omission or those of others.

    This guarantee, which supersedes and cancels all guarantees heretofore given to Customer by _____, is a continuing guarantee, which will remain in effect until it is terminated upon ten (10) days' prior written notice by _____ to Customer.

                      _____

           By _____
                          (Name and title)

REQUEST FOR EVIDENCE OF INSURANCE—VENDORS

In keeping with our corporate policy, we request that you supply us with information regarding your general liability/products liability, workers compensation, and umbrella insurance programs.

The information we require can be provided on a standard insurance certificate form, with which your insurance representative will be familiar. The minimum insurance requirements are:

1. Limits of general liability/products liability coverage of not less than $2,000,000 per occurrence and annual aggregate providing coverage for both bodily injury and property damage.
2. Coverage will include a vendor's liability endorsement and the insurance certificate will specifically state that _____ and its subsidiaries are an additional insured on the policy.
3. Coverage will be written on an occurrence basis and will be written on the 1986 ISO commercial general liability policy form or equivalent.
4. Worker's compensation insurance evidencing coverage and employers' liability insurance with minimum limits of $500,000.
5. Umbrella coverage written with minimum limits of $3,000,000.
6. Thirty days notice of cancellation, nonrenewal, or material change will be provided to _____ at _____.
7. All policies shall be written by an insurer with a rating of A–X or better by A. M. Best.
8. The name and address of the additional insured/certificate holder should be as follows: _____

    Attn: _____

    _____

    _____

You are reminded that these are the minimum requirements, and they will not limit your liability for claims or suits in excess of these limits. This policy is strictly enforced. Thank you for your cooperation and prompt compliance.

Supplier Risk Categorization Sheet

Supplier Name	Location	Item	Item Number	Risk Category	Survey	Audit

INGREDIENT SUPPLIER AND CO-PACKER QUALITY AND SAFETY SURVEY

To Our Current or Potential Supplier/Co-Packer:

The _____ is committed to providing products that are recognized by the consumer as high quality, safe, and wholesome. To this end, we have always considered our suppliers and co-packers as an extension of our own company; thus, require our vendors to achieve the high standards set by ourselves and ultimately by our customers. We must ensure that all our vendors comply with fundamental criteria and provide a vehicle to further strengthen our relationships.

In order to maintain a proactive Quality Control Program, we require your support in completing the attached food safety and quality survey. This information is essential to meeting our Quality Control goals. All information will be kept strictly confidential.

For both of our conveniences, we would prefer to send and receive this information form electronically. Therefore, if you can provide us with your email address, we will be happy to send the survey to you electronically. Otherwise, complete the attachments and mail/fax them to _____.

Telephone: _____

Fax: _____

E-mail: _____

Please return the completed survey by _____
(date)

to: _____

Contact _____

Address _____

City, State, Zip _____

Telephone _____

Fax _____

E-mail _____

INGREDIENT SUPPLIER AND CO-PACKER
QUALITY AND SAFETY SURVEY

**Please fill out a separate survey for each plant where you produce.
**For Brokers/Distributors/Copackers: Please have any Manufacturer that produces or stores product, complete this survey as well.

Please attach to this form:

1. Table of Contents of your documented quality system
2. Copy of GMP Policy
3. Explanation of how to read your lot codes
4. A copy of your HACCP documentation pertaining to the specific product(s) supplied.

Company Name:_____ Date_____

Respondent: _____ Title:_____

Phone #: _____ Fax #:_____

Name of Manufacturer:_____

Manufacturing Plant Address: _____

_____

Plant(s) supplied by this facility: _____

Name of the product(s) you supply: _____

- Technical Support Contact _____ Phone Number _____
  Title   _____ Fax   _____
                                             E-mail   _____

- Emergency Recall Contact _____ Business Number _____
  Primary   _____ Home Number   _____
                                             Fax   _____
                                             E-mail   _____

- Emergency Recall Contact _____ Business Number _____
  Secondary   _____ Home Number   _____
                                             Fax   _____
                                             E-mail   _____

I. PRODUCT SAFETY/REGULATORY COMPLIANCE

1. Safety devices in-line:

	Yes	No	Type/size and Location
Magnets?			
Screens/sieve?			
Filters?			
Metal detector?			

Sensitivity: Ferrous _____ mm   Non-ferrous _____ mm   Stainless _____ mm

2. Are inspection results of safety devices documented? | Yes | No |

3. Briefly describe how you ensure the safety of the water supply.

    _____

4. Does an outside inspection service conduct periodic quality / sanitation audits for your facility? | Yes | No |

    Name of auditing service(s) _____

    Frequency _____   Latest Result _____

II. PEST CONTROL

1. Is there an ongoing pest management program for insect and rodent control? Describe program, including types of devices used and frequency of checks. | Yes | No |

    _____
    _____

2. Does a state-licensed pest control operator maintain this program? | | |

III. PERSONNEL PRACTICES

1. Personnel Training

	Yes	No	Frequency of Training
HACCP			
GMP			
Allergen			
Safety			
Others (Please describe)			

GMP CHECKLIST

Please provide an explanation at the end of each section for any question answered no.

I. EXTERIOR MAINTENANCE

	Yes	No

1. The building perimeter is free of weeds and tall grass.
2. There is a paved/gravel perimeter of at least two feet around building exterior.
3. The surrounding premises are free of standing water.
4. There are no signs of tunneling or other rodent activity.
5. Scrap, pallets, equipment, etc. are not stored along the building walls.
6. The surrounding area is free of paper, trash, and litter.
7. The area around the docks and dock plates is kept clean.
8. All dock levels are gasketed and tight to exclude rodents.

Supplier Certification Program

9. All dock doors have dock seals.
10. Windows and doors have screens in good condition if open.
11. All doors are tight fitting to preclude pest entry. There is no clearance greater than ¼".
12. The outside waste compactor is clean and properly maintained.

Comments: _____
_____
_____

II. PEST CONTROL

	Yes	No

13. There is a program in place for controlling birds in and around the plant
14. Pest control locations are 30 feet apart inside and 40 feet outside.
15. There are pest control stations around all doors.
16. Pest control stations are examined at least bi-weekly.
17. Insectocutors have catch pans and bulbs are changed yearly.
18. A diagram is maintained indicating the location of all pest control locations.
19. All pesticides are kept under lock and key.
20. Records of all pesticide applications and pest inspections are kept on file.

Comments: _____
_____
_____

III. QUALITY RECORD KEEPING

	Yes	No

21. Facility has an up-to-date Quality Manual with lab procedures. ISO?_____
22. The Quality Program has procedures for conformance to specification, GMP's, Pest Control, Sanitation, Weight Control, Laboratory Control, Equipment Calibration.
23. Records are retained for 3 years.
24. Procedures exist to identify out of spec. material and then how to dispose of it. This consists of quarantine, corrective action, and record retention.
25. There is a written recall program that is up to date and is tested at least annually for Finished Good and Raw Material.
26. There is complete lot traceability.
27. There is a plan in place for Customer Complaint follow-up.

Comments: _____
_____
_____

IV. SUPPLIER QUALITY MANAGEMENT

	Yes	No

28. An approved supplier list is used and supplier audits maintained.

Comments: _____
_____
_____

V. PERSONNEL PRACTICES

	Yes	No

29. Restrooms are on a daily cleaning schedule.
30. There are signs in restrooms to remind people to wash hands before returning to work.
31. There are enough hand washing/sanitizing stations around the operation to ensure use.
32. Employees sanitize hands and/or gloves properly when handling product directly.
33. Employees have uniforms that are clean, without buttons on their shirts or items in shirt pockets.
34. Jewelry (earrings, watches, bracelets, necklaces, rings, etc.) is not worn in production areas.
35. Effective hair and beard restraints are worn to prevent contamination of the product.
36. If bandages are worn, they are metal-detectable and covered when possible.
37. Eating, drinking, smoking, etc., is limited to designated areas away from production.
38. Nail polish or false nails are not worn in production areas.

Comments: _____
_____
_____

VI. RAW MATERIALS

	Yes	No

39. Critical raw materials are checked upon receipt to ensure that they comply with specification and have COA's sent in to be checked and filed.
40. All Raw Materials are dated upon receipt.
41. All ingredients are tracked through production by recording lot numbers.
42. There is a program to ensure damaged raw materials are properly handled upon receipt.
43. All bagged ingredients are cleaned and/or outer layers stripped before using.
44. All Raw Materials that are identified not to be in specification are properly quarantined.
45. Vendor shall notify _____ of any regulatory action or product retrieval related to _____.

Comments: _____
_____
_____

VII. EQUIPMENT CONDITIONS

	Yes	No

46. Lighting has covers to prevent bulbs from breaking on Raw Materials and into the product.
47. There is adequate lighting to perform tasks.

Supplier Certification Program

48. All equipment is smooth and noncorrosive where there is direct contact with the product.
49. All equipment is free of oil leaks and grease build-up on bearings. Drip pans are used to collect drips.
50. All equipment is constructed for easy cleaning.
51. Conveyor belts do not have anything loose on them.
52. Brooms, squeegees, and mops have hangers and are kept off the floor.
53. Idle and obsolete equipment is removed from the processing area.

Comments: _____

VIII. PROCESS CONTROLS — Yes / No

54. Appropriate line covers are provided to protect the product from contamination.
55. Current and accurate formulas, plus processing parameters (mix times, temperatures, etc.) are documented and maintained.
56. All rework can be traced.
57. All meters and scales are routinely calibrated with the weights or measuring equipment of known accuracy. Frequency = _____
58. All thermometers and temperature recording devices are calibrated with a thermometer of known accuracy. Frequency = _____
59. Records of these calibrations are maintained on file for review.
60. Plant has an effective program to identify, monitor, and control potential product/ingredient quality deviations during processing.
61. Plant maintains complete, orderly, and easy to read records of all critical control points checks and analysis. There is documentation of all corrective actions.
62. Plant takes samples throughout processing in sufficient quantities and frequency to effectively monitor and control all in-process specifications.
63. All food containers are kept off the floor.

Comments: _____

IX. SANITATION — Yes / No

64. Plant has a schedule of written sanitation procedures, which include complete breakdown and cleaning. Records are kept of all cleanings.
65. Sanitation Audits are done at least monthly by the Q.C. Department. Preoperational Inspections are performed.
66. Micro swabbing of equipment and the environment (floors, drains, and air) is utilized to assess risk areas, and corrective actions are taken and documented.
67. ATP bioluminescence test equipment is used at least weekly to identify the presence of organic material on equipment.

68. Entry into the processing/packaging areas is controlled to reduce the potential to transfer microorganisms into the plant, utilizing footbaths/hand sanitizing as needed.
69. All sanitizers are EPA approved and MSDS are available.
70. Floor/wall junctions in production areas are sealed and easy to keep clean.
71. All walkways over product lines or stored product have solid bases and kick lips.
72. Waste containers are maintained and emptied regularly.
73. Overhead lines, ceilings, and walls are cleaned regularly to remove dust, food, and debris.
74. All walls are smooth and easily cleanable, free of cracks and holes. All electrical boxes are mounted flush to the wall, caulked, in good repair, and free of rust and flaking paint.
75. All ceilings are constructed of a smooth and easily cleanable material. Insulation line material is in good repair. No string, rope, wire, or tape is used as pipe or support.
76. Floors are well drained, smooth, and clean with no cracks.
77. Equipment which has undergone repairs or maintenance is cleaned and sanitized before using.

Comments: _____
_____
_____

XI. **STORAGE/SHIPPING AREAS**   Yes   No

84. All stock is stored at least four inches off the floor (six inches if on racks).
85. All stock is stored in a clean and dry environment.
86. All coolers and freezers have recording thermometers with a program for monitoring
87. Finished products, raw materials, and quarantined damaged goods are stored separately.
88. All stock is stored in an orderly manner and properly stacked to prevent damage.
89. All pallets, racks, and shelving are clean, in good repair, located 18" away from the wall.
90. All damaged finished products are disposed of in a timely manner.
91. All stock is rotated on a first in first out basis.
92. Finished product, packaging material, equipment, or ingredients are not stored in close proximity to any chemical, cleaning compounds, pesticides, or odorous material.
93. All carriers are inspected before loading for cleanliness and odor. This is documented.

Comments: _____
_____
_____

Supplier Certification Program

XII.	**FINISHED PRODUCT**	Yes	No
	94. Plant maintains a statistically sound sampling program, which ensures that all finished products either meet or exceed specifications.		
	95. All finished products are properly coded. This code will be complete and legible.		
	96. All finished products are coded on the outside of the container.		

Comments: _____
_____
_____

X.	**SECURITY**	Yes	No
	78. The points of entry locations are limited, secured, and locked after normal business hours and on weekends.		
	79. Water reservoirs, storage tanks, and unloading lines are secured, capped, and locked.		
	80. There is fencing around exterior storage areas and functional equipment outside the building at ground level.		
	81. The facility has visitor procedures that include badges and escorts.		
	82. All boxes, cases, drums, and bulk tanks close, tie, or seal and have tamper-evident closures.		
	83. All inbound and outbound full truck load shipments are sealed and seal number noted and crosschecked for tamper evidency.		

Comments: _____
_____
_____

XIII.	**WAREHOUSING**	Yes	No
	97. Outside warehouses are used to store product.		
	98. An outside inspection service conducts periodic product safety/ sanitation audits of those facilities.		

Comments: _____
_____
_____

FOOD SAFETY AUDIT REPORT

for:

Report Date:

Audit by:

This document sets forth the guidelines we require of those with whom we purchase raw materials for food production, those who manufacture products for _____ and those involved in the distribution of our products. The food distributors and processors with whom _____ performs business must at all times adhere to the guidelines listed within this document.
This document is also to be utilized as an audit report for determining a facility's compliance to _____ standards and requirements as set forth in this document and other supporting documents.

_____ sets forth our findings and recommendations on the enclosed report as of the date herein. _____ does not assume any responsibility for the programs and/or facility being audited nor for events or actions occurring prior or subsequent to this audit. _____ does not accept any responsibility or liability as to whether or not the plant carries out the recommendations, if any, as contained in this report. _____does not purport that this audit assesses the adequacy of the HACCP plan used by this facility nor does its brief review of some aspects of the HACCP plan as part of section I, food safety systems, represent a complete HACCP verification audit.

This report is furnished solely for the benefit of the above named associate in connection with the auditing services indicated, and this report may not be reproduced or published in full or in part, altered, amended, made available to or relied upon by any other person, firm or entity without the prior written consent of _____ .

The name of _____ or its affiliates or any of its employees may not be used in connection with any marketing or promotion of the products or in any publication concerning or relating to the aforementioned and/or their products without the prior written consent of _____ .

Company Information

Company:		Audit date:	
Plant address:		Auditor:	
Plant phone & fax numbers: E-mail:		Company associate(s) accompanying auditor (name & title):	
Customer audit was completed for:		Products produced by plant:	

Audit Summary

Audit score:		Rating:	
Date of last audit:		Score of last audit:	
Follow-up audit required:		Timing for follow-up audit:	
Reason:			

Audit Review

Company associate(s) with whom audit findings were reviewed:	

Auditor Signature: _____

Summary of Audit Findings

I. Issues/Areas Requiring Significant Improvement (core food safety issues)

Category	Findings	Improvement Path	Date to Be Completed

II. Other Opportunities for Improvement

Category	Findings

Food Safety/GMP Rating Analysis

	Category	# Points Received	# Possible Points	Percentage (%)
I	FOOD SAFETY SYSTEMS		165	
II	QUALITY SYSTEMS		345	
III	GROUNDS, BUILDING, AND EQUIPMENT		135	
IV	PEST CONTROL		65	
V	EMPLOYEE PRACTICES		50	
VI	RECEIVING, STORAGE, AND SHIPPING		105	
VII	PLANT SANITATION		65	
VIII	PROCESSING		70	
	OVERALL SCORE		1000	

I. Food Safety Systems

(Assessed by observation and review of records)

A. HACCP

		Rating
1.	A HACCP team, comprised of members representing Operations, Quality Control, R & D, and other functional areas, has been established and meets on a routine basis. The team includes a person trained in HACCP.	2
2.	A documented HACCP program, detailing the 7 principles, is established, up-to-date, and available. A hazard analysis has been completed.	10
3.	Each product has been described, and current process flow diagrams are available.	5
4.	Critical control points have been identified and are listed on the product flow sheets.	10
5.	Critical limits have been scientifically established and are documented.	10
6.	CCP's are monitored at regularly scheduled intervals. Monitoring procedures are documented and monitoring records are maintained.	10
7.	Corrective action procedures have been identified, are taken, and are documented, when critical limits are not met. Corrective action records are maintained. Product disposition is documented.	5
8.	Appropriate verification procedures have been established. They are documented, and verification records are maintained.	5
9.	All records are appropriately signed/initialed and dated.	3
10.	Audits of the HACCP plan are performed on a regular basis by someone independent of the development of the plan. Audit results are maintained. The last HACCP program audit was _____	5

B. Food Safety Practices

1.	Proper employee and equipment traffic flows are used to minimize contamination between raw products and finished products. Food processing areas are organized to minimize the risk of cross-contamination through adequate separation of raw materials, finished product, and storage and distribution areas.	3
2.	Employees with obvious sores, infected wounds, or other infectious illnesses shall not be allowed to have direct contact with exposed food products or production/storage areas.	3
3.	Employees are observed washing their hands after activities that may have contaminated them. Activities can include, but are not limited to using the restrooms, after breaks, prior to entering production and product packaging areas, prior to handling product, prior to touching product contact and non-food contact surfaces, after handling garbage.	5

Supplier Certification Program

C.	**Product Contamination**	
1.	No actual product contamination is observed. Product contamination must not occur, including the possible cross-contamination of organic with non-organic, Kosher with non-Kosher, etc. Steps must be taken to ensure that this does not occur and documentation of these steps must be maintained.	5
2.	No equipment used has the potential to contribute to the contamination/adulteration of product with physical, chemical, allergenic, or microbial contaminants.	5
3.	No sanitation practices are observed which could potentially cause product contamination. All food, food-contact surfaces and packaging are adequately protected from contamination during clean-ups.	5
4.	The use of hoses during production or mid-shift clean-ups is accomplished without contaminating food, food-contact surfaces, and packaging materials with water droplets and aerosols and without direct contact. High-pressure hoses are not used during production or where food is stored.	5
5.	**Glass/Plastic/Wood Program:** A written program outlining the management of glass, hard plastic, and wood is available and enforced. Glass is excluded; hard plastic and wood are managed.	5

D.	**Allergen Program**	
1.	A listing of food allergens present in the plant has been developed and is documented. Controls are in place to ensure that allergens are properly labeled, segregated and used only in properly labeled products.	10
2.	The proper testing of inbound materials and/or COA documentation of allergens is conducted and documented.	7
3.	Rework controls are applied to prevent cross-contact with unlabeled allergens. Controls are also applied during changeovers. A verification program has been established to ensure compliance with the allergen control procedures. (List as NA: if not applicable; if the allergen control program is only an ingredient labeling program; if an allergen program is applicable but not in place and has been appropriately scored in question D1.)	7

E.	**Food Safety Training**	
1.	Procedures for conducting ongoing food safety and GMP training for all employees, including new employees, have been established and are documented. Responsibility for the food safety and GMP training programs is assigned. The programs include evaluation criteria for knowledge learned. Completion of this food safety and GMP training is documented as to date(s) given and is a part of the employee's records.	5
2.	Employees have been trained and are aware of the HACCP-related activities in their immediate work areas. This training is documented as to date(s) given and is a part of the employee's records.	5

3.	Procedures for conducting ongoing training on cleaning and sanitation procedures for sanitation employees, including new sanitation employees, have been established and are documented. Responsibility for the sanitation training program is assigned. The program includes evaluation criteria for knowledge learned. Completion of this sanitation training is documented as to date(s) given and is a part of the employee's records. (Applicable even if the facility uses an outside service for sanitation. The sanitation company must provide appropriate documentation.)	5
4.	Refresher training programs are provided to all employees at least annually. Completion of this training is documented as to date(s) given and is a part of the employee's records.	5

F.   Bio-Security Program

1.	A written program to manage all security concerns relating to the building is available and supported by management.	5	
2.	All visitors must show identification and sign in before gaining access to the facility. All non-employees must be escorted within the facility at all times.	5	
3.	Access to the facility is contained through the use of double doors, card readers, and/or other means.	5	
4.	The facility parameter is adequately contained through the use of fences or other control devices.	5	

G.   GMO

1.	Do any of your ingredients and/or products contain, or have the potential to contain, genetically modified organisms (GMOs)? If yes, which ones?	Yes	No

Possible points   165

Actual points   _____

Percentage   _____

Comments:

CERTIFICATIONS:
(List all that are current)

II.   Quality Systems
(Assessed by review of records)

A.   QA/QC Program   Rating

1.	**Program:** A written program, which details policies and procedures for the operation of the QA/QC programs, is established, organized, and maintained.	5
2.	**Org. Chart:** A current organization chart is available.	2
3.	**Product Specifications:** There are written standards and specifications for raw and finished food products and packaging materials that come in contact with food. Any rework used in products must be defined.	5

Supplier Certification Program

4.	**Defective Material Program:** Procedures and criteria have been established for all hold and release programs. Documentation and records are maintained.	10
5.	**Positive Release Program:** A positive release system is used for all products produced.	5
6.	**Record Retention Program:** Records of all QC results and actions are documented and initialed. There is a record retention policy for all QC results.	5
7.	**QA Program Verification Program:** Self-audits are performed at least monthly. Copies are maintained for at least 12 months. The audits include all areas of the facility. Personnel from all departments participate. Corrective actions include what is to be done, when, and by whom.	5
8.	**Shipping/Receiving Program:** A documented shipping and receiving program is available. This includes inspection, adequate labeling, documentation, verification, and record retention.	5
9.	**Weight Control Program:** There is a documented system of weight or volume control for all products and lines.	5
10.	**Regulatory Inspection Program:** there is a documented program regarding how to handle regulatory inspections. This will include responsibility, samples, photographs, supplier or customer notification, and other procedures.	5
11.	**Internal Inspection Program:** There is a documented program outlining quality, maintenance, GMP, and sanitation issues within the plant. This program will include responsibility, frequency, and follow-up. These records are available for viewing.	10

B.	**Good Manufacturing Practices (GMP's)**	
1.	A documented GMP program has been established. It complies with all applicable regulations. Responsibility for managing the program is specified.	5
2.	GMP's include a section on employee hygiene practices. GMP's include a procedure for uniforms and outer garments. Appropriate signage on GMP's is posted for employees and visitors. Visitors are given a copy of the facility's GMP's.	4
3.	Corrective action procedures have been established for deviations, and appropriate records are maintained.	5
4.	GMP's are audited on a routine basis. These audits are documented.	5

C.	**Pest Control Program**	
1.	A written pest control program has been established. It must include a designated pest control operator (internal or an outside service), scheduled frequency of service, types of pesticides used and location of application, and a current map showing the location and type of all devices (internal and external). All records and documents must be readily accessible.	20

2.	The pest control book includes copies of all business licenses, proof of indemnity insurance, and certification for all pest control operators. The file is accurate, up-to-date, and complete.	5
3.	All pesticides, chemicals, and compounds used meet applicable regulations and approvals (EPA, USDA, OSHA, etc.). Labels and MSDS must be provided for all products used.	5
4.	Service reports must be up-to-date and available for review. They must show the service performed, chemicals used and amount, signs of activity, and applicable follow-up actions.	5

D.	**Cleaning and Sanitation Program**	
1.	A written master cleaning schedule lists all areas in the plant that require cleaning (including processing and non-processing areas and equipment) and the frequency of cleaning. Documentation of compliance is maintained, and these records are available for review.	10
2.	Current MSDS and labels are on file for all cleaners and sanitizers being used in the facility. Facility maintains the system so that it is organized, accessible, and easy to use.	2
3.	Sanitation SOP's are established, documented, and implemented. All necessary content, including responsibility, task to be performed, chemicals and equipment used, required signatures/sign-off is included. Kosher sanitation procedures are included and documented as necessary.	5
4.	A pre-operational sanitation inspection program is established. A visual inspection is conducted to assess sanitation prior to the start of production. Environmental monitoring using rapid methods and/or microbiological swabbing is used to verify sanitation on a pre-defined basis.	15
5.	Corrective action procedures are established and documented for incomplete or inadequate sanitation.	3
6.	Appropriate records are kept to verify SSOP's, including pre-operational and operational SSOP assessments. CIP records are accurate and retained.	10
7.	The water and plumbing are adequate for the intended operations, and the water is from a potable source. Potability is checked at least annually with a sample taken at the plant location, and proper records are maintained.	2

E.	**Supplier Certification Program**	
1.	A documented system has been established for approving and monitoring suppliers of raw materials and packaging.	5
2.	A documented inbound inspection program is required for all materials. Appropriate procedures for monitoring methods are used to document trailer condition and to examine incoming materials for evidence of contamination (pest, microbiological, chemical, and physical), temperature abuse, damage, quality and condition, and conformance to specifications and standards. Letters of guarantee are required where appropriate. Inspection records are documented and filed, including disposition of any rejected product.	5

Supplier Certification Program

3.	A system for identifying and labeling all ingredients and materials, including lot and date codes, has been established for traceability.	5

F. Process Control Measures for Achieving Product Quality

1.	Process control points have been identified. These match the HACCP program. There are written procedures for the monitoring of these control points and appropriate records are kept.	10
2.	Corrective action procedures have been established. There are adequate back-up systems in the event equipment in place does not operate properly.	5
3.	All measurement equipment for monitoring process control points (e.g., thermometers) is calibrated according to a schedule. The results are documented.	5
4.	A metal detector is on each line and is working. Procedures are in place and documented for testing whether the metal detector is operating and has compliant sensitivity.	15
5.	Procedures are documented on how to handle product when the metal detector is non-working or non-compliant.	5
6.	Procedures on how to handle product rejected by the metal detector, and calibration of the detector, must be established. Records of calibration checks are maintained.	
7.	Sensitivity of metal detectors is: SS_____, FE_____, non-FE_____	5

G. Maintenance

1.	A written program exists for the proper preventive maintenance of all equipment and appropriate areas of the facility in accordance with an established schedule.	5
2.	A program exists for employees to identify items that need maintenance. A system for reconciliation that maintenance has been completed is in place.	5
3.	Only approved food-grade lubricants are used, and they are appropriately stored.	5
4.	Equipment, which has undergone repairs, maintenance, or re-assembly is cleaned and sanitized before being used in processing. Responsibility for monitoring this process is assigned. Documentation of the sanitation is required, and records are maintained.	5
5.	Procedures are in place to ensure product protection in all maintenance activities. Activities are specific to include repairs when product is exposed versus repairs when product is not exposed.	5
6.	Procedures are established to ensure tool and part control when repairs are taking place. These should include proper placement of nuts and bolts, tools used in raw areas versus finished product areas, etc.	5

H.	**Good Laboratory Practices**	
	Check here if no internal lab exists _____ and rate this section as NA. Only labs with micro or chemical analysis should be evaluated.	
1.	A documented GLP program has been established. It includes written SOPs, including internal calibration and control procedures, for all tests or analyses performed. Lab results are documented and initialed. Responsibility for managing the program is specified. There is a documented verification program for internal proficiency, and records are available for review.	5
2.	All appropriate lab equipment is calibrated as scheduled or as necessary and is functioning properly on a continuing basis. The calibration results are documented. Verification procedures of calibration have been established.	5
3.	If the laboratory is testing for pathogens, the laboratory must be physically isolated from production areas. Controls to prevent pathogen contamination need to be in place. There must be a program for running positive controls/cultures with documented records for all analyses.	5

I.	**Customer Complaint, Product Tracking and Recall Programs**	
1.	**Lot Code Program:** A documented product recovery program that can trace the distribution of specific production lots and the source of all raw materials and ingredients used therein has been established and is maintained. Contact lists for responsible employees and customers are current. Responsibility for managing the recall program is assigned. (Collect an emergency contact list.)	10
2.	**Recall Program:** Mock recalls are conducted at least every 6 months to assess the effectiveness of the program. Mock recalls are performed using lot numbers from ingredient vendors as well as finished goods. The results of the mock recall are on file and available for review. Date of last mock recall _____	7
3.	**Product Tracking:** All finished products shall be properly coded for traceability. Lot or batch number records of ingredients can be linked to the finished products, including traceability for reworked product. Coding system for finished product is date of mfg. _____ or use by/best by date _____. (Check which system is used.) (Collect an example and an explanation of coding system.)	5
4.	**Customer Complaint Program:** A documented, formal program on how to evaluate customer complaints, especially those related to food safety, has been established. Food safety/QA personnel are notified of applicable customer complaints. Investigation and documentation are part of the program.	10

J.	**Product Testing**	
1.	All finished goods have a written specification. This is available in the production area and is used to determine if the product during processing is being manufactured correctly.	5
2.	Finished goods and intermediate products are tested to determine if they conform to written specifications. The results of these tests are documented and reviewed by upper management before the product is released.	15

Supplier Certification Program

3.	Are microbiological tests are required on the product? If so, is there a written microbiological program which includes sampling method, sample size, tests conducted exception, procedures, and responsibility. Are the results kept in a central location?	5
4.	Products which require microbiological clearance are held in a separate area until clearance.	10
5.	All outside laboratories are certified. A copy of the certification is available.	5

K.	Record Keeping	
1.	All records for a day's production are kept together in a central, easily accessible location. These include both quality and production records.	10

	Possible points	345
	Actual points	
	Percentage	

Comments:

III. Grounds, Building and Equipment
(Assessed by observation and review of records)

A.	Plant Grounds	Rating
1.	Roads, yards, grounds, and parking lots are maintained in neat and good condition, free of trash and litter. Grass and weeds are cut to minimize harborage areas for pests and are not within 20 feet of the building.	5
2.	Plant grounds have adequate drainage to prevent pooling water which can serve as source of contamination by seepage and foot-borne filth, or provide a breeding place for pests.	5
3.	Equipment stored on plant grounds is at least 20 feet away from the buildings, at least 6 inches above the ground, and in an organized manner to prevent breeding areas and harborage for pests. Any pipes within 20 feet of the building must have closed ends.	5
4.	Litter and waste are properly stored in enclosed containers. All waste is removed from the premises at appropriate intervals and in such a manner as to prevent spillage and litter. The dumpster areas are cleaned on a regularly scheduled basis and clear of debris and spilled product.	5
5.	The loading dock areas are clear of debris and spilled products. All bumpers, levelers, and shelters are in good repair and clean.	5

B.	Plant Facilities	
1.	Plant buildings and roofs are suitable in construction and design to facilitate maintenance and sanitary operations. There are no roof leaks.	5
2.	Interior floors, walls, and ceilings are constructed of materials which can be adequately cleaned and maintained in good repair. Adequate floor drainage is provided in all areas to prevent contamination.	5

3.	Adequate screening or other protection is provided as necessary for protection against pests. Doors and windows should be closed or screened with no gaps greater than ¼ inch. Cracks and crevices have been sealed to prevent entrance or harborage of pests. Drains protruding from outer building walls must be screened.	5	
4.	Processing areas must be free of overhead condensation.	5	
5.	Aisles or workspaces between processing equipment and walls are unobstructed and of adequate width to permit employees to perform their duties and protect against contamination.	5	
6.	All lights in receiving, shipping, production, and storage areas of the facility are to be shielded or protected against breakage.	5	
7.	There is adequate lighting in all areas of the facility, including processing, storage, receiving, shipping, locker rooms, restrooms, and breakrooms	5	
8.	Adequate ventilation or control equipment is in place to minimize odors and vapors. Fans and other air-blowing equipment are operated in a manner that minimizes the potential for contaminating food, equipment, or materials.	5	
9.	All water lines are protected against backflow or cross-connections between potable and waste water systems.	5	
10.	Hand wash/sanitize stations are appropriately located in the processing areas. Hand washing stations have hands-free operations and are provided with antibacterial soaps, hot water, and sanitary towels or suitable drying devices at all times. Signs in the appropriate languages direct employees to wash and sanitize their hands before they start work, after each absence from their workstation, and at any time their hands may become soiled or contaminated.	10	
11.	Break areas, locker rooms, and restrooms are maintained in a clean and sanitary condition. They are equipped with proper ventilation, self-closing doors, and sinks that minimize contamination of employees' hands and garments. Drains function properly and are free of standing water. Lunchrooms are separated from the food processing areas and are free of plant garments, aprons, etc. Ladies' restrooms must have covered trash receptacles. Hand-wash signage is posted in all of these areas.	10	
12.	Ladders and walkways over exposed product lines are protected to prevent potential contamination. Appropriate kick plates are installed as necessary.	5	

C.	**Equipment**		
1.	Equipment design: All plant equipment and utensils are designed and constructed to prevent contamination of food products. Food contact surfaces and seams are smoothly bonded. Wooden equipment and wooden food surfaces are not used in food processing areas.	10	
2.	Equipment maintenance: Equipment is in good repair and is being used for the task for which it was intended. Contact surfaces are corrosion resistant and able to withstand their processing environment. No mold or rust is observed on equipment.	10	

Supplier Certification Program

3.	Temporary repairs of equipment will not inhibit proper sanitation or be made with materials that contribute in any way to the contamination of the product or environment.	10
4.	Soiled or broken pallets are not used. Empty pallets are not stored near raw material, in food processing, or in food storage areas.	5
5.	Vehicles and equipment used for moving raw materials, finished products, and packaging throughout the facility are cleaned and maintained in good condition.	5
	Possible points	135
	Actual points	
	Percentage	

Comments:

IV. Pest Control
(Assessed by observation)

Rating

1.	There are an adequate number of interior pest control devices, spaced at consistent intervals (typically 20–30 ft.) around the interior perimeter of the facility, including mechanical stations within 4 ft. of both sides of doors leading to the exterior, including dock doors. Pest control devices must also be used in dry storage areas, coolers, locker rooms, restrooms, and lunchrooms. These devices must be located so that they do not contaminate product, packaging, or equipment. The proper number and color code to correspond with the master identification map should be used. Neither toxic bait nor glue boards are to be used inside the facility.	10
2.	There are an adequate number of tamper-resistant exterior pest control stations spaced at appropriate intervals (usually 30-50 ft.) around the building's exterior perimeter. Stations are secured in place next to the building, closed, and a key, or another tool (e.g., Allen wrench) is required to open. Bait must be anchored inside the stations to avoid being removed by a rodent or floating away during heavy rains. These devices must be located so that they do not contaminate product, packaging, or equipment. The proper number and color code to correspond with the master identification map should be used.	10
3.	Adequate inspection of all pest control devices: Live catch devices are checked at least monthly by an outside contractor and twice monthly by an in-house inspector. Exterior bait stations are checked at least monthly. The pest control operator must sign and date the labels on all traps. These labels should be on the inside of the devices.	10
4.	All pest control devices are functioning properly, i.e., are properly wound, have bait as appropriate, are of sound construction and working as intended. Bait in the stations has a fresh appearance.	5
5.	There is no evidence of decomposed rodents inside or outside pest control devices.	5

6.	There is no evidence of insects, flies, rodents, or birds inside the facility. There is no evidence of rodents or birds around the exterior perimeter of the facility.	5
7.	There is no evidence of insects, flies, rodents or birds on or in any food products.	5
8.	Insect control devices may be used, as needed at all exterior entrances. They must be at least 10 ft. from covered/protected products or packaging and at least 30 ft. from exposed product, packaging, or equipment. They must be cleaned and maintained on a scheduled basis. They must be located to prevent attraction of insects from outside of the building.	5
9.	Avicides are prohibited inside the plant's facilities and are used appropriately (as designated) if used on the exterior.	5
10.	All pesticides, chemicals and other compounds stored on site for pest control are properly labeled and kept in locked, secured areas away from any food storage or processing areas.	5

                                       Possible points   65

                                         Actual points

                                             Percentage

Comments:

V.   Employee Practices
(Assessed by observation)

                                                                                                Rating

1.	Employees follow written programs on employee hygiene practices. Employees maintain personal cleanliness and hygienic practices are followed.	5
2.	Exposed jewelry, other than a plain wedding band, and other objects that might contaminate a product, such as artificial nails, are not worn. Objects, such as pens or thermometers which might fall into food, equipment, or containers are not carried in above-the-waist pockets.	5
3.	Hairnets or other appropriate restraints are properly worn in food processing areas.	5
4.	All employees with facial hair who work in production areas must wear beard covers.	5
5.	Outer garments (aprons, smocks, lab coats, etc.) are clean and suitable to the operation.	5
6.	Uniforms do not contribute to potential product contamination. They are not worn outside the facility. Employees adhere to traffic flows when moving through the facility by changing uniforms or smocks to minimize cross-contamination.	5

Supplier Certification Program

7.	Gloves worn in the food processing areas are maintained in intact, clean, and good condition. Gloves must be used where there is direct hand contact with ready-to-eat products. Procedures for the proper handling and usage of gloves are established, implemented, and verified where required.	5
8.	Eating, chewing gum, drinking, and smoking are confined to designated areas outside the processing areas.	5
9.	Employees have a separate area away from the processing areas for storing their personal items. This area is kept in a neat and clean condition and is well maintained.	5
10.	Food should not be stored in lockers or consumed in locker rooms.	5

Possible points   50
Actual points   ____
Percentage   ____

Comments:

VI. Receiving, Storage, and Shipping
(Assessed by observation and review of records)

A. Receiving and Shipping   Rating

1.	All ingredients and materials are observed to be properly identified and labeled, including date of receipt and lot and/or date codes for traceability. All materials are to be labeled and separated to prevent the possible cross-contamination of Kosher with non-Kosher products, organic with non-organic products and/or allergen contamination.	5
2.	Products are not stored in the shipping and receiving areas, unless proper controls are used to prevent temperature degradation of the products. Products are maintained in their appropriate temperature range.	5
3.	Shipping and receiving areas are clean, organized, and free of debris and spilled products.	5
4.	Temperatures of refrigerated and frozen products are documented at the time of receiving and shipping.	5
5.	Transport vehicles used (incoming or shipping) are clean and free of any pest contamination. They are inspected before and after unloading. They are capable of maintaining proper product temperatures and preventing any product contamination. They must be pre-cooled, if applicable.	5
6.	If ingredients are received in bulk (tanker, rail, etc.), transfer procedures must protect the product from contamination. Hoses must be clean, capped, and stored off the ground, and connection ports into the building must be capped when not in use.	5
7.	A receiving log is kept at the receiving area and is used as a tool for receipt documentation.	10

8.	Inbound containers are required to be sealed. These seal numbers are recorded on the receiving log.	5
9.	Outbound containers are required to be sealed. These seal numbers are recorded on the bill of lading	5

B.   Storage

1.	Sufficient space (typically 18") is maintained along all walls to permit proper cleaning and inspection for pest activity. This space is clearly marked and typically painted white. No materials are stored within this space. All materials are stored at an adequate height (6" or pallet height) above the floor. Easy access to all areas around the walls for cleaning and inspections is provided.	7
2.	All materials are to be labeled and separated to prevent the possible cross-contamination of Kosher with non-Kosher products and/or organic with non-organic products.	4
3.	First in/first out (FIFO) rotation practices are used and documented for all raw materials, in-process materials, finished products, and packaging. Products which have been partially utilized are labeled as such and are properly rotated to the front of the inventory for immediate use. Products which have been exposed to warm temperatures and then returned to storage are appropriately labeled and rotated for immediate use.	5
4.	All stored ingredients, materials, and packaging are clean, dry, intact, and properly packaged or covered to prevent contamination of other products. They are in good condition and free from contamination and spoilage. They are stored under appropriate conditions (e.g., dry, cooler, freezer).	5
5.	Any damaged cases or packages are immediately segregated and repackaged or properly discarded. All materials rejected or on hold are properly identified, adequately segregated, clearly tagged with product name and original lot number, and protected from contamination. Hold areas are clearly identified.	5
6.	Ingredient containers are not reused, unless they are adequately sanitized or have protective liners. Single-use containers from microbiologically sensitive products are not reused.	5
7.	Dry storage areas are maintained in a clean and sanitary manner. All spills are immediately cleaned up; i.e., the floors and racks are not dirty and there is no evidence of spills, trash, or other litter.	5
8.	Restricted chemicals for use in processing or as an ingredient are stored in separate, locked areas away from food and packaging supplies.	5
9.	Floors, walls, and ceilings of coolers and freezers are maintained in a clean and sanitary condition. There is no evidence of spills, trash, or clutter. Floors are kept dry and free of ice.	5
10.	Coolers and freezers show no sign of condensation. Freezers have no ice. Products stored in freezers and coolers should be free from condensation and ice.	4

11.	Temperatures of coolers and freezers are maintained below the maximum levels. Monitoring systems include checking temperatures manually twice a day or via continuous recording devices.	5

Possible points	105
Actual points	
Percentage	

Comments:

VII. Plant Sanitation
(Assessed by observation and review of records)

A. Cleaning Equipment and Chemicals — Rating

1.	All chemicals used for cleaning and sanitizing have EPA/USDA/FDA approval and are properly labeled, used for their intended purposes, and stored in secure, locked areas away from any food processing or other storage. Chemicals that are connected to dilution devices do not have to be in a locked area if their location does not pose a contamination risk to food, packaging, or equipment.	5
2.	Test kits or sanitizer strength strips are routinely used to monitor chemical concentration in sanitizing hand dips, foot baths, and solutions. Procedures for these checks have been established, and records of the checks are documented.	5
3.	Containers, brushes, and applicators used for cleaning and sanitizing are color coded or labeled to properly identify them for their intended use. If a color coding system is used, appropriate signage on use of the containers and equipment is posted.	5
4.	Cleaning equipment is properly stored (when not in use) and is not stored in food processing areas.	5
5.	Cleaning equipment is non-porous and in good repair.	5

B. Cleaning, Sanitation, and Housekeeping Procedures

1.	Cleanliness is maintained in all non-processing and non-food contact areas.	5
2.	Cleanliness is maintained on all food contact surfaces.	5
3.	The cleanup of spills and accumulation of materials is conducted on a continuing basis during production.	5
4.	Proper cleaning and sanitizing procedures are followed. Equipment is disassembled as necessary for thorough cleaning.	5
5.	No gross product build-up is present during production.	5
6.	Excess moisture and pools of water and condensation are removed from equipment and the environment prior to the start of operations.	5
7.	Knives, saws, trimmers, and other tools used in processing are adequately cleaned and sanitized during processing.	5

8.	Equipment not used for 4 hours is sanitized again prior to use in wet processing environments.	5

	Possible points	65
	Actual points	
	Percentage	

Comments:

VIII. Processing
(Assessed by observation and review of records)

A. Raw Materials and Other Ingredients   Rating

1.	Water reused for washing, rinsing, or conveying food does not increase the level of contamination of the food.	5
2.	Thawing or tempering of frozen materials is done under controlled conditions (e.g., under refrigeration) and is monitored to ensure proper temperature controls are maintained. Thawing procedures and verification checks of compliance must be documented.	5

B. Process Control

1.	Appropriate process control points and limits are monitored on a regular basis. Results are monitored and corrective actions are being taken and documented on line.	5
2.	All processing operations (such as washing, trimming, sorting and inspection, shredding, extruding, and forming) are performed to protect the food against contamination, including adequate physical protection from contaminants that could drip, drain, or be drawn into the food. Product contamination must not occur and this includes the possible cross contamination of organic with non-organic, Kosher with non-Kosher, and/or allergen contamination, etc. Steps must be taken to ensure that this does not occur and documentation of these steps must be maintained.	5
3.	Heated ingredients and finished products are rapidly cooled to prevent the growth of harmful bacteria or are used in subsequent processing steps without delay.	5
4.	Breakdowns or line shutdowns are monitored to ensure that time delays, temperature fluctuations, and other factors do not contribute to contamination or decomposition of the food.	5
5.	All perishable product processing rooms have an easy to read, calibrated thermometer to monitor ambient temperature. The temperatures of products being processed or ingredients to be used in the process are maintained in their appropriate ranges.	5
6.	Ingredient containers are properly labeled and/or color coded and covered as appropriate. If a color coding system is used for labeling ingredient containers, appropriate signage on use of the containers and equipment is posted. This includes steps to prevent the possible cross-contamination of organic with non-organic, Kosher with non-Kosher, and/or allergen contamination, etc.	5

Supplier Certification Program

7.	An adequate kill step is utilized in the processing area. This is monitored and documented, and employees are trained to watch for, correct, and document any deviations.	15
8.	Packaging materials are kept clean, dry, and free from contamination during processing.	2
9.	Magnets, screens, sieves, etc., are used in the processing lines as necessary and are inspected on a routine basis to ensure proper performance. Inspection records are documented and maintained.	5
10.	Any compressed air (sanitary air in the freezer) or other gases (e.g., carbon dioxide, nitrogen) used in processing, packaging, or cleaning are treated in such a way to prevent contamination.	3
11.	Floors are free of standing water.	2
12.	Maintenance tools, gloves, rags, and other miscellaneous materials are not found on or near processing equipment to prevent potential contamination.	3
	Possible points	**70**
	Actual points	
	Percentage	

Comments:

IX. Other Programs

A. Other Program Information and Questions

1.	Does the plant have a documented and maintained Kosher program? If yes, who is the Kosherization body? _____	Yes	No
2.	Does the plant have a documented and maintained organic program? If yes, who is the certifying body? _____	Yes	No
3.	Has the plant had a third party audit within the last year? If yes, by whom: _____ Date: _____ Score: _____	Yes	No
4.	Has the plant registered under the Bioterrorism Act of 2004? If yes, what is the registration number? _____	Yes	No
5.	Does the plant conduct environmental microbial testing? If yes, for what organisms? _____ Is there a written program? _____ Are product contact surfaces tested? _____	Yes	No
6.	Does the plant maintain halal certification? If yes, who is the certifying body? _____	Yes	No
7.	Does the maintain vegan certification? If yes, who is the certifying body? _____	Yes	No

Food Safety and GMP Assessment Rating System

This rating system describes a food plant's level of compliance with recognized food safety and Good Manufacturing Practices. The point system and definitions are objective guidelines for evaluating the plant's compliance with the assessed standards and are intended to assure consistency in rating.

This rating system is an objective guideline. Auditors may use their discretion regarding scoring considering the severity of food safety issues and numbers of observations of issue noted.

Each plant will receive a total overall score based on the ratings of the individual standards in the audit form. *The minimum acceptable numerical score may vary depending upon the company requiring the audit.*

The base score is prorated based on the types of products manufactured, and systems needed to produce them in accordance to Federal regulations.

Rating	Numerical Score
Excellent	95% or higher
Good	90–94%
Fair	85–89%
Poor	< 85%

SUPPLIER CERTIFICATION PROGRAM

GENERAL OVERVIEW

Section A: General

In accordance with good manufacturing practices, _____ shall certify all suppliers of ingredients and packaging as key components of the hazard exclusion program. This program involves a combination of data and document collection and physical auditing.

Section B: Document Collection

As the underlying precursor to the hazard analysis component of the HACCP plan, each supplier shall be contacted prior to approval of the ingredient or package for use in the process. Using the supplier data sheet as a guide, a specification sheet, nutritional information, continuing food guarantee, biosecurity act registration verification, emergency contact sheet, and allergen statement are requested. In addition, a kosher certificate, organic certificate, halal certificate, material safety data sheet, and a non-GMO statement are requested when appropriate.

Section C: Document Processing

After collecting all of the required documents, as part of the hazard analysis, each ingredient is examined to determine if there are any chemical, physical, microbiological, or allergen hazards. When this examination is complete and no hazards are found, the ingredient is tentatively approved for use.

If the ingredient is a major component of the final product, if the supplier supplies numerous products, or if the ingredient has the potential to contain a hazard, then a supplier audit is warranted.

Section D: Supplier Audit

For all critical suppliers, a supplier audit shall be conducted. This audit shall be done at the supplier's manufacturing facility and focuses on food safety systems, quality systems, facility integrity, maintenance, and employee responsibility. The supplier will be rated on a 1000-point scale, with greater than 85% being the minimum qualification level. Upon conducting a supplier audit, a schedule of supplier improvement shall be developed in conjunction with the supplier as a means of continual improvement.

Approved by: _____ Date: _____

SUPPLIER DATA SHEET

Item Number: _____

Ingredient Name: _____

Supplier Name: _____ Vendor Item Number _____
Address: _____

Sales Contact:
  1 _____ Telephone Number: _____
  2 _____ Fax Number: _____
                                Email: _____
                                Cell Number: _____

Emergency Contact:
  1 _____ Telephone/cell Number: _____
  2 _____ Telephone/cell Number: _____
  3 _____ Telephone/cell Number: _____

Items to be in folder: Effective Date

Specification Sheet			Storage	_____
Ingredients			Shelf Life	_____
Nutritional				

	QA Verified	Certification	Expires Date
Kosher Certification			
Organic Certification			Lot Code Example
Vegan Certification			
Halal Certification			Pack Size to be purchased
Continuing Guaranty			
MSDS			Pallet Configuration
Non-GMO verification			
Biosecurity Act Verification			

Allergen	Contains (Y/N)	QA Verified		Contains (Y/N)	QA Verified
Milk			Celery Sds		
Egg			Mustard Sds		
Peanuts			Sesame Sds		
Wheat			Colors: ↓		
Cereal Gluten					
Soy					
Fish					
Shellfish					
Tree Nut					
Sulfites					
MSG					
Molusks					

22 Hold/Defective Material Program

TYPE OF PROGRAM: REQUIRED

THEORY

No matter how complete the quality systems and process controls are within a company, it is inevitable that an ingredient, packaging material, or intermediate or finished good will be considered out of specification. This might be for any possible reason, varying from food safety issues such as physical, microbiological, chemical, or allergenic contamination to just a quality issue such as granulation, color, or appearance. 21CFR110.80(b)9 states that "food, raw materials, and other ingredients that are adulterated within the meaning of the act shall be disposed of in a manner that protects against the contamination of other food." Additionally, section 110.80(a)5 says that "material scheduled for rework shall be identified as such." The way a company meets these legal requirements and handles these substandard products correctly is outlined in its hold/defective material program. The three basic parts to an effective hold/defective are the hold process, the release process, and the destruction process—each of which is supported by adequate record keeping.

When it has been identified that a raw material or intermediate or finished good is out of specification, it should be segregated and placed "on hold" pending further analysis leading to disposition. Normally, this process includes moving the component to a separate area or space within the facility as a means to prevent its inclusion in future products, but in some instances it is physically impossible to do so. In all cases, however, it is imperative that the nonconforming component be clearly labeled as "on hold." This is typically done with a large, brightly colored label indicating to all that there is a potential problem with the component.

Next, a formal hold notification should be completed and distributed. This document is the foundational form that notifies all parts of the company that there is a problem. It should be sent to all parts of the company that may be affected. In most companies this includes but is not limited to

- quality, so that this department may inspect the product to assist in the disposition determination;
- operations, so that this department can assist with the disposition, if it is rework; destruction; or return to the supplier;
- accounting, so that this department can bill the supplier if it is the supplier's responsibility;

purchasing, so that this department can replace the defective material if necessary; and

inventory control, so that this department can remove the defective material from active inventory.

A hold notification should include an assigned hold number, the product name and description, the item number, the lot code, the quantity, the reason for its being placed on hold, and a proposed disposition. The hold number should be assigned by taking the next number available from the hold log. CD (Book 2_Quality Control Manual\Section_18_Defective Material Program:Hold Notification) and at the end of the chapter is a customizable hold notification form.

The second part of the hold/defective material program is the release notification. This notification is a form that indicates to all parties how the defective material will be handled. It is an authorization to remove the defective material from the hold (or do not use) status and modify it as directed. In some cases this modification is just a rework at a low level and in some cases it could be a relabeling or repackaging; in other cases it could be the destruction or return of the material to a supplier. The release notification includes all of the information of the hold notice including a release number and a firm disposition. The release notification number should be assigned by taking the next number available from the release log.

Upon receiving a release notice, the designated party responsible for carrying out the disposition should schedule the appropriate action. This might be completed all at once or in stages. In any case, when the defective material is reworked, the hold sticker or label should be removed and attached to the release notice. When the entire lot has been reworked, the release notification and *all* of the hold stickers or labels should be returned to quality control for verification and document retention.

Destruction notices are handled in the third part of the hold/defective material program. This is due to the sensitive nature of returns and destructions and the required document trail. When it is necessary for a defective material to be destroyed or returned to a supplier, a separate destruction notice should be issued in conjunction with the release notification. This form contains most of the information from the hold and release forms, but gives specific information regarding the destruction method, timing, and documentation needed and has a separate destruction log number. This number is obtained by taking the next number from the destruction log. When the destruction notice is received with the release notice, it is completed as requested and the supporting materials are returned to quality control along with the release with the hold stickers or labels attached. These documents are reviewed and retained for future reference as needed.

Record keeping for the hold/defective material program is of primary concern. To support all of the documents needed and provide an easy method for keeping track of all holds, releases, and destructions, a notebook should be developed that contains each of the logs and copies of each hold, release, and destruction form. A notebook of this nature allows for easy reference for quality and operational personnel as they handle myriad defective material challenges.

APPLICATION

The first step in developing the hold/defective material program is to create the hold/defective material notebook for record retention. To do this, it is necessary to obtain a three-ring binder and separate it into three sections: holds, releases, and destructions, in that order. A copy of the respective log sheet sould be placed in each section. Label the front and spine of the book for easy reference. Templates for the company to utilize within the sections of the hold/defective material notebook can be found on the CD (Book 2_Quality Control Manual\Section 18_Defective Material Program:Hold Log; Book 2_Quality Control Manual\Section 18_Defective Material Program:Release Log; Book 2_Quality Control Manual\Section 18_Defective Material Program:Destruction Log) and at the end of this chapter. Forms on the CD (Book 2_Quality Control Manual\Section 18_Defective Material Program:Hold Defective Material Program Notebook Cover) and at the end of this chapter can be inserted into the cover and spine of the notebook for easy reference.

The second step is to develop a program document that explains why the company has the program, its key components, documentation and record retention, and responsibility. Within the key component section, the procedures used for putting defective materials on hold, releasing, and destroying them should be detailed. Because it is a program document, it should be signed by the quality control manager and inserted into the quality control manual along with copies of the hold, release, and destruction notices and logs and a copy of the hold stickers or labels to be used. The general overview on the CD (Book 2_Quality Control Manual\Section 18_Defective Material Program:General Overview) and at the end of this chapter can be customized by the company, signed, and then placed in the hold/defective material section of the quality control manual.

Finally, copies of the hold label should be printed out in preparation for labeling defective material. Most companies print this label onto peel-and-stick label stock to attach to the outsides of the pallets or cases. Sometimes, however, it is easier to print them on regular paper. As a means of making the label readily noticeable, the labels should be a minimum of 8.5 × 11 inches and printed on a color that gets attention. With the availability of an unlimited color pallet, most companies utilize fluorescent pink, green, red, or orange. These labels are to be filled out as needed based on the information provided on the hold form. CD (Book 2_Quality Control Manual\Section_18_Defective Material Program:Hold Sticker) and at the end of the chapter may be used to label pallets put on hold.

SUPPLEMENTAL MATERIALS

Hold/Defective Material Program

HOLD NOTIFICATION

To:

From:

Subject:

Date:

CC:
_____ ,

    Please let this memo serve as notification that the following product and lot code is to be placed on hold immediately for the reason(s) stated.

Hold Number: _____

Product: _____

Item Number: _____

Lot Code: _____

Quantity Produced: _____

Reason for Hold: _____

Specification: _____ Actual: _____

Other: _____

_____

Disposition: _____

_____

Thank you for your immediate attention to this matter.

RELEASE NOTIFICATION

To:

From:

Subject:

Date:

CC:

_____ ,

    Please let this memo serve as notification that the following product and lot code is to be released from hold immediately for disposition.

Release Number: _____

Hold Number: _____

Product: _____

Item Number: _____

Lot Code: _____

Quantity Produced: _____

Disposition: _____

_____

_____

Thank you for your immediate attention to this matter.

DESTRUCTION NOTIFICATION

To:

From:

Subject:

Date:

CC:
_____ ,

    Please let this memo serve as notification that the following product and lot code is to be destroyed immediately due to the reason stated below. Upon proper destruction, please provide documentation verifying that destruction has occurred.

Destruction Number: _____

Associated Hold Number: _____

Product: _____ Item Number: _____

Lot Code: _____ Quantity Produced: _____

Disposition: ☐ Garbage ☐ Donate ☐ Return to supplier

             ☐ Other: _____

Destroy by: _____

Documents needed: ☐ Pictures ☐ Shipping bill of lading

                  ☐ Other: _____

Thank you for your immediate attention to this matter.

Hold Log

Hold #	Date	Item Number	Product Description	Lot Code	Quantity	U/M	Ref.	Reason for Hold or Disposition Method	Tech

Hold/Defective Material Program

Release Log

Release #	Hold #	Date	Item Number	Product Description	Lot Code	Quantity	U/M	Ref.	Reason for Hold or Dispostion Method	Tech

Destruction Log

Destruction #	Hold #	Date	Item Number	Product Description	Lot Code	Quantity	U/M	Ref.	Destruction Method	Tech

HOLD/DEFECTIVE MATERIAL PROGRAM

HOLD DEFECTIVE MATERIAL PROGRAM

Hold/Defective Material Program

MANUFACTURING DEVIATION REPORT

_____ **Ingredient**   _____ **Packaging Material**   _____ **Finished Product**

Supplier: _____   Mfg. Location: _____

Date: _____

<u>Finished Product Deviation</u>

Prod No.: _____   Item Description: _____

Code Date: _____   Quantity: _____

Description of deviation from specification

_____

_____

<u>Inbound Material Deviation (Packaging and Ingredients)</u>

Vehicle No.: _____   Lot No.: _____

Date of Receipt: _____   Bill of Lading No.: _____

Deviation Report Originated by: _____

Investigation: _____

_____

_____

_____

_____

_____

Disposition		
Reject and return to supplier ☐	Plant Manager	Date
Waiver of specification for use ☐		
Destroy ☐	QA Manager	Date
Donate ☐		
Restack and use_____ ☐		

Comments: _____

_____

_____

HOLD/DEFECTIVE-MATERIAL PROGRAM

GENERAL OVERVIEW

Section A: General

In accordance with 21CFR7.110, _____ will have an effective means by which to identify, isolate, and handle defective materials in an appropriate manner. This will include handling of defective raw, intermediate, and finished goods, appropriate labeling, and final disposition.

Section B: Holds

All raw materials, intermediates, and finished goods that are deemed out of specification will be placed on hold. Where possible, these will be placed in a segregated area and labeled with a hold sticker or label. For each particular lot of product, a hold form will be completed and distributed to other departments within the company as needed. Each hold will be assigned a unique number. This is then logged in under the hold section of the hold/defective material notebook on the hold log. No product that is on hold will be used until a disposition is determined and the appropriate release issued.

Section C: Releases

When a disposition is determined for an item on hold, a release notification will be issued. For each release, a unique number will be assigned from the release log located in the release section of the hold/defective material notebook. When the release is issued, it will be completed according to the disposition, and the hold stickers or labels will be attached to the release and returned to the quality control department for retention and future reference.

Section D: Destruction

When the best possible avenue for handling a defective product is that of donation, destruction, or return to vendor, a destruction notice will be completed and accompany the release notice. The person responsible for donating the product, destroying the product, or returning it to a supplier will perform the required action as listed on the destruction notice. He or she will collect the hold stickers or labels, attach them to the release notice, and then attach any photos or shipping documentation to the destruction notice. All of the collected documentation will be returned to the quality control department for retention and future reference.

Section E: Records

To facilitate easy reference, a notebook containing a hold log, release lot, and destruction log will be developed and maintained in the quality control department.

Section F: Responsibility

The quality control manager will be responsible for developing and maintaining the hold/defective material program.

Approved by: _____ Date: _____

Hold Number: ☐

HOLD

LOT NUMBER: _____

PALLET #: _____

QC TECH: _____

PRODUCT: _____

DATE: _____

CS: _____   **LBS/CS:** _____

REASON: _____
_____
_____

RELEASED BY: _____

REWORKED BY: _____

23 Glass, Hard Plastic, and Wood Program

TYPE OF PROGRAM: REQUIRED

THEORY

In accordance with 21CFR110 good manufacturing policy (GMP), every food manufacturer or processor must establish and enforce regulations and policies that preclude any foreign object, including glass, plastic, metal, or wood, from contaminating or adulterating any product. Of these four contaminants, the easiest with which to deal is metal. Due to the nature of its properties, the technology to detect metal in the product stream, a metal detector, is readily available. This is not the case for wood, plastic, or glass. These foreign substances are not readily detectable and the current x-ray technology is too expensive for most companies and is easily confused by matrix effects. It is thus incumbent on the company to take the necessary steps to prevent these materials from entering the food stream in the first place. These preventative measures take two avenues: up-front prevention and in-plant prevention.

Up-front prevention comes as part of the risk prevention program that the company undertakes. As part of the HACCP plan, the company utilizes the hazard analysis portion to determine possible microbiological, physical, chemical, and allergen risks that might cause contamination. Under the physical contamination risk analysis, the company considers each product's source, manufacturing method, and the integrity of the supplier and then determines the possibility of any wood, plastic, or glass having gotten into the ingredient. The company then utilizes steps such as auditing and certificates of analysis to document that the initial ingredient provided is clean. Unfortunately, this approach only goes so far because the product can get contaminated during distribution. To manage these unforeseen avenues of glass, wood, and plastic contamination, the company should institute a program to eliminate and/or monitor them within the operation. This program should utilize the basic approach of exclusion, which includes removing these possible contaminants from the production and warehouse areas. When this approach is taken, it is first necessary to understand where glass, wood, and plastic reside within the operation.

The most common places in a facility to find glass are windows, light bulbs, forklift lights, gauge covers, mirrors, and insect lights. Although many of these are easy to replace at limited cost, others are not; however, all glass should be evaluated for removal if possible. Some considerations when removing glass include: windows in the production room should be replaced or covered with screens or other materials; light bulbs in all areas are to be completely covered or should be replaced with covered sodium lamps; forklift lights should be screened or covered even if they do not enter the production area; gauge covers can be replaced with hard plastic; mirrors should have a metallic surface, and insect lights should be properly contained within an approved insect light trap.

Wood contamination in the distribution chain, warehouse, and production room generally comes from contact with wooden pallets by the equipment moving them. This can occur by the forklift catching a runner and a piece of wood getting shoved through a container wall, setting a wood pallet on top of another pallet and getting pieces on top of product or containers, or putting wood pallets in racks and having pieces fall on the pallet below. Each of these scenarios has the potential to create little pieces of wood that might get into the product; although excluding wood pallets from the facility altogether would be nice, it is not entirely possible. Therefore, a more balanced approach is used that involves excluding wood from the production facility; requiring suppliers to use only top-quality pallets, such as #1 hardwood; and cleaning the tops of containers entering the production room.

Plastic in the operation is often the most difficult to exclude because it is commonly used as the medium for belts, guards, diverters, and rollers. In some cases it can be replaced with stainless pieces, but in many cases it cannot due to the physical and design flexibility of plastic and its cost effectiveness. Cases in which it is neither practicable nor cost effective to replace the plastic should be included in the glass, plastic, and wood inspection program.

This program is designed as a failsafe program to ensure that those glass and plastic pieces that cannot be replaced are inspected in a timely manner and verified that they are intact at all times. It involves identifying all glass and plastic in the facility and summarizing them on a check list. At a regularly scheduled time, an employee carefully inspects each of the identified components and verifies that they are intact and have not been damaged. When the employee finds that damage has occurred, operations, maintenance, and quality control are to be notified to fix or replace the damaged piece and determine if it might have contaminated any product.

The final component to the glass, wood, and plastic program is that of training. In most companies, these contaminants are caught by employees noticing that something is broken, damaged, or dirty, so it is imperative that all employees receive training regarding this issue. This training should occur as part of the new employee orientation, be documented, and placed in the employee's personnel file.

APPLICATION

The development of the glass, wood, and plastic control program within most companies has five parts: identification, exclusion, training, inspection, and document development. Identification involves a physical inspection of the facility to identify the kinds and locations of all glass, wood, and plastic. This list should include even the smallest use of any of these materials. Often overlooked are eye glasses, fire extinguisher and hose cabinet covers, emergency lights, and control panel covers. When this list is completed, it should be reviewed by the operations, maintenance, and quality managers to determine what could be replaced or removed.

The removal of glass, wood, and plastic from the operation is part two. The remaining items should be listed on a glass/plastic inspection chart. The form on the CD (Book 2_Quality Control Manual\Section 19_Glass_Hard Plastic_Wood Program:Glass/Glass Plastic Inspection Chart) and at the end of this chapter may be used to tabulate those glass and plastic items that cannot be replaced or removed.

Part three of this program is to train all employees to recognize the threat to the food safety posed by glass, wood, and plastic and to make sure that they understand that they are individually responsible to monitor and report concerns. The form on the CD (Book 2_Quality Control Manual\Section 19_Glass_Hard Plastic_Wood Program:Glass_Wood_Plastic Training Verification) and at the end of this chapter can be adapted by the company and used to verify that each employee has read, understands, and will adhere to the policy. These should be signed by all employees and placed in their personnel files. Training should be completed upon hiring and on an annual basis.

Part four involves performing an inspection of all the remaining glass and hard plastic. The list made after removing all of the possible glass, wood, and plastic from the facility may be used as the inspection document. Inspections should occur on a regular basis, whether daily, weekly, or monthly; the more often they are completed, the less liability there is if damage occurs and a recall is necessary. The glass/plastic inspection chart is an accountability document and should be filled out by the inspector as well as signed by the person repairing those items found to be unacceptable. This document should be turned in with the daily paperwork when completed.

The final part of this program involves document development. A program document for the company to utilize can be found on the CD (Book 2_Quality Control Manual\Section 19_Glass_Hard Plastic_Wood Program:General Overview) and at the end of this chapter. It should be customized to represent the strategy used by the company to exclude and inspect glass, wood, and hard plastic within the facility. When it is complete, it is signed and placed in the quality control manual with copies of the inspection sheet and training verification.

SUPPLEMENTAL MATERIALS

Glass / Plastic Inspection Chart

Date of Inspection: _____   Inspector _____

Location of Glass or Plastic	Condition		How Damaged	Person Notified	Repair Verified
_____	ok	not ok			
_____	ok	not ok			
_____	ok	not ok			
_____	ok	not ok			
_____	ok	not ok			
_____	ok	not ok			
_____	ok	not ok			
_____	ok	not ok			
_____	ok	not ok			
_____	ok	not ok			
_____	ok	not ok			
_____	ok	not ok			
_____	ok	not ok			
_____	ok	not ok			
_____	ok	not ok			
_____	ok	not ok			
_____	ok	not ok			
_____	ok	not ok			
_____	ok	not ok			
_____	ok	not ok			
_____	ok	not ok			
_____	ok	not ok			

VERIFICATION DOCUMENTATION

Glass, Plastic, and Wood Training

I, _____, have read and understand the section on glass, plastic, and wood control at _____. I will be on the alert for any nonapproved glass, plastic, or wood in the production area and will notify my supervisor if I notice any damage. If I have questions or concerns regarding this policy, I will immediately discuss them with my supervisor.

Date: _____

Name: _____

Instructor: _____

GLASS, HARD PLASTIC, AND WOOD CONTROL PROGRAM

GENERAL OVERVIEW

Section A: General

In accordance with 21CFR110 good manufacturing practice (GMP), _____ will establish and enforce regulations and policies that preclude any foreign object, including glass, hard plastic, metal, or wood, from contaminating or adulterating any product. These regulations and policies will be an integral component of training during the hiring process as well as during reviews.

Section B: Glass

There will be no glass allowed in any production area or warehouse. Exceptions to this rule are gauge fronts, lights, emergency lights, and forklift flood lights. Production, warehouse, and emergency lights will be covered and sealed. Forklift lights will be screened and inspected on a weekly basis for chips or other damage. Damaged lights will be replaced immediately. Due to the high-risk nature of having any glass in the production or warehouse areas, a weekly inspection of all glass elements will be made. A conscious effort to eliminate all glass will be made.

Section C: Plastic

All plastic allowed in the production area will be tethered or attached to a stationary surface. It will be made of hard plastic and be part of the internal inspection program.

Section D: Wood

Wood will be confined to the warehouse area. If pallets must be taken into the production area, they will be clean   number 1 hardwood pallets in good repair and will be kept on the floor. At no time will wood ever be allowed above products.

Section E: Prevention

All measures will be taken through training and/or mechanical means for the prevention of foreign objects falling into the product.

Approved by: _____ Date: _____

24 Loose-Material Program

PROGRAM TYPE: REQUIRED

THEORY

The federal government recognizes that it is easy to contaminate food during warehousing, manufacturing, and distribution. Many regulations are written to address microbial, physical, and chemical contamination. A key area among these three that many companies struggle to control is that of physical contamination. Specifically, this contamination is not from material brought into the facility, but rather from cross-contamination within the facility. 21CFR110.80(b) addresses this type of contamination:

> (5) Work-in-process shall be handled in a manner that protects against contamination. (6) Effective measures shall be taken to protect finished food from contamination by raw materials, other ingredients or refuse,.... (7) Equipment, containers, and utensils used to convey, hold, or store raw materials, work-in-process, rework, or food shall be constructed, handled, and maintained during manufacturing or storage in a manner that protects against contamination.

As a means to control this internal contamination stream, many companies develop a loose-material program. This program includes coding, labeling, and training.

Loose material in a food manufacturing or processing plant comes in many forms, including rework, waste on the floor, sanitation chemicals, and basic trash. With the high turnover of employees and lack of training in most companies, it is easy for these streams to get mixed and end up contaminating a finished good. Color coding of the containers or tools used for these materials provides immediate feedback that a potential contamination situation might be happening.

When preparing to develop the loose-material program, a company identifies the sources of cross-contamination and then assigns a color to that stream. The company then systematically purchases the tools and containers in that color for use within the plant. Some examples of this strategy would be the following:

> Floor (color: brown): All brushes, brooms, squeegees, and other tools that are used to sweep or clean the floor would be brown.
> Nonproduct waste (color: red): All garbage barrels, bins, and holders where miscellaneous packaging, Plastic, general trash, and nonproduct are put would be red. If the company is not collecting and measuring product waste, then this would go in the red barrels also.

Product waste (color: yellow): All products to be thrown away would be placed in yellow barrels, buckets, and containers and weighed or measured prior to disposal.

Rework (color: gray): This color would be used for all bins, barrels, scoops, and containers in which rework is stored. This includes containers used just to move product within the production room during production.

Drains (color: black): These would be used for brushes and tools that are used to clean drains.

An added reason to use a color scheme is to identify and separate those ingredients, intermediates, and finished goods that contain allergens. For example, if the plant produced peanuts in various forms and wanted to produce tree nuts also, this would be introducing an added allergen. One way to separate these is to designate anything used to process or store peanuts as one color and any container, brush, or tool used for tree nuts as another color. This would work just the same with other allergens such as dairy–nondairy, corn–soy, and oats–wheat. This type of color scheme would include the same colors for floors, nonproduct waste, product waste, and drains, but would segregate rework as follows:

Rework—peanuts (color: white): This color would be used for all bins, barrels, scoops, and containers in which rework is stored. This includes containers used just to move product within the production room during production.

Rework—tree nuts (color: blue): This color would be used for all bins, barrels, scoops, and containers in which rework is stored. This includes containers used just to move product within the production room during production.

or:

Rework—dairy (color: white): This color would be used for all bins, barrels, scoops, and containers in which rework is stored. This includes containers used just to move product within the production room during production.

Rework—soy (color: blue): This color would be used for all bins, barrels, scoops, and containers in which rework is stored. This includes containers used just to move product within the production room during production.

Colored brushes, scoops, garbage barrels, bins, and other containers can be purchased from various sources, including local restaurant supply stores, trade magazines, and on the Internet. Rubbermaid manufactures and distributes many of the colored tools and equipment needed for food manufacturers. In addition, industrial supply houses such as Grainger, Lab Safety Supply, and Rubbermaid are also good resources.

After identifying the possible cross-contamination streams and purchasing colored equipment, the next step is to post signs in conspicuous places within the facility for all employees to refer to during the daily operations. These signs should be large enough for the employees to see from a distance, in colors that match the color coding scheme, and printed with words that refer to the specific use. The signs should be made of a washable material, such as plastic or a laminated surface, when they are located in the production area.

The final part of the program is that of training. Due to the potentially devastating effects of cross-contamination from chemical, physical, microbiological, and allergen risks, each and every employee should be trained in what the color-coding scheme is, how it is implemented in the plant, and the potential consequences of not following it. As always, this training should be documented and kept in employee's file.

APPLICATION

Development of the loose-material program is prefaced by identifying the needed areas for separation. These should be evaluated based on possible cross-contamination from a physical, chemical, microbial, and allergen viewpoint. Once this is done, colors should be assigned to the designated areas and existing brushes, barrels, scrapers, containers, and other tools should be replaced to fit the color scheme.

During the purchasing process, the company can undertake the signage and employee training phase. A template for the development of wall signs is located on the CD (Book 2_Quality Control Manual\Section 20_Loose Material Program:Color Coding Scheme) and at the end of this chapter. Created in Microsoft PowerPoint, the colors and identification titles can be changed to fit the scenario of the company. This can be printed in color and laminated for small wall signs or it can be blown up and applied to a larger medium and mounted prominently on the wall. Office supply stores such as OfficeMax, Staples, and Kinkos can accomplish this relatively easily.

All operations and quality staff should be trained in the proper use of the designated colored tools and containers. The form on the CD (Book 2_Quality Control Manual\Section 20_Loose Material Program:Loose Material Training Verification) and at the end of this chapter should be used to document that the employee has been trained. The form should be signed and placed in the employee's file.

The final step in the development of the loose-material program is development of the program document. The form found on the CD (Book 2_Quality Control Manual\Section 20_Loose Material Program:General Overview) and at the end of this chapter can be adapted to fit the specifics of the company's color scheme. As a program document, it should be signed and placed in the Quality Control Manual with a copy of the training verification and color scheme.

SUPPLEMENTAL MATERIALS

Loose-Material Program

VERIFICATION DOCUMENTATION

LOOSE-MATERIAL CONTROL TRAINING

I, _____, have read and understand the section on loose-material control at _____. I have been trained in what the color scheme means and how it is to be applied within the production environment. Furthermore, I understand the contamination consequences if I do not use properly colored-coded tools and containers. If I have questions, I will immediately notify my supervisor.

Date: _____

Name: _____

Instructor: _____

LOOSE-MATERIAL PROGRAM

GENERAL OVERVIEW

Section A: General

Pursuant to 21CFR110.80(b)5–7, a program will be developed that dictates the separation of containers and tools used for different purposes within the manufacturing facility. Separation will be made as a means to prevent cross-contamination among tools used for cleaning, waste, rework, and drains. It will also be used as part of the allergen control program as a means to differentiate one allergen component from another.

Section B: Program

Tools, brushes, barrels, and containers will be purchased and color coded according to the following scheme:

garbage: red;
floor: brown;
rework: yellow;
drains: black;
peanuts: blue; and
tree nuts: white.

Section C: Signage

Signs that denote the color scheme will be made and posted in conspicuous places throughout the facility.

Section D: Training

All employees will be trained to understand what the color coding scheme is and the consequences of not adhering to it.

Section E: Responsibility

It will be the responsibility of operations to ensure that the color coding scheme is instituted and followed by all employees. Operations also will take responsibility for training employees. It is the responsibility of the quality control department to verify that this color code is being followed at all times.

Approved by: _____ Date: _____

25 MICROBIOLOGY PROGRAM

PROGRAM TYPE: OPTIONAL

THEORY

Foodborne illnesses have become a major public health issue in the past 30 years. Many agricultural and processed foods have been linked either directly or indirectly to contamination from microbiological sources. Recently, foods such as spinach, packaged lettuce, tomatoes, ground beef, cheese, and peanut butter have all been linked to major micro-based recalls. Each of these might have flown under the radar 20 years ago but due to today's instant communications through various media channels, news organizations, and the internet, the public has almost an instantaneous pulse on problems that occur within the food supply, whether domestic or imported. To ensure that food companies do not become today's news, it is important to understand the microbiological implications of the products they produce and the methods to control them.

Any time a food product is harvested, processed, packaged, or distributed there is a potential for contamination. Either directly or indirectly it may be contaminated by pathogenic organisms such as *E. coli, Salmonella, Listeria, Yersinia, Vibrio, Cyclospora, Shigella, Campylobacter, Staphylococcus* or *Norovirus* or non-pathogenic bacteria, yeasts, and molds. It is the job of the processor or manufacturer to analyze their components processes, finished products, and distribution chain to determine which organisms are naturally contained in the starting materials and which organisms may be introduced within the process. From this, a systematic program may be developed to reduce or remove the risk and develop and monitor the process to identify any contamination prior to it reaching the public.

Risk analysis is usually performed as a component of the Microbiological Risk Assessment portion of the HACCP plan. During this phase, the potential for the ingredients to contain a pathogen or microbial load are identified and the supplier worked with to either remove or reduce the presence of the unwanted bacteria. For those ingredients that are found to contain pathogens, the supplier should be required to remove them fully where possible, and provide a lot by lot certificate of analysis verifying that they have been removed. In the case of non-pathogenic bacteria and molds, these are usually controlled and a maximum presence limit set so that the supplier is required to certify prior to shipping the product. Certain ingredients, especially those that are raw agricultural commodities, will naturally have a bacterial load to some degree. This becomes a problem when they are minimally processed prior to consumer sale. To limit this, the processor must understand the farming practices of the supplier and conduct extensive testing on each raw material lot prior to use. After determining what the risks are, the manufacturer or processor must develop the plan to remove the organisms and then verify that the finished product poses no risks to the general public.

An effective program consists of three parts, raw material testing, process testing, and finished goods testing. Raw material testing occurs at the receiving area where samples are randomly drawn from different lots of the ingredient as they are received. Only ingredients that have been identified as possible risk containing should be tested. For those ingredients that have a certificate of analysis (C of A) for each lot, the lot numbers should be compared to the C of A prior to use. The drawn samples should be tested prior to use. If there is a concern about the perishability of the ingredient, then the ingredient should be processed and the product put on hold pending results of the raw material testing and any other tests conducted on the process or finished good. This is part of the positive release program.

Process testing involves sampling various points along the food contact surfaces. Normally this practice is done strictly as a means to locate potential microbial sources after a positive product test has been determined. Due to the time lag between sampling and results it is not a practicable exercise to test the process prior to each run.

Finished goods testing is used by companies as a final measure to verify that the run or lot is safe for human consumption. It requires the company to randomly sample the finished products during the run and then test them. When they are clear, the product is released for sale. For most companies this is a critical component of the positive release program.

After determining what, when, and where to sample, the next decision is what methodology to use for microbiological testing. Currently there are two basic approaches, in-house and contract laboratory. In-house microbiological testing may take the form of traditional plating or rapid testing methods. Traditional plating requires the company to invest in plates, spreaders, incubators, water baths, media, an autoclave, and other testing materials. It also requires an employee that has at least minimal lab and microbiological experience to prepare, run, and read the plates. While this method can be more specific and more accurate, it is also more material and labor intensive. For a small company this can be cost prohibitive.

In-house rapid testing utilizes 3M Petri films for organism identification and enumeration. These films are impregnated with a dry media that is hydrated with the sample and left to incubate for a set period of time prior to reading. The plates are available to identify various organisms including *E. coli*/coliform, aerobic plate count, yeast and mold, *enterobacteriaceae*, and *Staphylococcus aureus*, and are available at http://solutions9.3m.com/wps/portal/3M/en_US/Microbiology/Food-Safety/Products/Petrifilm-Plates/. A basic startup kit from 3M that contains the components needed to begin micro testing can be found at http://solutions9.3m.com/wps/portal/3M/en_US/Microbiology/FoodSafety/Products/Petrifilm-Accessories/Laboratory-Start-Up/. It includes a book of directions for plating samples, reading the plates, and interpreting the results.

The second approach companies utilize for micro-testing, is the use of a contract laboratory. Companies utilize contract laboratories when quality control time and labor resources are at a premium, when samples are taken from various locations, or when there is a potential to grow pathogens. Concerning the latter, it is critical that unless the correct handling procedures are in place, companies do not grow pathogens. Successful growth of pathogens might aimlessly contaminate the rest of the plant and products. Pathogen growth should be left to testing and growth by contract laboratories and/or trained professionals.

The final component of the company's microbiological program requires the development of the microbiological testing log and the program document. The testing log is the form used to document all of the non-environmental tests that are conducted by the company. It includes places for the test date, item number, description of the sample or location the sample was taken, the dilution of the sample, the tests performed, and the results determined, and finally, whether the product was released and by whom. This documentation should be placed in the testing notebook for easy access.

The program document is similar to other program documents in that it involves explaining why the program exists, what is tested, how often it is tested, the testing method used, who performs the tests, and who is responsible for the program. The program document should be stored under the Microbiology Program tab section of the Quality Manual. CD (Book 2_Quality Control Manual\Section 21_Microbiology Program:General Overview) and at the end of the chapter may be used to documet program.

APPLICATION

CD(Book2_QualityControlManual\Section21_MicrobiologyProgram:Microbiology Testing Log) and at the end of the chapter provides the company with a customizable microbiology testing log. The tests performed can be changed to reflect the actual tests conducted.

SUPPLEMENTAL MATERIALS

MICROBIOLOGY PROGRAM
GENERAL OVERVIEW

Section A: General

_____ shall develop and maintain a program that effectively eliminates the presence of pathogenic bacteria in all finished goods released for sale. This program has three parts, raw material testing, process testing and finished goods testing.

Section B: Raw Materials Testing

Representative samples will be taken from ingredients that possibly contain bacterial contamination. For those that a certificate of analysis is required it shall be obtained and its microbiological cleanliness verified.

Section C: Process Testing

When needed for poblem solving, micro samples shall be taken from various points within the food contact process. These shall be utilized to determine routes of contamination to the finished goods.

Section D: Finished Goods Testing

All finished goods shall have a representative set of samples during the production. These shall be tested and the product released when they are deemed clean.

Section E: Methodology

_____ all tests shall be conducted using 3M petrifilm. All pathogen testing shall be completed using contract laboratories.

Section F: Responsibility

The microbiology program shall be developed and administered by the quality control manager.

Approved by: _____ Date: _____

Microbiological Testing Log

Test Date	Item Number	Description or sample location	Dilution	APC		Yeast		Mold		E. coli / coliform		Staph		Released	By
				test	results	test	results	test	results	test	results	test	results		

26 Security/Biosecurity Program

PROGRAM TYPE: REQUIRED

THEORY

Recent events in world history have brought to everyone's attention the important role of security with regard to our food supply and its plethora of manufacturers, processors, transporters, distributors, importers, exporters, and sellers. Although the United States does have the safest food supply in the world, each company is left with the challenge of determining how to maintain this safety over and above the normal microbiological, chemical, physical, and allergen risks traditionally found in the food supply. Therefore, in order to provide a safe and secure environment for its employees, protect the safety of the food supply, and ensure compliance with the Public Health Security and Bioterrorism Preparedness Act of 2002, each company should develop a comprehensive biosecurity program. This program should be focused on all areas of the company where there is a potential for tampering or other malicious, criminal, or terrorist actions and is divided into five areas: management, staff, the public, facility, and operations.

MANAGEMENT

Company management and its attitude toward biosecurity are critical in establishing a solid front against any type of malevolent actions from personnel, customers, or other outside forces. As the foundation of the biosecurity program, management performs several time-consuming yet vital functions, beginning with designating an employee with the task of overseeing and implementing all aspects of the biosecurity program. Typically, this roll is assigned to the plant manager, although it may be assigned to the quality control manager or any other senior management member. The management member to whom this function is assigned should report directly to the president or an appropriate vice-president due to the potentially sensitive nature and wide scope of the responsibility.

After assigning responsibility for the biosecurity program at the facility, management should asses the risks associated with the staff, public, facility, and company operations. This risk assessment should be completed with a focus on preparing for and responding to tampering and other malicious, criminal, or terrorist actions (both threats and actual events), including identifying, segregating, and securing affected products. The outcropping of this risk assessment is the development or reinforcement of a formal and structured company recall program. This essential program is dealt with in an earlier chapter and includes the needed supporting materials and documents to develop and implement the program adequately. These can be found on the CD (Book 2_Quality Control Manual\Section 16_Recall Program:Recall Program Overview) and at the end of that chapter.

Next, management should develop and document a plan for emergency evacuations. Depending on where the facility is located and the types of emergencies the plant may face, the plan may be specific to single occurrences such as tornados or earthquakes or it might be generic and address all emergency types, including criminal employee actions such as fire, environmental contamination, or physical acts of violence. In all cases, however, the plan should address the people inside the facility, product inside and outside the facility, and the facility itself.

When considering the people inside the facility during an emergency, the decision of whether the employees should leave the facility immediately must be addressed. If this is the case, then an appropriate alarm or warning system must be installed and adequate directions given to the evacuating employees. This should include maps of the facility strategically placed near the exit doors and on employee bulletin boards showing employees the most direct routes to the exits in case of emergency. To facilitate a timely and safe evacuation of the premises, all employees should be trained. This training document should be placed in their personnel files.

During any type of evacuation, management must take care to understand and prepare for security breaches by other employees. These breaches include minor instances such as basic theft and major breaches such as malicious tampering or adulteration of the product. Generally, these types of situations occur in facilities where there is the potential for labor unrest, but they can be done by your basic disgruntled employee who thinks he or she will not be caught due to the confusion of the situation. One method to prevent these breaches is to assign a manager to monitor each exit door and other staff members to monitor specific groups of employees. This strategy provides the advantage of ensuring that all the employees and visitors have evacuated as well as acting as a deterrent for employees who want to perform malicious acts because they know they are being monitored by management.

Other things that members of management can do during the development of the biosecurity plan include:

Maintain a floor plan in an off-site location. This provides a reference for reconstruction in case of fire or other natural disaster and can be used by emergency personnel in situations where an employee is trapped or held hostage.
Understand the emergency response system within the community in which they live. Most cities and towns have emergency response systems that are activated during times of actual or potential natural, social, or environmental disasters. Knowing and understanding these systems help the company to protect its people and facility.
Establish a list of emergency numbers. This should include 24-hour contact information for local, state, and federal police, fire, rescue, health, and homeland security agencies. It should be made available to all management personnel and an abbreviated version posted on the employee bulletin board for reference.

- Educate personnel of whom in management they should alert about potential security problems. Employees are the first line of defense for most breaches of security. They see nonemployees outside the facility and normally enter the facility before management; they can take the "pulse" of disgruntled employees or potential labor unrest. Employees are the eyes and ears of the company and getting them involved with the biosecurity program acts as a front line defense.
- Educate personnel to alert the appropriate manager concerning any signs of tampering or other malicious, criminal, or terrorist actions or areas that may be vulnerable to such findings. Teaching the employees to be observant about what is abnormal behavior, the company's policy on family and other's visitations, and which employees are allowed in which areas is critical to having an informed, observant staff.
- Provide adequate management supervision at all times. When employees are not properly supervised, more opportunity is available to breach the company's security. This includes letting people into the plant after hours, bringing personal items into the plant, and providing unsupervised chances to contaminate or adulterate the product stream.
- Conduct routine security checks of the premises. This includes all outside doors, docks, computers, and utilities for signs of tampering or malicious acts.

HUMAN ELEMENT—STAFF

In today's modern society, with people traveling tens, hundreds, and sometimes thousands of miles to find employment, companies cannot simply continue to trust that the employees they hire have the same dedication to and concern for the company as in times past. Companies must also protect themselves from unscrupulous competition and outside international influence such as globalization and undocumented workers. To combat these types of problems, care should be taken to vet each potential employee properly and manage employee access to sensitive areas and information.

Under federal law, food establishments are required to verify the employment eligibility of all new hires, in accordance with the requirements of the Immigration and Nationality Act, by completing the INS employment eligibility form (INS form I-9). Completion of form I-9 for new hires is required by 8USC1324a and nondiscrimination provisions governing the verification process are set forth at 8USC1324b. To adhere to this standard, all employees should complete the I-9 form and undergo a background check prior to employment, including but not limited to reference checks, criminal background checks, credit checks, and drug checks where deemed appropriate for the security needed for the position applied for.

Furthermore, when temporary employees are used, they should also be required to undergo a similarly thorough background check, including drug and criminal history checks. To support this function, most staffing services can and will provide this type of background check prior to dispatching an employee to the jobsite. This is very important due to the transient nature of the workforce and the advent of smaller and easier methods of destruction and piracy.

Employees who leave voluntarily or involuntarily also should be debriefed for security reasons. Any keys, computers, telephones or other electronic devices, credit cards, uniforms, and miscellaneous documents, books, and files should be recovered during this exit interview. This is a critical step for security reasons, even if the employee is leaving on good terms, because sometimes the employee's new situation will change and he or she will look for ways to profit from previous knowledge and tools.

Management should identify those employees who require unlimited access to the facility to conduct the duties of their position properly. Access to areas where there are sensitive processes or information should be limited to those employees who require it. The company management should do a periodic review of access patterns and responsibilities with an eye on limiting open access to proprietary and confidential information.

As an aid in limiting access, all keys to doors, filing cabinets, computers, and machines should be maintained in a central repository and signed out to individuals as needed. A key log can be utilized to record which employees have which keys. The advantages of this system is that it assists in debriefing employees who are leaving the company, assists in the security of information and equipment, and acts as an aid for people to know to whom to go when a locked door, file, or office needs to be opened.

In addition to the preceding precautions, management should establish and enforce rules related to the types and storage location of personal items. These rules normally are part of enforcing good manufacturing procedures (GMPs) with employees and visitors. GMPs have been dealt with in a previous chapter and the supporting program and training documentation can be found on the CD (Book 2_Quality Control Manual\Section 3_GMP) and at the end of that chapter.

As a means of supporting the importance of biosecurity within the manufacturing facility, management needs to conduct yearly refresher training on security awareness, including how to prevent, detect, and respond to tampering or other malicious or criminal actions or threats. These training sessions might involve management presenting information on a biosecurity topic or inviting an outside expert in to cover a specific subject. In either case, this training should be documented and any training materials retained for further sessions or for new employee orientation.

Lastly, members of management need to be aware of what is happening within and outside their facility. They should watch for any unusual or suspicious behavior by employees. Some situations that management should be aware of that potentially could lead to a security risk are an employee who, without an identifiable purpose, stays late after the end of a shift or arrives unusually early; who accesses files, information, or areas of the facility outside the areas of his or her responsibility; who asks questions regarding sensitive or proprietary subjects; or who brings cameras to work. Each of these situations should raise red flags with management and the employees should be confronted to assure the company that there are no potential risks.

HUMAN ELEMENT—THE PUBLIC

Because no company exists in a vacuum, a company should be prepared for visitors at any time. A visitor is any person not on the company payroll who enters the plant by any means. This includes, but is not limited to, contractors, regulatory personnel, tours, vendors, suppliers, and family. Generally, these guests have not been vetted and have not passed the standard security screening, making them a higher security risk for the company. To minimize this risk, the company should take the following four steps:

1. Have all visitors sign in on a visitor log. This should be located at the entry to the facility (normally, with the receptionist).
2. Have all visitors show identification upon entering the facility. This can be accomplished with a business card but photo identification is better.
3. Limit access to nonreceptionist areas by having all visitors accompanied by a management representative or approved employee. To make it easy for all company employees to know that visitors are in the facility, some companies utilize a colored hat or a visitor badge system.
4. Limit admittance to areas with personal items to employees only. This includes locker rooms and private offices and helps to restrict access to items that can be stolen.

FACILITY

Open or unlimited entry into the facility is one of the most difficult aspects related to biosecurity that a company faces on a daily basis. This is due to the fact that most companies have shipping and receiving drivers, contractors, and suppliers on site daily. Combine this with the many combinations of exit and entry doors found in most buildings and facility security can become a sieve. To combat this openness, a company should keep all external doors locked at all times from the outside with the exception of the front door leading to the receptionist area. Entry into the facility should be done via key or employee key card. Front door access should be limited via a confined alcove or a two-door system. Shipping and receiving drivers should be limited in access to the facility by a bell system, a designated driver's area, or by not allowing them access to the dock. All rollup dock doors should be kept closed at all times.

Other considerations regarding facility security that management should consider include:

1. Secure all outside equipment and compressors to prevent tampering and interruption of power or service.
2. Install inside and outside lighting that is adequate to facilitate the detection of suspicious or unusual activities.
3. Parking lots should be separated from all food storage areas and, where appropriate, visitor and employee lots should be designated and separate.
4. Maintain chemicals and gasses under lock and key. Limit access to these to necessary personnel only.
5. Is the property completely fenced to prevent unwanted access? If it is, are employees required to show identification to enter?

Finally, facility security should be part of the monthly inspection program. This is discussed in a previous chapter and supporting material can be found on the CD (Book 2_Quality Control Manual\Section 7_Inspection Program) and at the end of that chapter.

OPERATIONS

When considering operations security, companies should evaluate their internal processes with an eye toward prohibiting any type of tampering or criminal activities. These might include:

1. Install an effective means of inbound and outbound carrier inspections. These programs are discussed in detail in earlier chapters with the supporting materials found on the CD (Book 2_Quality Control Manual\Section 13_Receiving Program and Book 2_Quality Control Manual\Section 14_Shipping Program) as well as at the end of the pertinent chapters.
2. Require all suppliers to take necessary precautions to prevent contamination of ingredients and packaging while in their possession or in transit. This can be done during the supplier certification program. This is discussed in an earlier chapter with the supporting material found on the CD (Book 2_Quality Control Manual\Section 17_Supplier Certification Program) and at the end of that chapter.
3. Restrict access to computers and the information contained therein. Utilize passwords for entry onto any given computer and cable tie downs to secure the physical unit to the facility.
4. Store all company records in an organized locked closet, file, or room. Limit access to these areas.
5. Back up all computer records on a nightly basis. Store a copy of these backups off site.
6. Ensure that all products manufactured and distributed can be traced by individual lot from their source ingredients through the process to the customer. This is done by a comprehensive lot tracking system and is discussed in detail in a previous chapter. The supporting forms for this program can be found on the CD (Book 2_Quality Control Manual\Section 11_Lot Coding Program) and at the end of that chapter.
7. To support the lot tracking system, develop a recall system that can trace and return, if necessary, all products produced. A recall program is discussed in a prior chapter and the supporting materials are found on the CD (Book 2_Quality Control Manual\Section 16_Recall Program) and at the end of that chapter.

APPLICATION

The first step involved in developing a security/biosecurity program is to conduct a survey of the current practices conducted within the company. This should involve exterior and interior door locking practices, employee traffic patterns and schedules, filing practices, computer locations, personal property locations, visitor handling, and shipping and receiving practices. After these elements are understood, the risk involved with each should be assessed and determination made as to how the company can sufficiently prevent any malicious or criminal behavior, acts of vandalism such as tampering or adulteration, and terrorist actions. The company's security plans should be documented on a security program general overview. A sample general overview can be found on the CD (Book 5_Other Programs\Section 1_Security_Bioterrorism:General Overview) and at the end of this chapter. This sample identifies the company policy with regard to the five general areas: management, staff, the public, the facility, and company operations. It can be modified to match exactly what the company is currently doing to ensure security.

Next, all employees, old and new, should be trained in the company's security and biosecurity programs. A training verification document should be used to document this training. An example of this is found on the CD (Book 5_Other Programs\Section 1_Security_Bioterrorism:Security Biosecurity Training Verification) and at the end of this chapter. These should be signed by the employee and placed in his or her personnel file.

When the company decides to regulate access to the facility, one of the initial steps is the development of a key log. This is just a tabulation of which employees have keys to doors, cabinets, and equipment. A sample of a key log can be found on the CD (Book 5_Other Programs\Section 1_Security_Bioterrorism:Key Check Out–Check In Log) and at the end of this chapter. This log should be maintained and all excess keys should be kept locked in a safe place.

Finally, a visitor log can be set up for all visitors to the facility to sign as a method to document their comings and goings. An example of this log is located on the CD (Book 5_Other Programs\Section 1_Security_Bioterrorism:Visitor Sign In Log) and at the end of this chapter. It can be placed on a clipboard or in a three-ring binder. If the latter approach is utilized, a binder cover and spine label are located on the CD (Book 5_Other Programs\Section 1_Security_Bioterrorism:Visitor Cover and Spine Labels) and at the end of this chapter.

SUPPLEMENTAL MATERIALS

SECURITY PROGRAM

GENERAL OVERVIEW

Section A: General

In order to provide a safe and secure environment for its employees, protect the safety of the food supply, and to raise awareness of security issues related to the promulgation of the Public Health Security and Bioterrorism Preparedness and Response Act of 2002, _____ will institute and maintain all measures necessary to ensure the safety of its food products. Such measures will be focused on all areas of the facility where there is a potential for tampering or other malicious, criminal, or terrorist actions and are divided into five areas: management, human element—staff, human element—the public, facility, and operations.

Section B: Management

1. The role of security will be assigned to the plant manager and the quality control manager reporting directly to the president and vice-president.
2. Management will:
   a. assess the risks associated with the staff, facility, public, and products;
   b. develop a recall program to prepare for and respond to tampering and other malicious, criminal, or terrorist actions, both threats and actual events, including identifying, segregating, and securing affected products;
   c. plan for emergency evacuation, including preventing security breaches during evacuation;
   d. maintain a floor plan in an off-site location;
   e. understand the emergency response system in the community;
   f. establish a list of emergency numbers including 24-hour contact information for local, state, and federal police, fire, rescue, health, and homeland security agencies and make it available to all management personnel;
   g. educate personnel of whom in management they should alert about potential security problems;
   h. educate personnel to be alert to any signs of tampering or other malicious, criminal, or terrorist actions or areas that may be vulnerable to such actions and to report any findings to the appropriate manager;
   i. provide adequate management supervision at all times; and
   j. conduct routine security checks of the premises, including outside doors, dock, computers, and utilities, for signs of tampering or malicious or terrorist actions.

Section C: Human Element—Staff

1. Under federal law, food establishments are required to verify the employment eligibility of all new hires, in accordance with the requirements of the Immigration and Nationality Act, by completing the INS Employment Eligibility Form (INS Form I-9). Completion of form I-9 for new hires is required by 8 USC 1324a and nondiscrimination provisions governing the

verification process are set forth at 8 USC 1324b. To adhere to this standard all employees will
   a. complete the I-9 form; and
   b. undergo a background check prior to employment, including but not limited to reference checks, appropriate to the security needed for the job applied for.
2. All temporary employees will also be required to undergo thorough background checks, including but not limited to criminal history.
3. When an employee leaves the company, for any reason, uniforms and keys will be obtained before the employee leaves the premises.
4. Management will identify staff members who require unlimited access to the facility. Individual employee access will be reassessed periodically.
5. All keys to the facility will be checked out on a key log when distributed and checked in when returned.
6. Management will establish and enforce rules related to the types and storage location of personal items.
7. Management will provide yearly refresher training on security awareness, including how to prevent, detect, and respond to tampering or other malicious or criminal actions or threats.
8. Management will watch for unusual or suspicious behavior by staff (for example, an employee who, without an identifiable purpose, stays late after the end of a shift; arrives unusually early; accesses files/information/areas of the facility outside the areas of his or her responsibility; removes documents from the facility; asks questions on sensitive subjects; or brings cameras to work.

Section D: Human Element—the Public

1. All visitors (contractors, supplier representatives, delivery drivers, customers, couriers, pest control representatives, regulators, tours, etc.) will sign in.
2. All visitors must provide some form of identification if they are not otherwise known.
3. Without prior approval, no visitor will have access to the production area or warehouse without accompaniment of a management representative.
4. No non-employee, unless otherwise approved, may have access to the locker room.

Section E: Facility

1. All doors will be locked at all times, except for the receiving door and the front door. The rollup doors are to be closed unless in use.
2. All compressors and outside equipment will be secured to prevent tampering and interruption of power and/or service.
3. Lighting will be adequate to facilitate detection of suspicious or unusual activities.
4. Employee parking and food storage areas will be separate to prevent unwanted activity.

5. All propane and kerosene are to be stored in locked cabinets. Access to keys will be limited to designated personnel.
6. Security will be a part of the monthly inspection program.

Section F: Operations

1. An effective means of inbound and outbound carrier inspection will be in place.
2. All suppliers will practice appropriate security measures to ensure the safety of the product/packaging that is purchased.
3. Truck seals will be inspected, verified, and the number recorded for inbound shipments.
4. Truck seals will be installed on full truckloads at the customer's discretion. This number will be recorded on the bill of lading.
5. Product that is suspect will be rejected before unloading.
6. Computers will have passwords to restrict entry to sensitive data.
7. All records will be stored appropriately in a secure place.
8. A nightly backup of the server data will take place.

Section G: Responsibility

All facets of this program will be administered by the plant manager and quality control manager.

Approved by: _____ Date: _____

VERIFICATION DOCUMENTATION

Security/Biosecurity Training

I, _____, have read and understand the section on security/biosecurity at _____. I understand that:

    all of my personal items are to be kept in the designated area;
    all outside doors are to be kept locked unless otherwise designated;
    no non-employee visitors are to be allowed in the facility without prior management approval;
    no equipment or records are to be taken from the facility without prior approval; and
    contamination and malicious behavior are serious with regard to _____ and will do everything possible to prevent it.

    Furthermore, if I see another employee violating any of _____'s security rules I will immediately notify a member of management. If I have questions regarding this policy, I will immediately notify my supervisor.

Date: _____

Name: _____

Instructor: _____

Key Check Out — Check In Log

Key #	Key	Date Out	Employee Name	Employee Signature	Date In

VISITOR SIGN IN SHEET

	Visitor Name	Company Name	Person to See	Badge Number	Badge Returned	Time In	Time Out
1							
2							
3							
4							
5							
6							
7							
8							
9							
10							
11							
12							
13							
14							
15							
16							
17							
18							
19							

VISITOR SIGN IN LOG

VISITOR SIGN IN LOG

27 Kosher Program

PROGRAM TYPE: OPTIONAL

THEORY

In today's complicated world of branded and generic products, niche and general categories, and consumer-specific products, many companies are turning toward finding ways to market their products to specific segments of the population. These segments can be split along various lines, including age, sex, eating habits, politics, race, and religion. One subgroup of the religious niche that increasingly is getting attention from large and small companies is the population that purchases based on the product having a kosher symbol. To attract this population to purchasing their products, companies pursue kosher certification.

Kosher certification means different things to various parts of this segment. For an observant Jew, it indicates that the product contained within the package was produced according to kashruth standards. Other religious groups, such as the Muslims and Seventh Day Adventists, also rely on the kosher symbol, as do some vegetarians and a portion of the non-Jewish population that believe that the kosher mark suggests that the product was manufactured in a cleaner or better manner than non-kosher products. Although the the products are manufactured kosher, the impression that they are cleaner or better is not necessarily so because products bearing a kosher mark are, by law, to be produced in the same food safety manner as those that do not.

In order for a company to obtain kosher certification it first needs to determine which of the more than 75 available kosher certifying bodies it wishes to utilize. In essence, it needs to ask which kosher hekhsher, or mark, it wants on its packages. This choice is usually dictated by where the company is located, the geographic regions in which it sells its products, and the customers to whom it sells its products. For instance, if the company sells its products in all states and internationally, it might want to contract with one of the more recognized agencies, such as Orthodox Jewish Congregations (OU), Star-K Kosher Certification, Organized Kashrus Laboratories (OK), or K-of-K Supervision. If, on the other hand, it sells to a regional market, then a smaller, more locally recognized certifying body might be used. For instance, if a company was from Chicago and sold mostly to Chicago and the surrounding area, then an appropriate choice might be the Chicago Rabbinical Council (cRc). A company selling to Southern California might use Kahillah Kosher (Heart-K), and a company selling in Atlanta might use the Atlanta Kashruth Commission.

The company's customers might also dictate the mark a company utilizes if the product is made exclusively for them. In this case it is easier sometimes for the customer to have the certifier certify the manufacturer, rather than to contract with a secondary certifier. In any case, the choice is quite varied.

Another question to be asked when considering the choice of certifiers is how well the mark is accepted by the target market. As with all brands, kosher marks are not all created equal; some certifiers adhere more stringently to the kashruth standards than others. When making the choice of certifiers, the company must take all these things into consideration prior to the final decision because changing to another certifier involves changing all the packaging and marketing materials and can therefore be time consuming and expensive.

Once the decision is made of which kosher certifier is to be used, the company contacts him or her and fills out an application. The certifier will usually at this point request some financial and sales data for the products to be certified. This is what the basic charge for certification is based on. This does not include the travel expenses the rabbi will submit for payment. Depending on the size of business and sales volume, this charge varies from hundreds to tens of thousands or even hundreds of thousands, but at any level the costs associated can make a company question whether it is worth the presumed sales increase. Only the company's management can decide this.

The next step in becoming certified is for the company to submit to the certifying body current kosher certificates for all of the ingredients used in its process. These are readily obtained from the ingredient manufacturer and can be faxed or digitally sent to the certifier. During this process, the company might find that some of the ingredients it uses are not from kosher-approved suppliers or that ingredients are not able to be kosher. In these cases, the company must find a new supplier or a new ingredient, or decide not to pursue kosher status on the finished products that contain these ingredients.

Upon approval of all the ingredients as certified kosher, a rabbi from the certifying agency will visit the plant and will ask questions regarding the cleaning methods, the process, the products manufactured at the plant, and the suppliers. The rabbi will also look at the packaging of the ingredients and verify that the kosher certificate supplied matches the ingredients and that, if it is to have a kosher mark, on the package it does. The purpose of this is not to "bless" the food, but rather to help the rabbi understand the products and processes to ensure that the product is manufactured in accordance with Jewish dietary laws. In some cases the rabbi may ask that a cleaning or processing method be changed or that a certain temperature be obtained; when this happens, try to accommodate the request in any reasonable manner.

After approval of the facility and the ingredients, the certifier will give the company approval to use the appropriate mark. Depending on the type of product, the process, and the ingredients, the mark might be pareve, meat, dairy, or dairy equipment. This should be added to the appropriate packaging and, when a proof is available, the artwork should be submitted to the rabbi for approval.

The final and continuing obligation of the company has two parts: certificate continuity and ingredient continuity. Certificate continuity involves gathering updated certificates from each ingredient when they expire and submitting them to the kosher agency for approval. Because most ingredients certificaitons expire on different dates during the year, this can be a tedious and labor-intensive task. Ingredient continuity requires setting up a system at the ingredient receiving station that double checks that the packaging bears the kosher mark as set forth by the supplier's certificate. One way to accomplish this is to develop a list of the ingredients with the approved source and the marks they must bear and have this located in the receiving area for reference.

APPLICATION

The decision for a company to apply for kosher certification begins with contacting a certifying agency. *Kashrus Magazine* (http://www.kashrusmagazine.com) is a great resource for finding a certifying agency and understanding kosher certification. On this Web site, the company can order a guide for kosher certification.

Once the certifying body is chosen, the certificates are collected and submitted, the manufacturing facility is inspected, and the company is approved, a master list should be created for the receiving area. The form on the CD (Book 5_Other Programs\Section 2_Kosher Certification:Approved Kosher List) and at the end of this chapter can be used as the basis for creating this list. To complete it, fill in the item number, ingredient name and the name of the supplier. Make sure the approved source is the manufacturer name that appears on the ingredient package. Then fill in the certifying agency. This should be the certifier's name on the kosher certificate supplied by the ingredient supplier. Finally, fill in the kosher symbol that should be on the package and any modifier. These also will be located on the ingredient package. Place the completed chart inside the receiving book for reference when ingredients are received as a method to ensure compliance with the agreed upon kosher program.

SUPPLEMENTAL MATERIALS

Receiving Kosher List

Item Number	Ingredient	Approved Supplier	Certifying Agency	Kosher Symbols Required	Dairy / Meat / Parve

28 Organic Program

PROGRAM TYPE: OPTIONAL

THEORY

The United States has one of the safest, cleanest food supplies in the world. Its consumers enjoy a vast variety of native and non-native, processed and unprocessed, and natural and non-natural products. This food supply has been significantly increased recently by the use of targeted and increased pesticide use and crops that have been genetically altered to increase or add certain desirable properties. With this increasing use of technology, many questions have arisen regarding how safe these products are for human consumption. This concern has spurred a backlash from the general population against this technology and has driven companies to undertake the manufacturing of products that do not contain chemicals and are not genetically modified.

The recognition by food companies of the desire for these types of clean products has led organic products to be one of the fastest growing segments of the food market. The organic movement is so strong that even larger companies are rethinking their product lines and introducing organic products. As such, these types of products are rapidly ceasing to be a niche and becoming more mainstream.

Companies that want to produce organic products that are labeled as "organic," "made with organic," or "100% organic" must follow all of the regulations as set forth by 7CFR205 of the National Organic Program (NOP) (http://ecfr.gpoaccess.gov/cgi/t/text/text-idx?c=ecfr&tpl=/ecfrbrowse/Title07/7cfr205_main_02.tpl).Failure to do so will result in civil penalties outlined in section of 7CFR3.91(b)(1)(xxxvii). Compliance with these regulations initially requires the company to evaluate how it cleans, the chemicals utilized for cleaning, and how pest control is conducted. These underlying programs and the chemicals they use dictate whether the company might be exposing the finished product to NOP noncompliant substances.

After reviewing these programs and deeming them to be in compliance, the company turns to ingredient and label compliance. The main portions of this governmental program for company compliance are laid out in compliance sections: section D—Labels, Labeling and Marketing Information and section G—Administrative. To comply with these sections, a company needs to gather ingredient information, calculate the percentage of organic material in the product, determine the organic statement, obtain certifier approval, and develop an organic plan.

Gathering ingredient information entails contacting each ingredient supplier and requesting a copy of its current organic certificate for the ingredient. This request is part of the supplier information gathering phase of the HACCP program development. During this phase, the ingredients may not have been reviewed with relation to their organic status. To assist manufacturers, the NOP provides guidance with regard to which ingredients can be approved as organic and those that cannot. This list is found in section 205.600 of the act; specifically, 205.605 lists those nonagricultural (nonorganic) substances that are allowed as ingredients in or on processed products labeled as "organic" or "made with organic" (specified ingredients or food groups) and 205.606 lists nonorganically produced agricultural products allowed as ingredients in or on processed products labeled as "organic."

When the certificates are received from the supplier, it is necessary to verify that the certificate is current and certified by a USDA-approved certifier. After obtaining all the current organic certificates for a given product, the percentage of organic material in the formula must be calculated based on one of the following methods:

Divide the total net weight (excluding water and salt) of combined organic ingredients at formulation by the total weight (excluding water and salt) of the finished product.

Divide the fluid volume of all organic ingredients (excluding water and salt) by the fluid volume of the finished product (excluding water and salt) if the product and ingredients are liquid. If the liquid product is identified on the principal display panel or information panel as being reconstituted from concentrates, the calculation should be made on the basis of single-strength concentrations of the ingredients and finished product.

For products containing organically produced ingredients in both solid and liquid form, dividing the combined weight of the solid ingredients and the weight of the liquid ingredients (excluding water and salt) by the total weight (excluding water and salt) of the finished product.

To arrive at the percentage of all organically produced ingredients in an agricultural product, it is necessary to round down to the nearest whole number. Based on the calculation, the company determines which of the following labeling categories the product is eligible for:

205.301a—100% organic: products sold, labeled, or represented as "100% organic." A raw or processed agricultural product sold, labeled, or represented as 100% organic must contain (by weight or fluid volume, excluding water and salt) 100% organically produced ingredients.

205.301b—organic: products sold, labeled, or represented as "organic." A raw or processed agricultural product sold, labeled, or represented as organic must contain (by weight or fluid volume, excluding water and salt) not less than 95% organically produced raw or processed agricultural products. Any remaining product ingredients must be organically produced, unless not commercially available in organic form, or must be nonagricultural substances or nonorganically produced agricultural products produced consistent with the national list in subpart G of this part.

205.301c—made with organic: products sold, labeled, or represented as made with organic (specified ingredients or food groups). Multi-ingredient agricultural products sold, labeled, or represented as made with organic (specified ingredients or food groups) must contain (by weight or fluid volume, excluding water and salt) at least 70% organically produced ingredients that are produced and handled pursuant to requirements in subpart C of this part.

205.301d—products with less than 70% organically produced ingredients. The organic ingredients in multi-ingredient agricultural products containing less than 70% organically produced ingredients (by weight or fluid volume, excluding water and salt) must be produced and handled pursuant to requirements in subpart C of this part.

After reviewing the ingredients, calculating the final organic percentage, and deciding what classification the product falls into, the company needs to contract with an approved certifier. A list of approved certifiers can be found at http://www.ams.usda.gov/NOP/CertifyingAgents/Accredited.html. Like other services contracted by the company, it is important to shop around to determine which organic certifier fits with the company's philosophy. Within this decision should be how closely the certifier adheres to NOP regulations, whether the certifier actively monitors developments related to organic products and regulation, and whether the certifier properly reviews each formula and inspects each facility. Although all certifiers are not created equal, by law they must recognize the certification of other approved certifiers and by default their certificates.

Having chosen a certifier, the company and certifier will sign a contract and confidentiality agreement prior to any formula sharing between the parties. The initial requirement of the certifier will be for the company to develop an organic system handling plan. This is a plan of management of an organic production or handling operation that has been agreed to by the producer or handler and the certifying agent and that includes written plans concerning all aspects of agricultural production or handling described in the act. During the development process of the plan, the certifier will contact the company and discuss the operation and ingredients. It might also request that the company complete an informational survey outlining its management structure and emergency contact information, cleaning procedures and chemicals, pest control practices, and suppliers' practices where important. From this information the company and certifying agent will develop a final plan that meets the standards of the NOP prior to inspection of the facility.

According to 205.403, the certifying agent must conduct an initial on-site inspection of each production unit, facility, and site that produces or handles organic products and that is included in an operation for which certification is requested. An on-site inspection should be conducted annually thereafter for each certified operation that produces or handles organic products for the purpose of determining whether to approve the request for certification or whether the certification of the operation should continue. This visit is to be conducted within a reasonable amount of time from when the certifier deems that the company can be in compliance, but it may be within up to 6 months. During this visit the certifier will verify the operation's compliance or capability to comply with the act; that the information, including the organic production or handling system plan provided, accurately reflects the practices used or to be used by the company for certification; and that the prohibited substances have not been or are not being applied to the operation. Means of ascertaining this are at the discretion of the certifying agent and may include the collection and testing of processed product samples.

Subsequent to plan approval and inspection, the company should submit the formula for each product for which it is requesting organic certification. This formula should be formatted based on organic percentage to the thousands place, with water and salt removed from the calculation. It should also be stated on this submission what claim—100% organic, organic, or made with organic—is to be made. Accompanying this submission should be a copy of the product ingredient statement and copies of each of the current certificates for the organic ingredients. The certifier will review all of the documentation and either grant approval for the use of the requested moniker or request more information. It is important for the company to obtain this approval in writing as a reference for review as needed at a later date.

After approval has been granted, the company can proceed to labeling the product in accordance with the requirements of the act based on the percentage of organic material contained in the product. If the product is 100% organic or organic (95% organic material), the company can (1) use the terms "100% organic" or "organic," as applicable, to modify the name of the product; (2) use the term "organic" to identify the organic ingredients in multi-ingredient products labeled "100% organic" or "organic"; and (3) utilize the USDA and the logo of the certifier on its package. If the product is utilizing a "made with organic" claim, then the company can (1) use the term "made with organic," (2) use the term "organic" to identify the organic ingredients in multi-ingredient products, and (3) utilize the logo of the certifier on its package. In products that contain less than 70% organic material a company can only use the term "organic" in the ingredient statement to describe those organic ingredients. The USDA logo or certifier logo may not be used. When a company makes organic claims on its packages, it is recommended that it consult its certifier to ensure compliance because, in all cases where a claim and/or a logo is to be used, there are font size constraints related to package design. The certifier can help the company navigate these regulatory waters.

The final step for a company developing an organic program is record keeping. This involves managing the organic plan, submission documents, and approvals. One easy method for record management is the use of a separate filing system. The first file in the system should contain the organic plan and the following files should be organized by product item number and house a copy of the submitted formula, ingredient certificates, and the documented approval from the certifier. It is important to keep these records separate from the actual formula files because they are open to inspection by both the certifier and regulatory officers.

APPLICATION

A company begins the development of the organic program by evaluating its cleaning and pest control procedures. These are discussed in earlier sections. During this evaluation, care must be taken to choose cleaning chemicals and a pest control system that are compliant with the organic regulations. The company should discuss its desire to manufacture organic products with its pest control contractor and sanitation chemical provider. Both of these companies will be able to direct any changes necessary to ensure full compliance with the NOP.

Next, the organic certificates should be collected for all the ingredients. This can be part of the initial supplier documentation done as part of the HACCP program development as a separate collection step. These should be placed in the ingredient files as they are obtained. After all of the certificates for a particular formula are accumulated, a formula sheet should be filled out for future submission to an organic certifier. The submission form on the CD (Book 5_Other Programs\Section 3_Organic Certification:Organic Formula Submission Form) and at the end of this chapter can be used to compile the formulas for submission to the certifier when contracted.

Contracting with a certifier involves determining which certifier the company wants to use. Normally this is a choice the company can make freely, although, in rare cases, a customer will dictate whom it requires. The list of USDA-approved certifiers should be consulted. It is located at http://www.ams.usda.gov/NOP/CertifyingAgents/Accredited.html and sorted by location of the certifier. This is not to say that the company must use a certifier in its state. The company should talk to several and consider the type of help they offer and whether they are aligned with the philosophy of the company. After a certifier is chosen, it should be asked to sign a company confidentiality agreement prior to receiving any of the company's documents. The supplier, in turn, will send the company a contract that outlines the services to be provided and the costs associated with certification. If the company is amenable, the contract should be signed and returned along with the formula sheets completed previously.

The company's contracted certifier will arrange an inspection of the facility, inspect and review the formulas, and work with the company on an approved organic plan. When the certifier deems that all programs and ingredients are in compliance, it will grant approval at the determined level of organic status for the formula. If not, it will suggest changes to company programs, ingredients, or suppliers. In this case, the company needs to comply with these and resubmit the necessary information. When this is approved, the company should make changes to the packaging and ingredient statement.

Finally, the approved organic plan, confidentiality agreement, and contract should be placed in a file. The rest of the submission forms should be filed in individual files labeled as organic, item number, and description. This file will be maintained as the program develops over time.

SUPPLEMENTAL MATERIALS

ORGANIC FORMULATION SUBMISSION FORM

Name of Product _____   **Item Number** _____

    Location product is to be made   _____

    Gross sales of product in last 12 months   _____

    New or existing product   _____

    Are all ingredients and processing aids non-genetically modified   _____

Proposed Organic Statement

    100 % Organic   ☐

    Organic   ☐

    Made with Organic   ☐

Percentage Calculation

Ingredient	Percentage	Supplier	Organic	Certifier

List any processing aids used in production that do not remain in the finished product?

_____

Submitted by: _____   Printed Name: _____

    Title: _____   Date: _____

29 Environmental Responsibility Program

PROGRAM TYPE: OPTIONAL

THEORY

Organic, *GMO* (genetically modified organism), *green*, *global warming*, *stewardship*, *buy local*, *recycle*—these words and concepts are becoming more commonplace in our world today. There is not a day that goes by that we are not bombarded with governments, organizations, friends, and family exhorting companies and society to be better stewards of the environment. As expected, these buzz words are being thrown around the food industry at a rapidly increasing pace. To some, these words mean a potential niche market to target; to others, they are symbols of being in harmony with the world in which we live. In an effort to align themselves with today's environmental movement, some companies choose to conduct business in a manner that has a minimal effect on the environment. Becoming a company that is environmentally friendly requires management commitment, employee time, and increased cost due to the amount of time necessary to plan and implement an environmental program, the centerpiece of which is the company's environmental policy.

A comprehensive environmental program involves considering how the company and those with whom it interacts protect the environment. The first step in development of this program requires the company to ask itself the following questions in the areas of management, hazardous materials, wastewater, hazard waste handling, air emissions, and waste management.

MANAGEMENT

1. Is the facility or all of the company facilities in compliance with environmental regulations as promulgated by the U.S. Environmental Protection Agency, state and local regulations, and Environment Canada, as appropriate?
2. Is the responsibility for environmental issues within the company assigned to a specific individual with the appropriate training and responsibility for environmental communication?
3. Are written guidelines or procedures for environmental activities prepared and regularly updated?
4. Are there procedures established to monitor developments in environmental regulations and legislation?

5. Does the company require its suppliers to conduct business in an environmentally friendly manner?
6. Does the company have internal policies that encourage employees to act in an environmentally friendly manner within their private lives?

Hazardous Materials

1. Are procedures in place that are practiced and followed for immediate notification or reporting of spills of hazardous materials?
2. Are notification forms for hazardous spills filled out in advance?
3. Are personnel assigned to handle cleanup of hazardous material spills and to notify appropriate agencies?

Wastewater

1. Is wastewater tested regularly to ensure compliance with local permit or pretreatment limits?
2. Does wastewater testing include biological oxygen demand (BOD), total soluble solids (TSS), oil and gas, and pH?

Hazardous Waste Handling

1. Are minimal quantities of hazardous waste generated and disposed of?
2. Does the facility generate less than 220 lb per month of hazardous waste?
3. Is hazardous waste properly stored, manifested, labeled, and transported with appropriate required documentation?
4. Do employees handling hazardous waste receive adequate training?

Air Emissions

1. Have permits been obtained for all sources of air emissions?
2. Are air emissions in compliance with permit limits?
3. Are copies of air emission permits retained on site?
4. Does the company actively pursue methods to reduce or eliminate emissions into the air?

Waste Management

1. Are programs in place for waste minimization, management of waste transportation and disposal liability, and identification of landfill disposal alternatives?
2. Is packaging designed in a manner in which waste is minimized at both the plant and consumer levels?

The answers to these questions lead the company to write an environmental policy for the company. This is simply a summary statement of the company's philosophy toward the environment and how the company will be environmentally friendly. It should include what the company believes is its relationship to the environment

Environmental Responsibility Program

and how the company fits in. It might also list specific goals, targets, or concepts the company uses in its manufacturing, waste management, and packaging.

After the company has established its environmental policy, it should develop methods to comply with it. This might entail establishing teams to carry out specific pieces of the policy, developing training tools to get employees excited and involved in being environmentally friendly, and contacting suppliers. It is important at this stage that members of management fully support the actions of the implementation teams; otherwise, the company will not embody the environmental philosophy that it professes.

APPLICATION

A company should begin the environmental program development process by forming a committee to develop the company's environmental policy. The committee should meet to examine the company's compliance with environmental regulations, its commitment to environmental stewardship, and employees' understanding of and support for environmental stewardship. From these meetings, a formal environmental policy should be written. A sample environmental policy can be found on the CD (Book 5_Other Programs\Section 4_Environmental Responsibility Program:Environmental Policy) and at the end of this chapter. If this is used, it should be carefully edited and customized to fit the actual positions and beliefs of the company.

When the environmental policy is complete, it should be discussed with all employees. This discussion should include the gathering of opportunities for the company and its employees to comply with the environmental policy. With the support of management, these ideas should systematically be instituted.

If part of the company environmental policy is to contact the suppliers to determine their compliance to environmental stewardship, a survey should be prepared. A basic supplier environmental check list that can be edited to reflect the company's stance on environmental compliance is located on the CD (Book 5_Other Programs\Section 4_Environmental Responsibility:Supplier Questionnaire) and at the end of this chapter. This may be used to help the company determine which suppliers are aligned with the company's environmental policy for the purpose of increasing their business.

SUPPLEMENTAL MATERIALS

ENVIRONMENTAL POLICY

_____ and its subsidiaries are committed to operating all facilities in accordance with company environmental policies and all applicable laws and regulations. At _____, we will conduct our business in ways that protect the environment and demonstrate good stewardship of our world's natural resources. Our commitment to protecting the environment will be demonstrated in the areas of development, measurement, and auditing. Specifically, we will:

- develop manufacturing facilities that utilize environmentally friendly design, promote energy conservation, minimize waste, and protect the air;
- utilize recycled products within packaging and where possible in other areas;
- minimize waste through source reduction and recycling;
- conserve natural resources through careful planning and efficient use;
- develop methods and technologies to measure resource reduction programs continually;
- train and encourage employees to live in an environmentally responsible manner;
- evaluate our partners' business practices and encourage them to conduct them in an environmentally friendly manner;
- institute environmental responsibility into all company job descriptions;
- report company and employee environmental performance to our customers; and
- review _____'s compliance with environmental stewardship on a quarterly basis.

ENVIRONMENTAL CHECKLIST

As part of _____'s commitment to environmental stewardship, we require that our suppliers conduct their operations, at a minimum, in compliance with all federal, state, and local regulations. We also value our suppliers' efforts not only to be in compliance but also to do business in an increasingly environmentally sensitive manner. Along these lines, _____ looks for suppliers whose environmental policy aligns with ours.

Management

1. Is the facility in compliance with all environmental regulations?

   ☐ Yes   ☐ Not applicable

   Comments: _____
   _____

2. Are procedures established to monitor developments in environmental regulations and legislation?

   ☐ Yes   ☐ Not applicable

   Comments: _____
   _____

3. Are written guidelines and procedures for environmental activities within the company prepared and updated?

   ☐ Yes   ☐ Not applicable

   Comments: _____
   _____

Waste

4. Does the company have a waste reduction program?

   ☐ Yes   ☐ Not applicable

   Comments: _____
   _____

5. Is packaging designed in a manner that minimizes waste for customers?

   ☐ Yes   ☐ Not applicable

   Comments: _____
   _____

6. Is wastewater tested regularly to ensure compliance with local regulations for biological oxygen demand, total soluble solids, oil and gas, and pH?

   ☐ Yes   ☐ Not applicable

   Comments: _____
   _____

Hazardous Materials

7. Are procedures in place to prevent hazardous material spills?

    ☐ Yes   ☐ Not applicable

    Comments: _____
    _____

8. Is a program in place for hazardous material usage reduction?

    ☐ Yes   ☐ Not applicable

    Comments: _____
    _____

9. Has any asbestos within the facility been removed or will it be removed?

    ☐ Yes   ☐ Not applicable

    Comments: _____
    _____

10. Is CFC usage limited? When use is necessary, are CFCs contained properly and the quantity monitored?

    ☐ Yes   ☐ Not applicable

    Comments: _____
    _____

11. Is CFC servicing and disposal done by certified technicians?

    ☐ Yes   ☐ Not applicable

    Comments: _____
    _____

Air Emissions

12. Have permits been obtained for all sources of air emissions?

    ☐ Yes   ☐ Not applicable

    Comments: _____
    _____

13. Are air emissions in compliance with permit levels?

    ☐ Yes   ☐ Not applicable

    Comments: _____
    _____

30 Environmental Testing Program

PROGRAM TYPE: OPTIONAL

THEORY

Monitoring of the environment inside a manufacturing facility is an important tool that is used to help determine if there are pathogens such as coliforms, *Listeria,* and *Salmonella* that might contaminate the products being produced. It is designed to provide a microbiological assessment of a plant's environmental control program. It is not in itself a control program, but rather a tool to provide information to improve environmental controls. Information collected should be used to correct problem areas before they pose a risk to products. Raw areas are not included in a monitoring program because they are considered positive at all times.

The underlying concept of an environmental testing program is to sample areas randomly throughout the facility and test the samples for pathogens. Sampling sites are divided into areas or zones as follow:

Zone 1: direct or indirect product contact surfaces with potential harborage and product buildup conditions. Examples are product conveyors and product discharge chutes, pipeline interior and storage hoppers to product fill, filler hoppers, nozzles, formers, cut-and-wrap equipment, and product scrapers and utensils.
Zone 2: non-product-contact areas adjacent to product contact areas. Examples are exterior of equipment, chill units, framework, equipment housing, phones, panel buttons, operator buttons, weight control buttons, and non-contact weight control scales.
Zone 3: non-product-contact areas within the processing room that are more remote from product contact surfaces. Examples are hand trucks, forklifts, walls, drains, and floors.
Zone 4: areas remote from product contact surfaces outside the processing room. Examples are coolers, floors, bathroom doors, lunchrooms, and halls.

Zone 1 sites should be sampled for indicator coliforms on a weekly basis. Zones 2, 3, and 4 should be sampled for *Listeria* genus and *Salmonella* or optional coliforms to indicate presence of *Salmonella*. Zones 2 and 3 should be sampled weekly and zone 4 sites should be sampled monthly.

Contact surfaces that fall into each of the zones are defined as the following:

Direct product contact surfaces are exposed to the product during the normal course of operation.

Indirect product contact surfaces are surfaces from which product may drain, drop, diffuse, or be drawn into the product or into the container, and surfaces that touch product contact surfaces of the container.

Non-product-contact surfaces are surfaces that, under normal operating procedures, do not contact the product or the product contact surfaces of the container.

Sampling should be done using either swabs or buffered sponges using aseptic techniques. All swabs should be taken during production at least 3–4 hours after startup. This tends to get the microbes because they are stirred up. When personnel safety is a concern, sampling may be done during a break period or at the end of production, prior to any cleanup. The time frame for taking swabs (shift, midweek, end of week, etc.) should be changed on a periodic basis. Swab site locations should be audited and changed on a periodic basis. The standard method for swabbing is to utilize sterile sponge swabs for sampling large areas when testing for *Listeria* or *Salmonella*. These are available from most microbiological and lab supply companies. Culturette swabs or 3M Quick Swabs should be used for coliform testing and they may be used for small or difficult to access areas for *Listeria*. If the presence of a sanitizer is used in the plant environment, a broth with a neutralizing agent such as D/E neutralizing broth or Letheen broth should be employed.

The technique for swabbing with a sponge or culturette is described next.

Sterile gloves should be worn when using the sponge and the appropriate procedure should be followed based on the type of surface to be swabbed. For large surfaces, an area no less than 40 square inches should be sponged. The sponge is then placed in a whirlpac bag for later testing. For irregular or hard to access surfaces, the entire area should be sampled as best as possible and then the sponge placed in the whirlpac bag for later testing. For small areas such as screw heads, small water points, thread surfaces, or interior corners of equipment, a culturette swab should be used and then the swab placed in the carrying tube. The swabs should be taken to the lab for testing or packed on refrigerated ice for shipping to an outside contract lab. If they are sent to an outside lab, a negative control sample should be added (for *Listeria* testing) with each group of weekly samples and all samples should be marked with a code number as opposed to the sampling site name. This acts as a baseline for the rest of the swabs and keeps the sampling locations as internal knowledge only.

Once the laboratory testing is completed, the company needs to analyze the results; if they are negative, no response is necessary; if they are positive, action should be based on the zone. For zone 1 coliform testing only, the action limit is <10 cfu/g per area swabbed. For ≥10, the following tasks should be performed:

- Review cleaning records and procedural documentation.
- Complete an audit of cleaning procedures and execution of same.
- Conduct teardown inspections of filling/packaging equipment involved.
- Conduct pre-op inspections.

- Review environmental data from zones 2 and 3 to see if this is a system-wide problem.
- Review environmental data from zones 1, 2, and 3 from the lines feeding the product into the line in question.

The day after exceeding limits and after the preceding action steps are completed, the line should be reswabbed at each site of the original swab. Samples should be taken at pre-op if available and/or each swabbing time (at least 3–4 hours into the shift) until three consecutive standard results are obtained. In the event that the corrective actions do not fix the problem, the company should institute preventative actions, including reinforcing general manufacturing procedure (GMP) training and improved attention to sanitation procedures, observing the area during operation for GMP violations, collecting finger plates from employees on the production line, rewriting equipment cleaning protocols, and redesigning equipment.

For zone 2 *Listeria* and *Salmonella* testing, the action limit is negative. If either test is positive, then the site should be examined and the potential causes should be investigated. The company should then perform the same action steps as for a zone 1 positive. The day after receiving a positive test and after completing the required action steps, the company should reswab the area of the original positive test and also include other sites on or near the equipment. Samples should be taken at least 3–4 hours into the shift for the next three swabbing times until three consecutive negative tests have been obtained. In the event that the company's corrective actions do not fix the problem, preventative actions should be taken to correct it. These might include the same action steps from zone 1 as well as interdictive equipment cleaning and examining adjacent areas.

The company should sample zone 3 for *Listeria* and *Salmonella* with an action limit of negative. If a positive is obtained in zone 3, the site should be examined and potential causes investigated. Action plans are different from those in zones 1 and 2 because a zone 3 positive, in the absence of zone 1 and 2 positives, is an early indicator of a control program that is not robust enough. Action plans are targeted to GMPs, period cleaning, environmental controls, and facility maintenance. In the event that corrective actions do not fix the problem, preventative actions will need to be taken, including reinforcing GMPs, eliminating water collection points, placing footbaths in entryways, limiting foot traffic, repairing damaged floors, restricting forklift movement, increasing cleaning frequency of adjacent zone 4 areas, and redirecting high-risk traffic from any adjacent areas. The day after receiving a positive in zone 3, the company should reswab the positive area and include other sites on or near the equipment. Samples should be taken 3–4 hours into the shift for the next three shifts until three consecutive negative results are achieved.

Zone 4 *Listeria* and *Salmonella* testing has an action limit of negative. Zone 4 sites are selected by the company to provide an indication of the microbial quality of the area. If a positive test is recorded, the site should be examined and potential causes investigated. Action plans should be used to examine period cleaning frequencies and traffic patterns. These areas are remote from the production and generally present less risk to products; however, they do provide information about the nonproduction environment and traffic flow.

Understanding and instituting an environmental testing program provide the company with an internal baseline for its microbial load. However, some argue that conducting the program may be harmful. The argument behind this is that if a company knows that it has a problem in the environment, it would have to test the product for the known pathogen. Thus, if the company does not test, it will not know and will not incur any additional costs related to testing. In the short term, this attitude might suffice; however, if a customer is sickened by a pathogen, this might increase the long-term liability of the company.

APPLICATION

When the company decides to implement an environmental testing program, the first step is to determine how and when testing will be conducted, what surfaces will be tested, what action steps will be taken when a positive result is found, and who will be responsible for the program. The form on the CD (Book 5_Other Programs\Section 5_Environmental Testing Program:General Overview) and at the end of this chapter can be customized to reflect the company's approach to environmental testing. A company can choose to test all four zones or it can sample from zones 2–4 as non-product-contact zones. The program can be implemented as a true indicator program of the company's environmental microbial load.

For ease of recording the testing results, a microbiological sample result log template can be found on the CD (Book 5_Other Programs\Section 5_Environmental Testing Program:Environmental Microbiological Sample Result Log) and at the end of this chapter. This document should be completed as samples are taken, making sure that the location and zone of where the sample was obtained are filled in. This sheet should be stored in a microbiological results log book.

SUPPLEMENTAL MATERIALS

ENVIRONMENTAL TESTING PROGRAM

GENERAL OVERVIEW

Section A: General

Environmental monitoring is designed to provide a microbiological assessment of a plant's environmental control program. It is not, in itself, a control program, but rather a tool to provide information to improve environmental controls. Information should be used to correct problem areas before they pose a risk to products. Raw areas are not included in a monitoring program because they are considered to be positive.

Section B: Sampling Sites

Sampling sites are divided into areas or zones. Zone 1 sites will be sampled for indicator coliforms weekly. (See zone 1 below for additional detail). Zone 2, 3, and 4 sites will be sampled for *Listeria* genus and *Salmonella* or optional coliforms to indicate *Salmonella* presence. Zones 2 and 3 will be sampled weekly and zone 4 sites will be sampled monthly. All swabs will be taken during production, at least 3–4 hours after startup. When personnel safety is a concern, sampling may be done during a break period or at the end of production, prior to any cleanup. The time frame for taking swabs (shift, midweek, end of week, etc.) should be changed on a periodic basis. Swab site locations should be audited and changed on a periodic basis. Zones are defined as follows:

> *Direct product contact surfaces* will mean all ready to eat (postprocessing/filling) surfaces exposed to the product during normal operation.
> *Indirect product contact surfaces* will mean all surfaces from which product may drain, drop, diffuse, or be drawn into the product or into the container, and surfaces that touch product-contact surfaces of the container.
> *Non-product-contact surfaces* will mean all surfaces that, under normal operating procedures, do not contact the product or the product-contact surfaces of the container.

- Zone 1: direct or indirect product contact surfaces with potential harborage and product buildup conditions (e.g., product conveyors and product discharge chutes, pipeline interior and storage hoppers to product fill, filler hoppers, nozzles, formers, cut-and-wrap equipment, product scrapers/utensils)
- Zone 2: non-product-contact areas adjacent to product (e.g., exterior of equipment, chill units, framework, equipment housing, phones, panel buttons, operator buttons, weight control data input, weight scales)
- Zone 3: non-product-contact areas within the processing room that are more remote from product contact surfaces (e.g., hand trucks, forklifts, walls, drains, floors)
- Zone 4: areas remote from product contact surfaces outside the processing room (e.g., coolers, floors, bathroom doors, cafeteria, halls)

Section C: Testing Protocol

Zone 1: Coliform testing only. Individual samples of product contact surfaces, direct and indirect (e.g., filling/packaging equipment surfaces) will be sampled weekly. Only product contact surfaces exposed to the environment should be included in the swabbing protocol. This assumes that product contact surfaces not accessible to swabbing without disassembly will not be included in the protocol.

- Action limit: <10 coliforms per area swabbed or finished product specification
- Action plan if limits are exceeded on a zone 1 swab may include some or all of the following:
  - Review cleaning records and procedural documentation.
  - Complete an audit of cleaning procedures and execution of same.
  - Conduct teardown inspections of filling/packaging equipment involved.
  - Conduct pre-op inspections.
  - Review environmental data from zones 2 and 3.
  - Review environmental data (zones 1, 2, and 3) from lines feeding product into the line in question.
- The day after exceeding limits and after completing the preceding action steps, the line will be reswabbed at each site of the original swab. Samples should be taken at pre-op if available and/or each swabbing time (at least 3–4 hours into shift) until three consecutive in-standard results are obtained.
- All corrective action plans will be developed by the plant and will be documented.
- In the event corrective actions do not fix the problem, preventative actions may need to be taken. Examples of preventative actions are:
  - Reinforce GMP training and provide additional attention to sanitation procedures.
  - Observe area during operation for GMP issues/practices.
  - Collect finger plates from employees on the production line.
  - Rewrite equipment cleaning protocol.
  - Redesign equipment.
- Repeat >10 coliform positives or finished product specification, whichever is greater.
- Pre-ops for >10 coliforms per area swabbed.
- Protocol for swabbing is a minimum of 40 square inches. Sponge swabs are preferred for *Listeria* and *Salmonella* and single tip culturette or 3M Quick Swabs are preferred for coliform environmental testing (see environmental sampling procedures section).
- Repeat positives and/or counts exceeding the action limits: *notify quality control manager.*

Zone 2: *Listeria* genus and *Salmonella* or coliform testing. Zone 2 sites will be sampled weekly. Specific sites will be selected that are adjacent to product contact surfaces. The type of zone 2 sites that should be selected are areas that, if not cleaned properly, may provide a risk to product or areas or sites that employees frequently handle that could lead to postprocess contamination (e.g., control panels, operator buttons, equipment exterior).

If a positive is obtained in Zone 2, the site will be examined and potential causes will be investigated. Action plans are similar to those for zone 1, due to the proximity to exposed product.

    Action limit: *Listeria* genus negative per area swabbed
    Salmonella negative or <100 coliforms per area swabbed or one log higher than the finished product
    Action plan if a positive is obtained on a zone 2 swab may include:
        Review cleaning records and procedural documentation.
        Audit of standard cleaning methods (SCMs)
        Conduct teardown inspections.
        Conduct pre-op inspections.
        Review environmental data (zones 1, 2, and 3) from lines feeding product into the line in question.
        Review same line environmental data from zones 1 and 3.
    The day after receiving a positive and/or exceeding limit results and after completing the preceding action steps, the area will be reswabbed to include the positive and/or exceeding limit site and other sites on or near the equipment. Samples will be taken (at least 3–4 hours into shift) for the next three swabbing times until three consecutive negatives and in-standard results are obtained. All corrective action plans will be developed by the plant and will be documented.
    In the event corrective actions do not fix the problem, preventative actions may need to be taken. Examples of preventative actions are:
        Reinforce employee GMP practices and provide additional attention to sanitation procedures.
        Eliminate water collection points.
        Rewrite equipment or period cleaning protocol.
        Redesign equipment/equipment maintenance and preventative maintenance.
        Conduct interdictive equipment cleaning.
        Examine equipment cleaning of adjacent areas. Investigative swabs may be required at pre-op to determine if the sanitation procedures were effective.
    Repeat positives and/or counts exceeding the action limits: *notify quality control manager.*

Zone 3: *Listeria* genus and *Salmonella* or coliform testing. Zone 3 sites will be sampled weekly. Specific sites will be selected that provide an indication of the microbial environment of the entire processing area. Floors will always be included in the zone 3 sampling sites. Other types of zone 3 sites that should be selected are walls, doors, fork truck handles, and drains.

If a positive is obtained in zone 3, the site will be examined and potential causes will be investigated. Action plans are different from those for zones 1 and 2 because a zone 3 positive, in the absence of zone 1 and 2 positives is an early indicator of a control program that is not robust enough. Action plans are targeted to GMPs, period cleaning, environmental controls, and facility maintenance.

In the event corrective actions do not fix the problem, preventative actions may need to be taken. Examples of preventative actions are:

Reinforce employee GMP practices and provide additional attention to sanitation activities.
Eliminate water collection points.
Place footbaths/foamers in entryways and seal doors and limit foot traffic (zones 3 and 4)
Rewrite period cleaning protocol.
Change period cleaning frequency.
Repair damaged floors.
Restrict fork truck movement.
Increase cleaning frequency of adjacent zone 4 areas.
Redirect high-risk traffic from adjacent areas.

The day after receiving a positive and/or exceeding limit results and after completing the preceding action steps, the area will be re-swabbed to include the positive and/or exceeding limit site and other sites on or near the equipment. Samples will be taken (at least 3–4 hours into shift) for the next three swabbing times until three consecutive negative and/or in-standard results are obtained. Investigative swabs may be required at pre-op to determine if the sanitation procedures were effective.

All corrective action plans will be developed by the plant and will be documented.

Action limits: *Listeria* genus negative per area swabbed
Salmonella negative or coliform: floor drains < 1000 per area swabbed, handtrucks, forklifts, and miscellaneous zone 3 equipment < 500 per area swabbed.
Repeat positives and/or counts exceeding the action limits: *notify quality control manager.*

Zone 4: *Listeria* genus and *Salmonella* or coliform testing. Zone 4 sites will be sampled monthly if immediately adjacent to an RTE area and quarterly in other areas. Sites will be selected that provide an indication of the microbial quality of the area (e.g., cafeteria, locker rooms, and hallways). The types of zone 4 sites that should be selected are walls, doors, fork truck handles, floors, and drains.

If a positive is obtained in zone 4, the site will be examined and potential causes will be investigated. Action plans should be used to examine period cleaning frequencies and traffic patterns. These areas are remote from production and generally present less risk to products; however, they do provide information about the nonproduction environment and traffic flow.

In the event corrective actions do not fix the problem, preventative actions may need to be taken. Examples of preventative actions include:

Reinforce employee GMP practices and provide additional attention to sanitation activities.
Place footbaths/foamers in entryways, seal doors, limit foot traffic.
Rewrite period cleaning protocol and change frequency.
Repair damaged floors.
Restrict fork truck movement.
Increase cleaning frequency in zone 4 outlining areas.
Redirect high-risk traffic from adjacent areas.

Samples will be taken (at least 3–4 hours into shift) weekly until three consecutive negatives and/or in-standard results are obtained.
All corrective action plans will be developed by the plant and will be documented.

Action limits: *Listeria* genus negative per area swabbed
Salmonella negative or coliform: floor drains < 1000 per area swabbed; handtrucks, forklifts, and miscellaneous zone 3 equipment < 500 per area swabbed.
Repeat positives and/or counts exceeding the action limits: *notify quality control manager.*

Section D: Environmental Sampling Procedures

The swabbing procedures and methods are consistent with the standard methods regarding microbiological examination of food contact or other surfaces. Sterile sponge swabs are the most effective for sampling large areas for *Listeria* and/or *Salmonella* testing. Sponge swabs are available from International Bio-Products (IBP). Culturette swabs or 3M Quick Swabs should be used for coliform testing and they may be used for small or difficult to access areas for *Listeria*. If the presence of a sanitizer is used in the plant environment, a broth with a neutralizing agent such as D/E Neutralizing broth or Letheen broth should be employed.

Sponge Technique and Procedures

1. Using sterile gloves, remove the sponge and then follow the appropriate sampling procedure as described:
    (a) Large surfaces: sponge an area no less than 40 in.^{2}. Replace the sponge in the whirlpac bag.
    (b) Irregular or hard to access surfaces: sample the entire area as indicated by the surface description. Replace the sponge in the bag.
    (c) Small areas: Certain areas may be more appropriately sampled using a culturette swab (e.g., head screws, small water collection points, screw holes, threaded surfaces, or interior corners of equipment). Swab the entire area as indicated by the surface description. Replace the swab in the culturette tube.
    (d) Place completed swabs in a container separate from the one holding the swabs to be taken. Other disposable materials (gloves, tear strips, etc.) should be placed in a separate container from the one holding the swabs. It is also recommended to use an alcohol-based hand sanitizer during the swabbing process to prevent cross-contamination.

2. After sampling, return the samples to the lab and refrigerate until they are tested or shipped to the outside testing laboratory. Samples should be properly identified, packaged with ice packs, and shipped within 24 hours after sampling; 48 hours should be the maximum time frame for receipt at the external lab for testing.
3. A negative control sample (for *Listeria* testing) should be included in each group of weekly swabs.
4. Composite sampling (maximum six sites) should only be composited from the same zone. Composite sampling is discouraged for potential harborage areas.

Note: It is recommended that samples submitted for external testing be identified by a code and not include the sampling location name.

Approved by: _____ Date: _____

31 Outside Audits

PROGRAM TYPE: OPTIONAL

THEORY

As a company grows and becomes a larger player in the marketplace, chances are that it will encounter customers or suppliers that require its facility to be audited. For those companies that have the required programs in place and functioning, this can be an enlightening experience as the quality control systems come under scrutiny. For those companies whose systems are not in place, it can be a terrifying experience. In general, outside audits come in two forms: customer audits and third-party audits.

Customer audits are audits that involve the customer visiting the facility. During this visit the auditor will ask to view records and quality system documents pertaining to products. Care should be taken to prevent the sharing of information that might belong to other customers due to confidentiality agreements. Third-party audits are conducted by a company not affiliated with the customer, but hired by the customer to examine the quality systems of the company. Although they are directed from different points of view, both types of auditors try to understand the fundamental workings of the company's quality control program and utilize the same basic questions to get there. Prior to auditing a facility, the inspectors will contact the company and arrange an inspection time. Depending on the size and complexity of the facility, the quality manager should set aside at least 4 hours for the inspection; however, in many cases it will take all day. The determining factors for the length of the audit are how clean the facility is, how complete and well documented the quality program is, whether the auditor has another facility to audit on the same day, and when the auditor's plane leaves. In short, the length of audits varies, but auditors generally follow the same basic format after they arrive.

The format of the audit generally has four parts: walk-through, questions and answers, document and system review, and wrap-up. Prior to the arrival of the auditor, the sanitation manager, plant manager, and quality control manager should conduct a walk-through of the facility to evaluate and correct any areas that can be improved. This will make the audit proceed more smoothly and informs other managers of the impending audit so that they may notify their staffs and prepare. During this preinspection walk-through, the managers should focus on sanitation, maintenance, and good manufacturing issues. Any concerns found should be corrected prior to the arrival of the inspector.

Upon arrival, the inspector should be treated as any visitor would be, be given the GMP document, and sign in. When the guide—usually the quality control manager—arrives, the inspector should be taken to a conference room where introductions can be made. At this time the inspector will provide a card of identification and possibly a copy of the inspection format and discuss the plan of how he or she will conduct the audit. This leads to the first phase—the walk through.

The inspector will ask to see the facility and the guide should be prepared to conduct a tour of the plant. Depending on the inspector, this might be a very thorough tour where a lot of questions are asked or just a quick "look-see." If the inspector has been to the facility before, he or she will normally just take a quick look around. If this is a first visit, the auditor will look carefully at sanitation, maintenance, pest control, equipment storage, and other good manufacturing practices. Often the inspector will want to see inside freezers, refrigerators, closets, and storerooms and even walk around the outside of the facility. During the walk-through, expect that the auditor will take notes in preparation for phase two: the question and answer session.

When the walk-through is complete, the guide and auditor will return to the conference room for the question and answer session. Generally, the walk-through will spur some questions regarding the company, its products, and its personnel. This is usually a casual exchange of information, although some inspectors will use it to determine the qualifications, knowledge, and experience of the quality control manager as a gauge for how well the quality program has been implemented. Do not let this be a bother; just answer the questions rapidly and move on to phase three—document and system review.

In preparation for this phase, the quality control manager should have the HACCP Manual, the Quality Control Manual, Pest Control Books 1 and 2, the Other Programs Manual, and the Shipping/Receiving Manual ready for evaluation. The inspector will determine the course of this session, but will work through all the manuals and ask questions as he or she does so. This is the meat of the audit because it allows the auditor to determine if all of the required systems are in place, fully documented and compliant, and routinely audited by management. Be prepared to answer all questions in full but reserve the temptation to expand on your answers because, in many instances, the additional information will lead to other questions. Also, do not volunteer information about any other part of the company or its products, customers, or personnel. Keep in mind that the sooner the question and answer phase is complete, the sooner the wrap-up phase can begin.

Upon conclusion of the question and answer phase, it is normal for the auditor to ask for some quiet time to fill out the internal survey form. In today's technological age, it is becoming increasingly common for the auditor to input all findings into a portable computer and then print the company a copy when finished. In preparation for this, the company should inquire as to whether the auditor will need an outlet for the computer and/or paper for the printer. During this quiet period, the guide may attend to other duties as long as he or she is in the same proximity and available to answer the auditor's questions and complete the wrap-up.

After the auditor has completed the report, he or she will request a quick wrap-up meeting. Sometimes other management will be requested to join the meeting, but it is up to the company as to who is involved in this closing meeting. When all parties are ready, the auditor will present a short summary of findings both verbally and with a copy of the audit report. If the audit is a scored audit, the auditor will usually provide some indication of what the score is and how it relates to whether the company passed, did not pass, or conditionally passed. If the audit was a conditional

Outside Audits

pass, the company should take all steps to understand what the conditions for passing are before the auditor leaves.

After completion of the wrap-up phase, the auditor will leave and allow the company to further digest any recommendations and implement a strategy for correcting any deficiencies found. It should be noted that the company should consider each and every recommendation that an auditor makes during an audit—whether it is a cleanliness, maintenance, GMP, or system change. However, it is very common for two or more auditors to see the same thing differently and want the company to change its program to fit their audits. Resist this urge to comply for compliance's sake—especially when programs are involved. If the company has the basic program in place and functioning, then do not feel the need to change it to fit an auditor's specific directions. Doing so will be cumbersome and lead to multiple unnecessary changes.

32 Social Responsibility Program

PROGRAM TYPE: OPTIONAL

THEORY

Food manufacturers are a major part of the globalization of the marketplace today. Ingredients are available from most countries and all corners of the world, and most companies are looking to sell their products within an ever expanding marketplace. As its impact in the marketplace increases, it is increasingly important that the company think globally and act locally. One way to bring a company to focus on its actions within society is to develop a social responsibility program. This involves developing a statement of beliefs and then integrating them into the company's daily actions.

Developing a statement of beliefs regarding the company's social responsibility can be a very time-consuming and arduous task. To begin, management needs to evaluate how the company fits in locally, nationally, and internationally. Some basic and more obvious questions it should ask include:

1. How involved should the company be with the community in which it is located? Does the company donate money to various charities and organizations to promote the community?
2. Does the company encourage its employees to contribute their personal time and money within the local community? Does the company offer special arrangements for time off for employees to participate in community activities?
3. Does the company recognize employees' social, environmental, and volunteer activities?
4. Does the company actively promote being a good steward of the environment? (This position lends itself to the company's environmental program discussed earlier.) Does the company actively recycle, utilize second-generation materials, and conserve energy?
5. Does the company encourage its employees to be good environmental stewards?
6. Does the company assist employees to better their lives through training and education?
7. Does the company actively employ those that need physical or financial assistance?
8. Does the company participate in or sponsor activities within the community that promote community spirit?
9. Does the company conduct business with other socially and environmentally responsible companies?

10. Does the company partner with other companies to facilitate better social and environmental stewardship?
11. Does the company set goals regarding its social impact?

By answering these questions and others, depending on the situation of the company, management can begin to develop a corporate social responsibility statement. During the course of this exploratory period, the company should work to solicit employee involvement and support. The more the staff is involved, the greater are the chances for a successful and meaningful program. While engaging the staff, the management should take particular note of interests and ideas that the employees have regarding their social responsibility because these can be the nexus of the company's activities.

When management and the company are ready, the social responsibility statement should be developed. Like other company position statements, such as the company's vision and environmental policy, this statement outlines what the company believes its role within society is and how it will interact with that society. The statement will then outline specifically what the company will do and how its actions will be measured. The final statement needs to be disseminated to each of the company's employees and added to the new employee training packet and the employee manual.

The final and ongoing step in development of the social responsibility program is that of implementation. This involves the company evaluating its newly developed social responsibility statement and, together with the employees, determining strategies, programs, and measures for the company and employees to implement. Some possible foundational steps the company might take are to create teams for the development of ideas, appoint individuals to lead efforts within the community, and seek marketing opportunities for community interaction. Whatever direction implementation takes, the company must seek a long-lasting commitment by investing in the community and its employees.

APPLICATION

A customizable statement that the company can use as a basis for its own social responsibility statement is located on the CD (Book 5_Other Programs\Section 7_Social Responsibility:Social Responsibility Statement) and at the end of this chapter.

SUPPLEMENTAL MATERIALS

SOCIAL RESPONSIBILITY STATEMENT

At _____, we are committed to conducting our business in a socially responsible and ethical manner. We recognize our responsibility to contribute to the community that supports us. As such we donate a portion of our profits to local nonprofits and community partners. Some of our favorites are _____, _____, and _____.

At _____, we believe that employees are one of our most important assets. We recognize our responsibility to provide opportunities to improve their lives through continuing education and community involvement.

We encourage our employees to take active roles in the community and support this with flexibility in time and funds.

We are proud to provide a work environment where employees are appreciated, valued, and given regular feedback. Our management promotes an "open door" policy for concerns of, thoughts of, and feedback from staff. We promote based on merits and skill and do not discriminate based on age, color, creed, or relationship orientation.

We recognize parental and family obligations of our employees and ensure that all staff members are supported by being flexible with their hours or allowing them personal time on short notice.

We provide a creative working environment where all of our staff members are encouraged to develop their skills in their current position as well as in the work environment at large.

At _____, we strive to be a good steward of the environment. We strive to recycle, conserve energy, and reuse materials where possible. It is our goal to reduce our ecological footprint. Some ways we are doing this are _____, _____, and _____.

33 Continuing Food Guarantee Program

PROGRAM TYPE: REQUIRED

THEORY

Most manufacturers of foodstuffs in the United States, Canada, and Western Europe are bound by standards for food safety and hygiene. These are promulgated by international, national, and local governmental bodies and dictate to manufacturers the safety measures that they must take to produce a product safe for human consumption. Unfortunately, not all companies, even in these more regulated countries—and especially in developing countries, produce products that are hygienically safe. In such cases the federal government outlines some legal assurances that companies can obtain from suppliers in the form of a continuing food guarantee.

In accordance with section 303c(2) of the code of Federal Food, Drug and Cosmetic Act, food manufacturers need to provide a legal guarantee to customers that those products they sell comply with federal, state, and local laws. This guarantee usually takes the form of a letter or document that outlines compliance with one or more of the following laws:

the Federal Food, Drug and Cosmetic Act and its amendments;
California State Proposition 65;
state laws; and
local laws.

Its verbiage should include the statement that, as of the date of each shipment, the products, articles, or shipments supplied by the company are not, when shipped, adulterated or misbranded within the meaning of the Federal Food, Drug and Cosmetic Act. The supplier further should guarantee that it will not introduce a product that is adulterated or misbranded into interstate commerce.

As a required document, the company should be prepared to provide this document to its customers as part of the initial information packet. It should be on company letterhead and signed by a corporate officer. Copies of all continuing food guarantees should be placed in the customer's file.

APPLICATION

The continuing food guarantee form found on the CD (Book 5_Other Programs\ Section 8_Continuing Food Guarantee:Continuing Food Guarantee) and at the end of this chapter can be customized by the company when a customer requests that it send a continuing food guarantee for its products. This should be placed on company letterhead and signed by a corporate officer.

The form to be used with the supplier located on the CD (Book 5_Other Programs\Section 8_Continuing Food Guarantee:Continuing Food Guarantee for Supplier) and at the end of this chapter can be utilized by the company as a template to send its suppliers that do not have a standard continuing food guarantee of their own. The company name and address should be placed in the upper right-hand corner and in paragraph (a). The supplier's name and address are placed in the first paragraph. The supplier should place this on its letterhead, sign and date the bottom, and return it to the company for inspection.

SUPPLEMENTAL MATERIALS

To: _____
    _____
    _____

For the purpose of Section 303c(2) of the Federal Food, Drug, and Cosmetic Act ("the Act"), the California Health and Safety Codes, Section 25249.6 of Proposition 65, and all local codes, and for no other purpose, _____ (Supplier) hereby guarantees that, as of the date of each shipment by it to you of any product labeled "food grade," such product is not, when shipped, adulterated or misbranded within the meaning of the Act.

The Supplier further guarantees that, as of the date of each such shipment, no such product is an article that may not, under the provision of Section 404 of the Act, be introduced into interstate commerce. This guarantee shall also apply under substantially identical state or municipal regulations.

This guarantee shall, however, be void and of no effect in any instance where the particular use by you or your customer of any product to which this guarantee would otherwise apply is a use that is not in accordance with the requirements of the Act or applicable state laws.

By the acceptance of this guarantee, you agree to notify the Supplier promptly in writing of any demand, complaint, or proceeding within your knowledge for claimed violation of the Act resulting from or in any manner arising out of any such shipment, including the name and address of the complainant and the name of the product involved.

This guarantee shall continue in effect until such date as you shall receive from the Supplier written notice of the revocation of the guarantee contained herein.

Date: _____   By: _____

                        Printed Name: _____

                                Title: _____

                              Address: _____

                                       _____

                                       _____

Company Name
Address
Address
Telephone

FOOD AND DRUG GUARANTEE

The undersigned, _____ ("Seller"), with principal offices at _____, by its duly elected officer herby certifies that:

(a) The article making up each shipment of other delivery thereafter made by the seller to, or in the order of, *Company Name, Company Address* is hereby guaranteed, as of the date of such shipment or delivery, to be, on such date (1) not adulterated or misbranded within the meaning of the Federal Food, Drug and Cosmetic ("the Act"), 21 U.S.C 301 et seq.; (2) not an article which may not, under the provisions of Section 404, 505, or 512 of the Act, be introduced into interstate commerce; and (3) not adulterated or misbranded within the meaning of the state or local laws, regulations or ordinances, the adulteration and misbranding provisions of which are the same as those found in the Act.

(b) This guarantee shall continue to be effective until revoked by the Seller at any time but with not less than ten (10) days' notice to Buyer.

Company: _____   Date: _____

By: _____

Title: _____

34 Contract Laboratory Testing Program

PROGRAM TYPE: OPTIONAL

THEORY

During the course of normal business, food manufacturers conduct various tests on ingredients, finished goods, and the environment. Some of these tests are conducted to establish a quality base line; others are done to refute claims or determine if problems exist. The types of testing required can be as varied as microbiological testing, physical testing, or wet chemistry testing and may be either destructive or nondestructive. A significant challenge that companies large and small encounter is how to deal with performing a test that requires equipment and methodologies not currently available in house. Although some companies will ante up and purchase equipment and hire technicians to conduct the tests, many companies today are seeking an easier solution: the contract laboratory.

Located in most major cities, contract laboratories are businesses that have the equipment and staff to perform most microbiological, physical, and chemical testing necessary for the food industry. From the normal proximate analysis to specific chemical and organism identification, they can identify the specific BAM, AOAC, Codex, or other approved methods for the test required and complete it using the highest levels of quality assurance. This provides the customer three advantages: cost, timeliness, and confidentiality.

When deciding to contract with an outside lab, the customer should consider the tests required, the lab's location, fees, turnaround time, and most importantly, quality control. If the company needs a government-required test, then an analytical laboratory is appropriate. These are laboratories that specialize in a limited number of tests directed toward a particular industry. An example of an analytical laboratory is a state department of agriculture lab that conducts export tests for companies sending products overseas. For companies that require a varied and ever changing set of testing and consulting requirements, a full-service lab is required. These are labs that perform a range of lab-related services, such as HACCP program evaluation, proximate analysis and nutritional labeling development, shelf life studies, microbiological testing, chemical analysis, and foreign material evaluation. Full-service labs generally provide problem-solving and data analysis services as well as a variety of analytical testing. They will also provide expert witness testimony in court if necessary. This is especially important with the combination of global sourcing, tightening margins, and an increasingly litigious society. When deciding if the lab can conduct the tests required, a company should verify its expertise in conducting the tests, that it conducts the tests itself and does not contract them out to another laboratory, and that both the company and lab agree on the testing method to be used. This method should be from a verifiable and approved source such as BAM, AOAC or CODEX.

Location of the contract laboratory is important to the company when expediency of the testing is crucial. If the nature of the sample degrades with time or the timeliness of results is important, then having a testing laboratory close may be critical. With the current availability of overnight freight services such as FedEx or DHL, this is becoming less relevant; however, these services are not always reliable due to weather and other factors.

The third factor for the company to evaluate when choosing a contract laboratory is that of the fees charged per test. Depending on the test, the costs can vary greatly between labs for the same test. This is partly due to the expertise of the lab and partly due to the cost of equipment needed to conduct the test and the number of tests run on the particular piece of equipment. Fees can be an important factor for most companies because they add up quickly; however, they should be only one factor.

Turnaround time is another important issue to consider because timeliness can be a significant factor, especially as a component of the company's positive release program. With the pressures of today's HACCP programs, companies need and want to be proactive and a large component of that is having timely results for clearing products. Turnaround time comes into play especially during seasonal testing times.

The final requirement for a contract lab is that of quality control. This involves several components, including whether the lab has a good reputation in the industry, is qualified and accredited to ISO17025 standards, conducts and provides proficiency studies, and has an internal quality control program. To find out about different labs, the company might consult its suppliers and inspectors, who generally know what labs are proficient conducting the required tests and whether they have sufficient quality systems. Another way to determine if the lab should be used is to conduct a quality audit similar to what the company does with ingredient and packaging suppliers.

After choosing a lab, the company prepares the samples and sends them for analysis. Accompanying all samples should be a chain-of-custody document. This states who sent the sample, where it is going, a description of the samples, a sample number, and the tests requested. The company should retain a copy of the chain of custody for its records.

APPLICATION

A customizable chain-of-custody form can be found on the CD (Book 5_Other Programs\Section 10_Contract Labs:Contract Labs) and at the end of this chapter. This should be altered to reflect the name, address, phone, and fax of the company sending the samples and of the laboratory receiving them. The description of the sample and a sample number should be filled in for each sample. It is recommended that the number be the lot code representative of its source. Although some companies apply random numbers to samples sent to contract labs, this forces the company to develop a cross-reference system. After completion, a copy should be retained by the company and the original inserted in the package with the sample. Some labs will also request that the completed chain of custody be faxed to them prior to the sample arrival. This can be done with the retained copy.

SUPPLEMENTAL MATERIALS

CHAIN OF CUSTODY

Company Submitting Samples: _____
Contact Address: _____

Send Report To: _____
Fax Number: _____
Phone Number: _____

Date Samples Sent: _____
Date Report Faxed: _____

Send Samples To: _____

Lab Name: _____
Address: _____
Address: _____

Phone: _____
Fax: _____

Date Results Faxed: _____
Date Samples Received: _____

Sample Description	Sample #	Salmonella	Listeria	Aerobic Plate Count	E. coli	Coliform	Y & M	Other	Other	Other

35 Record Keeping

PROGRAM TYPE: OPTIONAL

THEORY

Every company that produces products creates reams and reams of documents that need to be retained. From daily production documents, quality control paperwork, customer paperwork, and receiving and shipping documents, the quantity of these documents can be quite overwhelming. Although some companies store documents in various departmental locations, this can be very cumbersome and limits the speed needed to track products in the event of a recall. Therefore, to handle this plethora of paper, companies must create a central filing system, the cornerstone of which is the Julian date file.

A Julian date file is 366 manila folders labeled with consecutive Julian date stickers. Because the Julian date without the year notation is the same each year, the files can be labeled as 001–366 and reused each year. Placed in a file cabinet or two in numerical order, the files act as repositories for all documents pertaining to each individual day's activity. The production and quality control documents generated on that particular day should be placed in this file, as well as product inspection documents; microbiological and other lab testing records; quality control e-mail; hold, release, and destruction notices; customer complaints; shipping and receiving documents and inspections; and other papers. Any record that pertains to the particular day should be placed in this file as a central repository for storage and future reference.

Typically, in January, the first 6 months of the previous year, or roughly 001–185, should be archived. To do this, everything in a particular date file can be stapled together and placed in order in a box. Bankers' boxes are an easy box for record retention because they hold an ample number of records and are easy to transport. Both ends of the box should be labeled with the year and the date range of the records contained therein. The cases should be stored in a safe, locked, cool room for a period of no less than 3 years. The file folders should be purged every 6 months so that there are always open files for record retention.

APPLICATION

To create a functioning record repository, 366 consecutive five-tab manila folders should be purchased. On these, a label dated from 001 to 366 is added. Book 5_Other Programs\Section 10_Record Keeping:File Labels on the CD can be used to print out the labels. They are set up for Avery 5167 standard labels and are four labels wide. The labels should be printed out and added to the file folders, and the manila folders are then put into hanging files in a filing cabinet.

36 Other Forms

PROGRAM TYPE: OPTIONAL

THEORY

During the course of normal business, the quality control manager will need to source documents that may, or may not be, outside the purview of his or her job description. This might include confidentiality and broker agreements and credit applications. This chapter provides many of these needed forms.

APPLICATION

These forms are all found on the CD (Book 5_Other Programs\Section 11_Misc Forms) and at the end of this chapter.

SUPPLEMENTAL MATERIALS

CONFIDENTIALITY AGREEMENT

This agreement, effective as of the _____ day of _____ 20__, is entered into by and between _____, Incorporated ("_____"), located at _____ and _____ ("the Company") located at _____.

WITNESSETH:

WHEREAS, _____ wishes to protect the confidentiality of its information which may be supplied to the Company as a result of the business relationships between _____ and the Company;

WHEREAS, the Company wishes to protect the confidentiality of its business information which may be supplied to _____ as a result of business relationships between the Company and _____; and

WHEREAS, the parties hereto desire and have agreed that Confidential Information (as hereafter defined) made available by one party (the "Disclosing Party") to the other party (the "Receiving Party") shall be kept confidential by the Receiving Party.

NOW, THEREFORE, in consideration of the mutual covenants and premises contained herein, _____ and the Company agree as follow:

1. As used in this Agreement, the term "Confidential Information" shall mean any information in any form emanating, directly or indirectly, from the Disclosing Party or any of its employees or agents, and any of its divisions and subsidiaries, including, but not limited to, trade secrets, customer lists, product lines, methods of business operation of the Disclosing Party, technical information, economic information data, specifications, know-how, process information, and methods of manufacture, distribution, and sale relating to the development and marketing of the Disclosing Party's product and general business operations. "Confidential Information" does not include any information which (a) at the time of disclosure is generally known by the public or thereafter becomes public knowledge through no act or omission of or on behalf of the Receiving Party; (b) is disclosed to the Receiving Party obligations of confidentiality in respect thereof; or (c) is known to the Receiving Party, as can be documented, prior to disclosure.
2. Each party hereto acknowledges that the other party has a proprietary interest in maintaining the confidentiality of its Confidential Information and further agrees not to, either during or after completion of the purpose for which the Confidential Information has been disclosed to it, disclose the Confidential Information (except in accordance with Paragraph 3) or use the Confidential Information for any purpose other than the purposes stated herein.
3. The Receiving Party shall
   a. limit the disclosure of the Confidential Information in its organization to those of its officers and employees to whom such disclosure is necessary to fulfill its obligations to the Disclosing Party;

b. ensure that such officers and employees acknowledge that the information is confidential before it is imparted to them and ensure that such officers and employees are bound by obligations restricting use and disclosure of the Confidential Information equivalent to those set out in this Agreement;
  c. use its best efforts to ensure that such officers and employees abide by such obligations; and
  d. accept full liability for and indemnify the Disclosing Party (and its officers, directors, employees, agents, and affiliates) against any wrongful disclosure or use of the Confidential Information by any of its officers and employees.
4. Each party shall assume full and exclusive liability for the acts and omissions of itself, and agent or employee ("Indemnifying Party"), and shall Indemnify and hold harmless the other party against all liability to third parties arising from or in connection with the negligence of the Indemnifying Party's employees, agents, or subcontractors.
5. Unless the Disclosing Party has agreed otherwise in writing, upon completion of the for which the Confidential Information has been disclosed, the Receiving Party shall return any and all materials which contain any Confidential Information including, but not limited to, all documents, plans, samples, drawings specifications, notebooks, computer software, and any other materials whatsoever and all copies made of them.
6. If the Receiving Party develops a product or a process which, in the opinion of the Disclosing Party, might have involved the use of any of the Confidential Information it shall, at the request of the Disclosing Party, promptly supply information reasonable necessary to establish that the Confidential Information has not been used or disclosed.
7. The Company shall not advertise its association with _____ or any _____ affiliate in any manner, written, verbal, or pictorial.
8. No work is to be performed or costs incurred without a purchase order number issued through _____ purchasing department.
9. This Agreement may not be changed or amended except in a writing signed by the party to be bound. This Agreement and legal relations between the parties shall be construed and determined in accordance with the laws of the State of _____ without regard to conflict of law principles.

IN WITNESS WHEREOF, the parties have caused this Agreement to be executed as of the date set forth above.

_____ Company Name

By: _____ By: _____

Title: _____ Title: _____

Dated: _____ Dated: _____

NONCONFORMING MATERIAL REPORT

Date: Person filling out form:
Record number:
Plant location: Date of run:
Supplier: Quantity:
U/M:
Supplier item number: Is the material usable?
Are samples pulled?
Lot no.:   Production line:
Date of manufacture: Category of complaint:
Exp. date:

Problem noted:

Finished goods dumped
Finished goods held
Extra labor used
Downtime
Is there a claim?

Purchase order number:

Date received:

Issue external: Issue internal:

Name of contact: Title:
Street address: Phone:
City: Cell phone:
State: Fax:
Zip   E-mail:

Credit to be issued:

Material to be returned:

All suppliers must provide a written statement on their findings and corrective actions. Please submit to the quality control manager.

Date of notification: _____   By: _____

OSHA's Form 300
Log of Work-Related Injuries and Illnesses

Year _____

U.S. Department of Labor
Occupational Safety and Health Administration

Form approved OMB no. 1218-0176

Attention: This form contains information relating to employee health and must be used in a manner that protects the confidentiality of employees to the extent possible while the information is being used for occupational safety and health purposes.

You must record information about every work-related injury or illness that involves loss of consciousness, restricted work activity or job transfer, days away from work, or medical treatment beyond first aid. You must also record significant work-related injuries and illnesses that are diagnosed by a physician or licensed health care professional. You must also record work-related injuries and illnesses that meet any of the specific recording criteria listed in 29 CFR 1904.8 through 1904.12. Feel free to use two lines for a single case if you need to. You must complete an injury and illness incident report (OSHA Form 301) or equivalent form for each injury or illness recorded on this form. If you're not sure whether a case is recordable, call your local OSHA office for help.

Establishment name _____

City _____ State _____

Identify the person				Describe the case		Classify the case							
(A) Case No.	(B) Employee's Name	(C) Job Title (e.g., Welder)	(D) Date of injury or onset of illness (mo./day)	(E) Where the event occurred (e.g. Loading dock north end)	(F) Describe injury or illness, parts of body affected, and object/substance that directly injured or made person ill (e.g. Second degree burns on right forearm from acetylene torch)	Using these categories, check ONLY the most serious result for each case				Enter the number of days the injured or ill worker was:		Check the "injury" column or choose one type of illness:	
						Death (G)	Days away from work (H)	Remained at work		On job transfer or restriction (days) (K)	Away from work (days) (L)	(M)	
								Job transfer or restriction (I)	Other recordable cases (J)			Injury (1)	Skin Disorder (2)	Respiratory Condition (3)	Poisoning (4)	All other illnesses (5)
Page totals						0	0	0	0	0	0	0	0	0	0	0

Be sure to transfer these totals to the Summary page (Form 300A) before you post it.

Injury (1) Skin Disorder (2) Respiratory Condition (3) Poisoning (4) All other illnesses (5)

Public reporting burden for this collection of information is estimated to average 14 minutes per response, including time to review the instruction, search and gather the data needed, and complete and review the collection of information. Persons are not required to respond to the collection of information unless it displays a currently valid OMB control number. If you have any comments about these estimates or any aspects of this data collection, contact: US Department of Labor, OSHA Office of Statistics, Room N-3644, 200 Constitution Ave, NW, Washington, DC 20210. Do not send the completed forms to this office.

Page 1 of 1

OSHA's Form 300A
Summary of Work-Related Injuries and Illnesses

Year _____

U.S. Department of Labor
Occupational Safety and Health Administration

Form approved OMB no. 1218-0176

All establishments covered by Part 1904 must complete this Summary page, even if no injuries or illnesses occurred during the year. Remember to review the Log to verify that the entries are complete and accurate before completing this summary.

Using the Log, count the individual entries you made for each category. Then write the totals below, making sure you've added the entries from every page of the log. If you had no cases write "0."

Employees, former employees, and their representatives have the right to review the OSHA Form 300 in its entirety. They also have limited access to the OSHA Form 301 or its equivalent. See 29 CFR 1904.35, in OSHA's Recordkeeping rule, for further details on the access provisions for these forms.

Number of Cases

Total number of deaths	Total number of cases with days away from work	Total number of cases with job transfer or restriction	Total number of other recordable cases
#REF!	#REF!	#REF!	#REF!
(G)	(H)	(I)	(J)

Number of Days

Total number of days away from work	Total number of days of job transfer or restriction
#REF!	#REF!
(K)	(L)

Injury and Illness Types

Total number of...
(M)
(1) Injury   #REF!   (4) Poisoning   #REF!
(2) Skin Disorder   #REF!   (5) All other illnesses #REF!
(3) Respiratory Condition #REF!

Post this Summary page from February 1 to April 30 of the year following the year covered by the form

Public reporting burden for this collection of information is estimated to average 50 minutes per response, including time to review the instruction, search and gather the data needed, and complete and review the collection of information. Persons are not required to respond to the collection of information unless it displays a currently valid OMB control number. If you have any comments about these estimates or any aspects of this data collection, contact: US Department of Labor, OSHA Office of Statistics, Room N-3644, 200 Constitution Ave. NW, Washington DC 20210. Do not send the completed forms to this office.

Establishment Information

Your establishment name _____

Street _____

City _____ State _____ Zip _____

Industry description (e.g., Manufacture of motor truck trailers) _____

Standard Industrial Classification (SIC), if known (e.g., SIC 3715) _____

Employment Information

Annual average number of employees _____

Total hours worked by all employees last year _____

Sign here

Knowingly falsifying this document may result in a fine.

I certify that I have examined this document and that to the best of my knowledge the entries are true, accurate, and complete.

_____ Company executive

_____ Phone   _____ Title   _____ Date

OSHA's Form 301
Injuries and Illnesses Incident Report

Attention: This form contains information relating to employee health and must be used in a manner that protects the confidentiality of employees to the extent possible while the information is being used for occupational safety and health purposes.

U.S. Department of Labor
Occupational Safety and Health Administration
Form approved OMB no. 1218-0176

This *Injury and Illness Incident Report* is one of the first forms you must fill out when a recordable work-related injury or illness has occurred. Together with the *Log of Work-Related Injuries and Illnesses* and the accompanying *Summary*, these forms help the employer and OSHA develop a picture of the extent and severity of work-related incidents.

Within 7 calendar days after you receive information that a recordable work-related injury or illness has occurred, you must fill out this form or an equivalent. Some state workers' compensation, insurance, or other reports may be acceptable substitutes. To be considered an equivalent form, any substitute must contain all the information asked for on this form.

According to Public Law 91-596 and 29 CFR 1904, OSHA's recordkeeping rule, you must keep this form on file for 5 years following the year to which it pertains

If you need additional copies of this form, you may photocopy and use as many as you need.

Completed by _____
Title _____
Phone _____ Date _____

Information about the employee

1) Full Name _____
2) Street _____
   City _____ State _____ Zip _____
3) Date of birth _____
4) Date hired _____
5) ☐ Male
   ☐ Female

Information about the physician or other health care professional

6) Name of physician or other health care professional _____
7) If treatment was given away from the worksite, where was it given?
   Facility _____
   Street _____
   City _____ State _____ Zip _____
8) Was employee treated in an emergency room?
   ☐ Yes
   ☐ No
9) Was employee hospitalized overnight as an in-patient?
   ☐ Yes
   ☐ No

Information about the case

10) Case number from the Log _____ *(Transfer the case number from the Log after you record the case.)*
11) Date of injury or illness _____
12) Time employee began work _____ AM/PM
13) Time of event _____ AM/PM ☐ Check if time cannot be determined
14) What was the employee doing just before the incident occurred? Describe the activity, as well as the tools, equipment or material the employee was using. Be specific. Examples: "climbing a ladder while carrying roofing materials"; "spraying chlorine from hand sprayer"; "daily computer key-entry."
15) What happened? Tell us how the injury occurred. Examples: "When ladder slipped on wet floor, worker fell 20 feet"; "Worker was sprayed with chlorine when gasket broke during replacement"; "Worker developed soreness in wrist over time."
16) What was the injury or illness? Tell us the part of the body that was affected and how it was affected; be more specific than "hurt", "pain", or "sore". Examples: "strained back"; "chemical burn, hand"; "carpal tunnel syndrome."
17) What object or substance directly harmed the employee? Examples: "concrete floor"; "chlorine"; "radial arm saw." If this question does not apply to the incident, leave it blank.
18) If the employee died, when did death occur? Date of death _____

Public reporting burden for this collection of information is estimated to average 22 minutes per response, including time for reviewing instructions, searching existing data sources, gathering and maintaining the data needed, and completing and reviewing the collection of information. Persons are not required to respond to the collection of information unless it displays a current valid OMB control number. If you have any comments about this estimate or any other aspects of this data collection, including suggestions for reducing this burden, contact: US Department of Labor, OSHA Office of Statistics, Room N-3644, 200 Constitution Ave. NW, Washington, DC 20210. Do not send the completed forms to this office.

Credit Application

Company name
DBA (if different)
Contact person
Address
Phone   Fax
Federal tax ID or Social Security number.
Type of business   No. of employees
Date business established
Types of products you will purchase
Amount of credit requested $

Are you a:
☐ CORPORATION
State of incorporation

Names, titles, and addresses of your three chief corporate officers

Name and address of your resident agent

☐ PARTNERSHIP
Names and addresses of the partners

☐ SOLE PROPRIETORSHIP
Are you sales tax exempt?   ☐ Yes   ☐ No
Have you ever had credit with us before? ☐ Yes   ☐ No
If yes, under what name?

Authorized purchasers

Purchase order required?   ☐ Yes   ☐ No

TRADE REFERENCES

Reference #1
- Name: _____
- Address: _____
- Phone: _____
- Fax: _____

Reference #2
- Name: _____
- Address: _____
- Phone: _____
- Fax: _____

Reference #3
- Name: _____
- Address: _____
- Phone: _____
- Fax: _____

BANK REFERENCES

Bank #1
- Account #: _____
- Phone: _____
- Fax: _____
- Contact person: _____
- Name of bank: _____
- Address: _____

Bank #2
- Account #: _____
- Phone: _____
- Fax: _____
- Contact person: _____
- Name of bank: _____
- Address: _____

I represent that the above information is true and is given to induce to extend credit to the applicant. My company and I authorize the requestor to make such credit investigation as sees fit, including contacting the above trade references and banks and obtaining credit reports. My company and I authorize all trade references, banks, and credit reporting agencies to disclose to the requestor any and all information concerning the financial and credit history of my company and myself.

I have read the terms and conditions stated below and agree to all of these terms and conditions.

Authorized signature: _____
Printed name: _____
Title: _____ Date: _____

GENERAL TERMS AND CONDITIONS AND PERSONAL GUARANTEE

1. Invoices are sent upon shipment and payment is due NET 30.

2. No additional credit will be extended to past due accounts unless satisfactory arrangements are made with our credit department.

3. PERSONAL GUARANTEE: If the credit customer is a corporation, then those signing this application, whether signing as an officer or not, personally guarantee payment for all items purchased on credit by the corporation.

CARRIER SEAL POLICY

FINISHED GOODS
(FULL TRUCKLOADS AND MULTISTOP LOADS)

Due to the current national security issues, _____ mandates that no trailers of finished goods are to be received without a seal or with an unsatisfactory seal, receiving personnel must examine the vehicle contents and determine whether the product is to be placed on MANDATORY 24-HOUR HOLD or MANDATORY LOCKOUT. The facility is to notify the quality control manager or plant manager of the seal issues.

MANDATORY 24-HOUR HOLD PROCEDURE

If any vehicle transporting finished goods is received:

- without a seal or with an unsatisfactory seal,
- but shows no evidence of product tampering or pilferage,

THEN

- contact the quality control manager or plant manager to place the product on mandatory 24-hour hold.

MANDATORY LOCKOUT HOLD PROCEDURE

If any vehicle transporting finished goods is received:

- without a seal or an unsatisfactory seal,
- and shows evidence of tampering or pilferage,

THEN

- contact the quality control manager or plant manager to place the product on mandatory 24-hour hold;
- contact logistics manager; and
- quality control manager should take digital pictures of the trailer seal area and areas of concern before the trailer is unloaded.

Shipment of Finished Goods

Because of the current national security concerns, _____'s quality control department mandates that all finished goods shipments must be properly sealed. The seal number should be printed on the bill of lading prior to shipment.

GMO INQUIRY LETTER

Supplier Product: _____

Company Item No.: _____

Supplier Name: _____

Please answer the following questions on the product you supply to us. Where available we have indicated your item number and description. If it is not listed, please fill in the information so we can update our databases.

1. Is it manufactured with ingredients containing or produced from genetically modified organisms? _____
2. If yes, what is the source? _____
3. If no, is it "identity preserved" (IP)? _____
4. If not IP, is it "PCR (Polymerase Chain Reaction) negative"? _____

If you would like to discuss this request in further detail, please do not hesitate to contact me at _____.

Thank you for your assistance.

Name
Title
Company

Extended Nutritional Information Inquiry Form

Supplier Product Number: _____

Company Item Number: _____

Supplier Name: _____

_____ requests as much of the following information as you can supply beyond the Nutritional Labeling and Education requirements.
Nutrients in bold; italic type are required to complete our regulatory requirements.

NUTRIENTS	Amount per 100(g) of ingredients	CHECK ONE		NUTRIENTS	Amount per 100(g) of ingredients	CHECK ONE	
		*by analysis	*by calculation			*by analysis	*by calculation
Total Solids							
Moisture				***Total Sugars (g)***			
Calories				Sucrose (g)			
Total Fat (g)				Dextrose (g)			
Saturated Fatty Acids (g)				Lactose (g)			
4:0				Glucose (g)			
6:0				Fructose (g)			
8:0				Maltose (g)			
10:0				Starch (g)			
12:0				***Protein (g)***			
14:0				***Polyols (g)***			
16:0				***Cholesterol (mg)***			
18:0				***Sodium (mg)***			
Polyunsaturated Fatty Acids (g)				***Vitamin A (IU)***			
18:2				Retinol (mcg)			
18:2 Trans				Beta Carotene (mcg)			
18:3				***Iron (mg)***			
18:4				***Calcium (mg)***			
20:4				***Vitamin C (mg)***			
Cis Poly Fatty Acids (g)				Potassium (mg)			
Trans Poly Fatty Acids (g)				Vitamin D (IU)			
Monounsaturated Fatty Acids (g)				Vitamin E (IU)			
14:1				Thiamin (mg)			
16:1				Riboflavin (mg)			
16:1 Trans				Niacin (mg)			
18:1				Vitamin B6 (mg)			
18:1 Trans				Folic Acid (mg)			
Cis Mono Fatty Acids (g)				Vitamin B12 (mcg)			
Trans Mono Fatty Acids (g)				Biotin (mg)			
Total Carbohydrate (g)				Phosphorus (mg)			
Total Dietary Fiber (g)				Magnesium (mg)			
Soluble Fiber (g)				Zinc (mg)			
Insoluble Fiber (g)				Alcohol (g)			
Sulfites				Caffeine (mg)			
Ash (g)				Pantothenic Acid (mg)			

_____
Name and Title of person completing the nutritional information

Date: _____

Date

Company Name
«Address Block»
«Address Block»

Dear Supplier:

_____ is applying for halal certification on several products from a nationally and internationally recognized body. The Islamic certifier requires a letter from our suppliers stating your ingredients do not contain alcohol, natural L-cysteine extracted from hair or feathers (synthetic L-cysteine is acceptable), animal fats and/or extracts, bloods of any origin, blood plasma, or pork and/or other meat by-products, and that alcohol is not used as a processing aid.

    If your product meets these requirements, please fax a letter to _____ at _____ stating that the above is met by your company.

    If you have any concerns or questions, please call us at _____.

    Thank you.

AFFIDAVIT FOR INGREDIENTS

Dear Supplier:

Due to _____'s policies for organic and nonorganic ingredients, it must be verified that all ingredients comply with all NOP regulations.
   Ingredients used in organic products **cannot** be:

1. from a GEO (genetically engineered organism) source;
2. irradiated (ionizing radiation); or
3. grown with the use of sewage sludge.

Company: _____

We will need to know the status of the ingredient/ product listed below:

Raw material ingredient number:	Name of ingredient:

Please check off below to describe the status of your ingredient.

To the best of our knowledge the ingredient we provide you:

	Does	Does not
Contain genetically engineered ingredients or carrier sources	_____	_____
Use ingredients that have been exposed to irradiation	_____	_____
Use ingredients that were grown on land exposed to sewage sludge	_____	_____
Use any processing aides declared or undeclared	_____	_____
Contain enzymes	_____	_____

   a. Identify source: rennet, pepsin, microbiological, etc. _____

   b. If microbiological, is it produced from GMO technology? ☐ Yes ☐ No

Is there backup material available for review by an approved authority concerning this statement? ☐ Yes ☐ No

Signed: _____

Name: _____

Company Affiliation: _____

Date: _____

BROKER AGREEMENT

THIS AGREEMENT is made as of the ___ day of _____, 20___, between _____ (hereinafter referred to as the Company) and _____ a _____ Corporation (hereinafter referred to as Broker).

1. Engagement.
   (a) Subject to the terms and conditions herein set forth, the Company contracts for the Broker, on a nonexclusive basis, to use its best efforts to sell all Products produced by the Company, without limitation or exclusion, to those companies listed in Exhibit A. From time to time, Broker will request that the Company supply its literature, goods, or services to other companies not listed in Exhibit A attached hereto. It is understood and agreed to that the recipient or addressee of such a request shall be treated by the Company as a customer of the Broker. Once established as a customer of the Broker, said company will be afforded and subjected to the same terms, stipulations, conditions, and protections set forth in this agreement as is granted to the company named in Exhibit B attached hereto. The Broker hereby covenants, warrants, and agrees to sell and cause to be sold the Products according to any and all laws, and all regulations and rules pertaining thereto, heretofore or hereafter issued by the Company. The terms of this agreement shall be in effect for ___ (_) years from the date of this agreement and will automatically renew for an additional ___ (_) years following review and approval from both parties unless either party gives written notice 60 days prior to the end of the term.
2. Representations and Warranties of the Broker. The Broker represents and warrants to the Company as follows:
   (a) Broker is duly organized, validly existing, and in good standing under the laws of the Territory with all requisite power and authority to enter into this Agreement and to carry out the obligations hereunder.
   (b) Broker shall comply with all Federal laws and the laws of the State of _____ pertaining to the sale of the Products.
3. Protocols—sampling, order placement, order acknowledgment. Company agrees to comply with all stipulations set forth in Exhibit C of this agreement.
4. Commissions to Broker. Broker shall receive payment of a commission calculated to be no less than ____ percent (_.0%) of Net Sales (as defined and adjusted below) on all orders received by the Company, either directly or indirectly, and for which Brokers' efforts, contacts, or relationships are determined to be the efficient procuring cause of such sale. It is further agreed that to the extent the Broker may be able to procure Orders for Products at a price higher than listed by the Company, the Broker shall have the authority to sell the Products at such increased price. For such sales at an increased price, the Broker shall be paid the commission rates set forth above for the list price of the products and an additional fifty percent (50%)

of the difference between the list price and the increased sales price. Company agrees to pay to Broker its earned commissions on a monthly basis, doing so within 15 days after the end of each calendar month based on actual Net Sales collected during the immediately preceding calendar month. "Net Sales" shall mean receipts collected by the Company from the sale of the Company's products less delivery costs if included in the price. Broker will continue to receive payments from the Company for any commission-earning sales in accordance with this schedule and without interruption for the duration of this contract. In the event the Company terminates this agreement for any reason other than for "cause," Company agrees to continue payment of commissions to the Broker for an additional period of ___ (_) months immediately following the agreement's termination.

5. Independent Contractor. Broker shall serve solely as an independent contractor, and no other relationship shall be created by this Agreement. Nothing in this Agreement shall be construed to give Broker any right to own or control Company's business, or to give Company any right to own or control Broker's business. Broker shall not have nor shall represent itself as being an agent of Company or having the power to make contracts or commitments in the name of or binding upon Company. Broker shall assume and pay all costs of its activities hereunder and shall not be entitled to reimbursement from Company for any expenses incurred by it in connection with its performance hereunder.

6. Termination. This Agreement may be terminated by either party hereto on sixty (60) days prior written notice to the other. Within sixty (60) days following receipt of such notice, Company shall pay Broker all commissions due on all orders received (a) during the 60 days immediately preceding the effective date of termination in addition to (b) the 60 days immediately following the effective date to terminate unless Broker is terminated "for cause" (as defined herein).

7. As used herein, "cause" shall include negligence or willful misconduct by Broker in the performance of its obligations hereunder, Broker's willful breach of its obligations under this Agreement, misappropriation of corporate opportunity of Company, and unauthorized dissemination of confidential information of Company.

8. Confidentiality. Company and Broker possess "Proprietary Information" such as formulations, manufacturing techniques and processes, customer lists, sales and marketing plans, and other information (including the terms and provisions relating to Commissions found herein) which are to be treated by each as confidential. Both parties agree not to do or omit to do anything which could result in the same being disclosed to the public or to any person, firm, or corporation (a "Prohibited Disclosure") unless disclosed under order of any court or in order to comply with any applicable law, regulation, rule, or ordinance. It is further understood and agreed to that Proprietary Information shall not include information which: (a) is or has become generally available to the public other than through a breach of this agreement, (b) which can be demonstrated to have been known to the

Broker prior to disclosure by the Company, (c) which is lawfully obtained by Broker from a source independent of the Company. Broker is permitted to disclose such information to its employees, partners, agents, and/or representatives as is necessary for it to carry out the purposes of this agreement. Both parties acknowledge that the other would be greatly injured by any breach of Paragraphs 7 and 8.

9. Injunctive Relief. Both parties specifically recognize that any breach of Paragraph 8 shall cause irreparable injury to both parties and that actual damages may be difficult to ascertain and, in any event, will be inadequate. Accordingly (and without limiting the availability of legal or equitable including injunctive remedies under any other provisions of this Agreement), both parties agree that in the event of any such breach, the party suffering the breach of confidentiality shall be entitled to injunctive relief, without the requirement of posting a bond or other security, in addition to such other legal and equitable remedies that may be available.

10. Notices. All notices or other communications required by this Agreement shall be in writing and shall, except as otherwise required by law or this Agreement, be deemed duly served and given when personally delivered, or instead of personal service, three (3) days after being deposited in the United States mail, certified, return receipt requested, first class postage prepaid, addressed to the address of the party to whom it is directed as shown at the end of this Agreement. Any party may change that party's address for the purpose of this paragraph by giving written notice of such change to the other party in the manner provided for in this paragraph.

11. Pricing. Broker shall quote all prices to customers as per Company's instructions but retains the right, and at its own discretion, to quote prices higher than those prices set forth by the Company, such quotes being subject to change without notice, and shall, in addition to the solicitation of orders, make such calls on existing accounts as are necessary to properly service such accounts for the preservation of existing business and the promotion of new business.

12. Waivers. All failures of either party to comply with any obligation, agreement, covenant, or condition set forth in this Agreement must be expressly waived in writing by the party intended to be bound by such waiver. However, any waiver or failure to insist upon strict compliance with any obligation, agreement, covenant, or condition shall not operate as a waiver of, or estoppel with respect to, any other failure to strictly comply with the terms of this Agreement.

13. Governing Law and Resolution of Disputes. This Agreement and the legal relations between the parties shall be governed by, and construed in accordance with, the laws of the State of _____. The parties further agree that all actions, proceedings, or disputes under this Agreement shall be resolved in the courts in the jurisdiction of the complaining party.

14. Amendment and Modification. Subject to applicable law, this Agreement may be amended, modified, and supplemented by mutual written consent of each party.

15. Severability. If any term, provision, covenant, or condition of this Agreement is held by a court of competent jurisdiction to be invalid, void, or unenforceable, the remainder of the provisions shall remain in full force and effect and shall in no way be affected, impaired, or invalidated.
16. Entire Agreement. This Agreement and the attached Exhibits A, B constitute the entire agreement between the parties with respect to the subject matter of this Agreement, superseding all prior agreements (whether written or oral) and understandings between the parties with respect to the subject matter.
17. Headings. The headings of the paragraphs and articles of this Agreement are inserted for convenience only and do not constitute part of this Agreement and shall not be used in its construction.
18. Attorneys' Fees. In the event that any party hereto institutes an action or proceeding for a declaration of the rights of the parties under this Agreement, for injunctive relief, for an alleged breach or default of, or any other action arising out of this Agreement, or the transactions contemplated hereby, or in the event any party is in default of its obligations pursuant thereto, whether or not suit is filed or prosecuted to final judgment, the non-defaulting party or prevailing party shall be entitled to its actual attorneys' fees and to any court or arbitrator's costs incurred, in addition to any other damages or relief awarded.
19. Mutual Contribution. This Agreement has been drafted on the basis of the parties' mutual contributions of language and it is not to be construed against any party as being the drafter (or causing the drafting) of this Agreement.
20. Counterparts. This Agreement may be executed in one or more counterparts and by the parties hereto on separate counterparts, each of which counterpart, when so executed and delivered, shall be deemed an original and all of which counterparts, taken together, shall constitute one agreement.
21. Cooperation. Each party agrees to perform any further acts and to execute and deliver any and all additional documents which may be reasonably necessary or appropriate to carry out the provisions of this Agreement and each party agrees that he will not act in any manner which will hinder, interfere with, or prohibit the performance of the parties under this Agreement.

IN WITNESS WHEREOF, the undersigned have executed this Agreement as of the date first written above.

Broker   Company

_____   _____

By: _____   By: _____
Print name: _____   Print name: _____
Title: _____   Title: _____
Address: Address:

_____   _____
_____   _____

Fax: _____   Fax: _____

EXHIBIT A

CUSTOMERS/TERRITORY

The following list of company names is to be considered as (a) current clients of _____, (b) soon to become clients of _____, or (c) those companies with whom _____ and it principals and/or associates have had some basis or degree of prior awareness and/or contact:

_____   _____

_____   _____

_____   _____

It is the sole responsibility of _____ to safeguard and protect _____'s rights hereunder as being the exclusive sales Broker for the Company at these accounts in addition to any new clients as they become identified and agreed to by Company and Broker. In this regard, all orders received either directly or indirectly by _____ directly from or on behalf of those companies referred to as "_____ Accounts," regardless of reason or cause will be credited to _____ as though receipt of said orders was the direct result of _____'s efforts.

EXHIBIT B

SAMPLING/ORDER PLACEMENT PROTOCOL

1. Sampling protocol/price quotation

    _____ will designate an employee to act as a liaison with _____ in fulfilling sample requests and purchase order placement. From time to time, _____ will contact this individual with information considered to be confidential in nature and as such to be used solely by _____ for the expressed purpose of sample selection or project initiation. Upon shipment of samples, a copy of packaging invoice will be mailed directly to _____ or via fax _____ or via e-mail _____.
    It is important to note that_____ will only provide pricing information at time of making sample shipment directly to _____. The Company is prohibited from providing pricing information directly to any _____ Customer, contract manufacturer, or other third party who might be representing a _____ Customer. It is understood that a final selling price can only be established by _____ with the Company receiving written pricing confirmation from _____ via fax or e-mail.

2. Order placement and acknowledgment

   _____ may receive written purchase orders directly from _____, directly from a _____ Customer, or from a contract manufacturer or other third party representing a _____ Customer. _____ will furnish _____ with copies of all orders received and will forward same directly to _____, or via fax _____ or via e-mail _____. _____ agrees to process these orders using the final pricing established by _____ unless advised otherwise in writing.

3. Order shipment

   Within twenty four (24) hours immediately following the shipment of any order for which _____ is to receive commission, _____ will send confirmation of said shipment directly to _____ via phone _____, fax _____, or e-mail _____.

Index

A

Adversarial relationship, between production, quality control, 15
Affidavit for ingredients, 493
Air emissions, environmental responsibility program, 440, 445
Allergen food hazards, risk assessment worksheet, 79
Allergen program, 189–204
    allergen inclusion chart, 198
    allergen information request form, 196
    allergen testing form, 200
    application, 192–194
    daily plant sanitation inspection form, 199
    documentation, 203
    egg, egg derivatives, 190
    fish, fish derivatives, 190
    gluten nonwheat, derivatives, 190
    initial training verification, 201
    manufacturing, 202–203
    milk, milk derivatives, 189
    mollusks, 190
    monosodium glutamate, 190
    overview, 202–203
    peanuts, peanuts derivatives, 190
    raw materials, 202
    required program, 189
    risk assessment worksheet, 195
    seed, seed derivatives, 190
    shellfish, shellfish derivatives, 190
    soy, soy derivatives, 190
    sulfites, derivatives, 190
    supplemental materials, 194–203
    testing, 203
    theory, 189–192
    tree nuts, tree nut derivatives, 190
    wheat, wheat derivatives, 190
    yellow #5, 190
    yellow #6, 190
Allergenic risk analysis, 55–57
Allergens food hazards, risk assessment worksheet, 93
Audit procedure/schedule, 154
Auditing, 64–65, 69
    good manufacturing practice, 155
Auditing meeting attendance, 82
Audits, outside, 459–462

B

Baked product, process flow for, 87
Baseline
    establishing, 11–12
    measuring compliance against, 12–13
Biological control, pest control program, 159
Biosecurity Act compliance, 42–43
Biosecurity program, 409–424
    application, 414–415
    check in log, 421
    facility, 413–414, 418–419
    general information, 417
    key check out, 421
    management, 409–411, 417
    operations, 414, 419
    overview, 417–419
    public, 412–413, 418
    required program, 409
    responsibility, 419
    staff, 411–412, 417–418
    supplemental materials, 416–424
    theory, 409–414
    verification documentation, security/biosecurity training, 420
    visitor sign in log
        cover, 423
        spine label, 424
    visitor sign in sheet, 422
Bird, pest control program, 166–167
Broker agreement, 494–499
Building, 359–361

C

Campylobacter jejuni, 50, 403
Carrier seal policy, 486
Ceilings
    sanitation program, 227
    sanitation standard, 134
Chain of custody, contract laboratory testing program, 476
Charts
    allergen inclusion, 198
    color scheme, 399
    glass inspection, 391
    organizational, 113–116
        application, 113–114
        required program, 113

supplemental materials, 115–116
theory, 113
plastic inspection, 391
preorganizational, 37
summary, 81, 97
weight control, 210, 212
Check in log, biosecurity program, 421
Chemical control, pest control program, 159–164
Chemical food hazards, risk assessment worksheet, 76, 90
Chemical risk analysis, 48–49
Chopped product, process flow for, 86
Clam meat, frozen, *Salmonella* in, 2
Clostridium botulinum, 50
Color scheme chart, 399
Company insurance, 38
Compartmentalized relationship, between production, quality control, 15
Compensation worksheet for sample company, pest control program, 168
Compliance, measuring against baseline, 12–13
Conciliatory team-building relationship, between production, quality control, 15–16
Confidentiality agreement, 481–482
Container, shipping, inspection log, 300
Continuing food guarantee program, 467–472
application, 467–468
required program, 467
supplemental materials, 469–471
theory, 467
Contract applicators sheets, 38
Contract laboratory testing program, 473–476
application, 474
chain of custody, 476
optional program, 473
supplemental materials, 475–476
theory, 473–474
Contract sample, pest control program, 169–170
Control limit establishment, 61–62
Conveyor 1 belt, sanitation program, 226
Conveyor 3 belt, sanitation program, 227
Conveyor 2 bucket, sanitation program, 227
COP tank, sanitation program, 227
Corrective action, 68
Corrective actions, 62
Cost of quality, 5–6
Critical control point determination, 57–60
Critical control point identification, 68
Critical limits, establishing, 68
Cultural control, pest control program, 159
Customer complaint program, 277–286
application, 280
customer complaint flow diagram, 283
customer complaint form, 282
customer complaint log, 284
documentation, 286
general information, 285

initiator, 285
marketing, 286
overview, 285–286
quality control, 285
required program, 277
sales, 286
supplemental materials, 281–286
theory, 277–279
Customer relationship, quality control and, 17
Cyclospora, 403

D

Daily inspection, 220
Daily line check sheet, 132
Daily non-production schedule, sanitation program, 236, 244
Daily plant sanitation inspection form, 222
allergen program, 199
Daily production schedule, sanitation program, 235, 243
Date of expiration, lot coding program, 274
Declaration of allergens, 43
Defective material program, 38
Defective product handling, 324–325
Department of Agriculture, regulation by, 1
Destruction, hold/defective material program, 384
Development of quality control systems
allergen program, 189–204
continuing food guarantee program, 467–472
contract laboratory testing program, 473–476
customer complaint program, 277–286
environmental responsibility program, 439–446
environmental testing program, 447–458
glass, hard plastic, wood program, 387–394
good manufacturing practices program, 117–156
Hazard Analysis Critical Control Point program, 39–98
hold/defective material program, 371–386
inspection program, 213–222
kosher program, 425–430
loose-material program, 395–402
lot coding program, 269–276
metal detection program, 247–260
microbiology program, 403–408
organic program, 431–438
organizational chart, 113–116
other forms, 479–500
outside audits, 459–462
overview, 19–22
pest control program, 157–188
quality control program overview, 99–112
recall program, 313–330
receiving program, 287–296

Index

record keeping, 477–478
regulatory inspection program, 261–268
sanitation program, 223–246
security/biosecurity program, 409–424
shipping program, 297–302
social responsibility program, 463–466
specification program, 303–312
supplier certification program, 331–370
weight control program, 205–212
Divisive relationship, between production, quality control, 15
Documentation, pest control program, 167
Doors, sanitation standard, 136

E

Egg, egg derivatives, allergen program, 190
Electrical boxes, sanitation standard, 133
Emergency recall notification list, 327
Environmental responsibility program, 439–446
    air emissions, 440, 445
    application, 441
    environmental checklist, 444–445
    environmental policy, 443
    hazardous materials, 440, 445
    hazardous waste handling, 440
    management, 439–440, 444
    optional program, 439
    supplemental materials, 442–445
    theory, 439–441
    waste, 444
    waste management, 440–441
    wastewater, 440
Environmental testing program, 447–458
    application, 450
    direct product contact surfaces, 452
    environmental sampling procedures, 456–457
    general information, 452
    indirect product contact surfaces, 452
    non-product-contact surfaces, 452
    optional program, 447
    overview, 452–457
    sampling sites, 452
    supplemental materials, 451–457
    testing protocol, 453–456
    theory, 447–450
Equipment, 359–361
    maintenance, design, installation, sanitation standard, 139
    sanitation, 122–123
    sanitation standard, 139
Equipment cleanliness, sanitation standard, 142
Escherichia coli, 2, 51, 403
Extended nutritional information inquiry form, 491

F

Facility map, 38
FDA. *See* Food and Drug Administration
Finished goods, 305
    full truckloads, multistop loads, 489
    specifications, 304–305
Finished product, specification, 309–310
Fish, fish derivatives, allergen program, 190
Floors
    sanitation program, 227
    sanitation standard, 135
Flow diagram, customer complaint, 283
Flow master list, 74
Food and Drug Administration, regulation by, 1
Food guarantee program, continuing, 467–472
Food safety audit report, 348–350
Food safety/good manufacturing practice assessment rating system, 368
Food safety/good manufacturing practice rating analysis, 351–367
Foreign material investigation log, 258
Frozen peas, *Staphylococcus* in, 2

G

Genetically modified organisms inquiry letter, 490
Giardia lamblia, 51
Glass, 387–394
    inspection chart, 391
    quality control program, 105
    verification documentation, 392
Gluten nonwheat, derivatives, allergen program, 190
Good manufacturing practices, 37, 103, 117–156
    application, 128–130
    audit procedure/schedule, 154
    ceilings, sanitation standard, 134
    checklist, 342–347
    controls, 123–126
    daily line check sheet, 132
    distribution, 126
    doors, sanitation standard, 136
    electrical boxes, sanitation standard, 133
    equipment, 122–123
        maintenance, design, installation, sanitation standard, 139
        sanitation standard, 139
    equipment cleanliness, sanitation standard, 142
    floors, sanitation standard, 135
    good manufacturing rules, 149–150
    grounds, 119–120
    hand-washing facilities, sanitation standard, 145
    installation of equipment, sanitation standard, 139

natural, unavoidable food defects, no health hazards, 126–128
odors, objectionable, sanitation standard, 141
overview, 155
plant, 119–120
processes, 123–126
product protection, sanitation standard, 143–144
required program, 117
sanitary facilities, controls, 121–122
sanitary operations, 120–121
standards, 128–130
stock arrangement, sanitation standard, 140
supplemental materials, 131–155
surroundings, sanitation standard, 147
theory, 117–128
toilet facilities, sanitation standard, 146
unessential items, sanitation standard, 138
utensils, 122–123
ventilation, sanitation standard, 137
verification documentation
    employee good manufacturing practice, annual review, 151
    good manufacturing practice, training, 148
visitors, good manufacturing rules, 152–153
walls, sanitation standard, 133
warehousing, 126
windows, sanitation standard, 133
Good manufacturing rules, 149–150
Governmental bodies, regulation by, 1

H

HACCP. *See* Hazard analysis critical control point
Halal certificate, 44
Hand-washing facilities, sanitation standard, 145
Hard plastic, 387–394
    application, 388–389
    overview, 393
    plastic, 393
    prevention, 393
    quality control program, 105
    supplemental materials, 390–394
    theory, 387–388
    verification documentation, training, 392
Hasson twin tube form filler, sanitation program, 227
Hazard Analysis Critical Control Point program, 39–98
    allergen food hazards, risk assessment worksheet, 79
    allergenic risk analysis, 55–57
    allergens food hazards, risk assessment worksheet, 93
    auditing, 64–65

auditing meeting attendance, 82
baked product, process flow for, 87
chemical food hazards, risk assessment worksheet, 76, 90
chemical risk analysis, 48–49
chopped product, process flow for, 86
control limit establishment, 61–62
corrective action, 68
corrective actions, 62
critical control point determination, 57–60
critical control point identification, 68
critical limits, establishing, 68
flow master list, 74
hazard analysis, 47–48
    risk assessment, 67
implementation team roster development, 70
ingredient specification sheet, 42
material safety data sheet, 43
microbiological food hazards, risk assessment worksheet, 77, 91
microbiological risk analysis, 50–53
monitoring procedures, 68
overview, 45–46, 67–68
physical food hazards, risk assessment worksheet, 78, 92
physical risk analysis, 54–55
planning implementation, 44–45
point evaluation worksheet, product/process control, 80, 94–96
process analysis, 68
process flows, 46
product description, 46–47
product description worksheet, 73, 88–89
program auditing, 69
program verification, 69
record keeping system, 69
required program, 39–41
summary chart, 81, 97
summary sheet, 62–63
supplemental materials, 66–83
supplier data sheet, 72, 85
team meeting attendance, 71
theory, program planning, 41–44
training, 65
    verification documentation, 83
verification, 63–64
Hazardous materials, environmental responsibility program, 440, 445
Hazardous waste handling, environmental responsibility program, 440
Health departments, regulation by, 1
Hold/defective material program, 371–386
    application, 373
    cover, 381
    destruction, 384
    destruction log, 380
    destruction notification, 377

Index

general information, 384
hold log, 378
hold notification, 375
hold tag, 386
holds, 384
manufacturing deviation report, 383
overview, 384–385
records, 384
release log, 379
release notification, 376
releases, 384
responsibility, 385
spine label, 382
supplemental materials, 374–386
theory, 371–372
Hold procedure, mandatory, 489

I

Implementation team roster development, 70
Ingredient specification, 307–308
    sheet, 42
Ingredient statements, 38
Ingredient supplier, co-packer quality, safety survey, 340–341
Ingredients, affidavit for, 493
Insect, pest control program, 166
Inspection, pest control, 167
Inspection program, 213–222
    application, 215–216
    daily inspection, 220
    daily plant sanitation inspection form, 222
    documentation, 220–221
    inspection team roster, 218
    monthly inspection, 220–221
    monthly inspection form, 219
    overview, 220–221
    regulatory, 261–268
        application, 264
        initial introduction, 266
        inspection, 266–267
        inspection conclusion, 267–268
        inspection follow-up, 268
        overview, 266–268
        required program, 261
        supplemental materials, 265–268
        theory, 261–264
    required program, 213
    responsibility, 220–221
    supplemental materials, 217–222
    theory, 213–215
Inspection team roster, 218
Installation of equipment, sanitation standard, 139
Insurance vendors, request for evidence of, 338
Ishida scale, sanitation program, 227

K

Key check out, 421
    biosecurity program, 421
Kosher certificates, 43
Kosher program, 425–430
    application, 427
    optional program, 425
    receiving kosher list, 429
    supplemental materials, 428–429
    theory, 425–427

L

Labels, returned products, issues regarding, 2
Listeria monocytogenes, 2, 51, 403
Lockout hold procedure, mandatory, 489
Loose-material program, 395–402
    application, 397
    color scheme chart, 399
    general information, 401
    overview, 401
    program, 401
    required program, 395
    responsibility, 401
    signage, 401
    supplemental materials, 398–401
    theory, 395–397
    training, 401
    verification documentation, 400
Loose tool/material, quality control program, 104
Lot coding, 102, 269–276
    application, 271–272
    basic code, 274
    date of expiration, 274
    general information, 274
    line code, defined, 270
    lot code explanation, 275
    overview, 274
    period code, defined, 270
    plant code, defined, 270
    printing, 274
    record keeping, 274
    required program, 269
    responsibility, 274
    shift code, defined, 270
    supplemental materials, 273–275
    theory, 269–271
    time, defined, 270

M

Management, role in quality control, 6–9
Mandatory lockout hold procedure, 489
Manufacturing, allergen program, 202–203
Marketing, customer complaint program, 286
Material safety data sheet, 43

Index

Mechanical control, pest control program, 158–159
Metal detection program, 37, 227, 247–260
    accountability, 254
    application, 250–251
    check sheet, 257
    foreign material investigation log, 258
    handling, 253
    metal detection, 253
    metal detector operation, 255–256
    overview, 253–254
    periodic testing, 255–256
    placement, 247–248
    raw materials, 253
    record keeping, 254, 256
    record storage, 254
    rejected product, 254
        action required for, 256
    rejection, 249–250
    required program, 247
    setup, sensitivity testing, 255
    size, 248
    supplemental materials, 252–259
    testing, 248–249
    testing fails, action required, 256
    theory, 247–250
    training, verification documentation, 259
Microbiological food hazards, risk assessment worksheet, 77, 91
Microbiological risk analysis, 50–53
Microbiology program, 403–408
    application, 405
    Escherichia coli, 403
    finished goods testing, 407
    general information, 407
    methodology, 407
    microbiological testing log, 408
    overview, 407
    process testing, 407
    raw materials testing, 407
    responsibility, 407
    supplemental materials, 406–408
    theory, 403–405
Milk, milk derivatives, allergen program, 189
Mollusks, allergen program, 190
Monitoring procedures, 68
Monosodium glutamate, allergen program, 190
Monthly inspection, 220–221
Monthly inspection form, 219
Monthly schedule, sanitation program, 238, 246
MSDS. *See* Material safety data sheet
MSG. *See* Monosodium glutamate

N

Natural, unavoidable food defects, no health hazards, 126–128
Non-Genetically Modified Organisms Program, 38
Noncompliance, reporting, 13–14
Nonconforming material report, 483
Nongenetically modified organisms statement, 44
Nutritional statement, 42

O

Odors, objectionable, sanitation standard, 141
Organic certificates, 43
Organic program, 431–438
    application, 434–435
    optional program, 431
    organic formulation submission form, 437
    supplemental materials, 436–437
    theory, 431–434
Organizational chart, 113–116
    application, 113–114
    required program, 113
    supplemental materials, 115–116
    theory, 113
OSHA Form 300, log of work-related injuries, illnesses, 484
OSHA Form 300A, summary of work-related injuries, illnesses, 485
OSHA Form 301, injury and illnesses incident report, 486
Other Programs Manual, 31, 38
Outside audits, 459–462

P

Peanuts, peanuts derivatives, allergen program, 190
Peas, frozen, *Staphylococcus* in, 2
Pest control program, 157–188
    application, 164
    biological control, 159
    chemical control, 159–164
    compensation worksheet for sample company, 168
    contract, sample, 169–170
    cultural control, 159
    examples, 164
    mechanical control, 158–159
    pest control service report, 187
    required program, 157
    rodent, pest control program, 166–167
        bird, 166–167
        documentation, 167
        general information, 166
        insect, 166
        inspection, 167
        rodent, 166
    rodent/pest inspection form, 185
    sample company inspection form, 184
    sanitation, 158
    supplemental materials, 165–187
    theory, 157–164

Index

Physical food hazards, risk assessment worksheet, 78, 92
Physical risk analysis, 54–55
Plastic, 387–394
    application, 388–389
    plastic, 393
    prevention, 393
    quality control program, 105
    supplemental materials, 390–394
    theory, 387–388
    verification documentation, training, 392
Plastic inspection chart, 391
Point evaluation worksheet, product/process control, 80, 94–96
Pre-operational check, quality control program, 103
Preorganizational chart, 37
Printing, lot coding program, 274
Process analysis, 68
Process flows, 46
Processing room, sanitation program, 226–227
Processing specifications, 304
Product descriptions, 37, 46–47
Product descriptions worksheet, 73, 88–89
Product protection, sanitation standard, 143–144
Product tracking, recall programs, 358
Production, quality control, relationship, 14–16
    conciliatory team-building relationship, 15–16
    divisive, adversarial relationship, 15
    secretive, compartmentalized relationship, 15

Q

Quality Control Manual
    cover, 25
    spine label, 26
Quality control program, 99–112
    allergens, 103
    application, 99–100
    customer complaint, 104
    defective material, 104
    documentation, 103
    environmental testing, 103
    examples, 100, 107–111
    general information, 102
    glass, 105
    good manufacturing practices, 103
    hard plastic, 105
    hazard analysis critical control point plan, 102
    loose tool/material, 104
    lot coding, 102
    metal detection, 103
    plastic, 105
    pre-operational check, 103
    quality control, 102
    quality monitoring scheme, 106, 108–111
    recall, 104
    receiving, 103
    regulatory inspection, 104
    responsibility, 105
    rodent/pest control, 102
    sanitation, 102
    shipping, 103
    supplemental material, 101–106
    supplier certification, 104
    theory, 99
    weight control, 104
    wood, 105

R

Raw material recall sheet, 328
Raw material specifications, 303–305
Raw materials, allergen program, 202
Recall, quality control program, 104
Recall program, 313–330
    application, 317–321
    communication plan, 324
    conducting recall, 317–321
    defective product handling, 324–325
    documentation, 325
    effectiveness checks, 325
    effectiveness of product recall sheet, 330
    emergency recall notification list, 327
    finished product recall sheet, 329
    general information, 323
    hazard evaluation, 323
    raw material recall sheet, 328
    recall classification, 323–324
    recall program overview, 323–326
    recall steps, 318–321
    record retention, 325–326
    required program, 313
    supplemental materials, 322–330
    theory, 313–317
Receiving, 37, 103, 287–296
    application, 289
    column instruction, 292
    damaged goods, 296
    general information, 292, 295
    ingredients, 295–296
    inspection, 295
    kosher, 296
    labeling, 295
    overview, 295–296
    quality, 295
    receiving log, 291
    receiving log directions, 292–293
    record keeping/documentation, 295
    record storage, 293
    required program, 287
    storage, 296

supplemental materials, 290–296
theory, 287–289
verification documentation, receiving
ingredient/packaging training, 294
Receiving Manual, 38
cover, 33
spine label, 34
Record keeping, 38, 69, 477–478
lot coding program, 274
Record retention, 325–326
Record storage
receiving program, 293
shipping program, 301–302
Regulatory agency relationship, quality control
and, 18
Regulatory inspection program, 37, 104, 261–268
application, 264
initial introduction, 266
inspection, 266–267
inspection conclusion, 267–268
inspection follow-up, 268
overview, 266–268
required program, 261
supplemental materials, 265–268
theory, 261–264
Rejected product, 254
action required for, 256
Rejection notification, shipping program, 302
Relationship between production, quality control,
14–16
conciliatory team-building relationship,
15–16
divisive, adversarial relationship, 15
secretive, compartmentalized relationship, 15
Release log, hold/defective material program, 379
Release notification, hold/defective material
program, 376
Reporting noncompliance, 13–14
Required program, 39–41
Responsibility, quality control program, 105
Returned products, issues regarding label, 2
Rodent/pest control, 102, 166–167
Rodent/pest inspection form, 185
Role of quality, 11–14

S

Sales, customer complaint program, 286
Salmonella, 2, 50, 403
Sampling/order placement protocol, 498–499
Sanitary facilities, controls, 121–122
Sanitary operations, 120–121
Sanitation, 102, 223–246
accountability, 242
application, 229–230
ceiling, 227
cleaning methods, 242

cleaning schedule, 242
conveyor 1 belt, 226
conveyor 3 belt, 227
conveyor 2 bucket, 227
COP tank, 227
daily non-production schedule, 236, 244
daily production schedule, 235, 243
floors, 227
general information, 242
Hasson twin tube form filler, 227
Ishida scale, 227
master sanitation schedule, 232–234
metal detector, 227
monthly schedule, 238, 246
postcleaning inspection, 242
processing room, 226–227
record storage, 242
required program, 223
sanitation procedure, 240
sanitation procedure training, 241
sanitation program overview, 242
screener, 227
supplemental materials, 231–246
tables, 227
tape machine, 227
theory, 223–229
Urschel RA-A chopper, 226
walls, 227
weekly schedule, 237, 245
yearly quarterly semi-yearly schedule, 239
Screener, sanitation program, 227
Secretive relationship, between production,
quality control, 15
Section labels, manual, 37–38
Security/biosecurity program, 409–424
application, 414–415
check in log, 421
facility, 413–414, 418–419
general information, 417
key check out, 421
management, 409–411, 417
operations, 414, 419
overview, 417–419
public, 412–413, 418
required program, 409
responsibility, 419
staff, 411–412, 417–418
supplemental materials, 416–424
theory, 409–414
verification documentation, security/
biosecurity training, 420
visitor sign in log
cover, 423
spine label, 424
visitor sign in sheet, 422
Seed, seed derivatives, allergen program, 190

Index

Shellfish, shellfish derivatives, allergen program, 190
Shigella, 403
Shipment of finished goods, 489
Shipping, quality control program, 103
Shipping Manual
    cover, 35
    spine label, 36
Shipping program, 297–302
    application, 298
    column instruction, 301
    documentation, 302
    general information, 301–302
    inspection, 302
    overview, 302
    record storage, 301–302
    rejection notification, 302
    required program, 297
    shipping container inspection log, 300
    shipping log directions, 301
    supplemental materials, 299–302
    theory, 297–298
Signage, 401
Social responsibility program, 38, 463–466
    application, 464
    optional program, 463
    social responsibility statement, 466
    supplemental materials, 465–466
    theory, 463–464
Soy, soy derivatives, allergen program, 190
Specification program, 303–312
    application, 305
    certificate of compliance, 311
    finished goods, 304–305
    finished product, 309–310
    ingredient specification, 307–308
    optional program, 303
    processing specifications, 304
    raw material, 303–305
    supplemental materials, 306–311
    theory, 303–305
Standards, establishment of, 11–12
Staphylococcus, 2, 50, 403
State departments of agriculture, regulation by, 1
Stock arrangement, sanitation standard, 140
Storage, receiving program, 296
Sulfites, derivatives, allergen program, 190
Summary chart, 81, 97
    HACCP plan, 97
Summary sheet, 62–63
Supplemental materials, 66–83
Supplier certification, 104, 331–370
    allergen program, 353
    application, 333–335
    bio-security program, 354
    customer complaint, product tracking, recall programs, 358
    equipment conditions, 344–345
    exterior maintenance, 342–343
    food safety audit report, 348–350
    food safety good manufacturing practice assessment rating system, 368
    food safety/good manufacturing practice rating analysis, 351–367
    food safety systems, 352–354
    genetically modified organisms, 354
    good laboratory practices, 358
    good manufacturing practice checklist, 342–347
    grounds, 359–361
    ingredient supplier, co-packer quality, safety survey, 340–341
    insurance vendors, request for evidence of, 338
    overview, 369
    pest control, 343, 361–362
    pest control program, 355–356
    plant facilities, 359–360
    plant grounds, 359
    product contamination, 353
    quality record keeping, 343
    raw materials, 344
    required program, 331
    shipping areas, 346
    supplemental materials, 336–370
    supplier data sheet, 370
    supplier quality management, 343
    supplier risk categorization sheet, 339
    theory, 331–333
    warehousing, 347
Supplier data sheet, 72, 85, 370
Supplier risk categorization sheet, 339
Surroundings, sanitation standard, 147

T

Tables, sanitation program, 227
Tape machine, sanitation program, 227
Team-building relationship, between production, quality control, 15–16
Team meeting attendance, 71
Testing fails, action required, 256
Theory, program planning, 41–44
Toilet facilities, sanitation standard, 146
Top-down quality management, 3–5
Training, 37, 65, 259, 401
    food safety, 353–354
    good manufacturing practice, 148
    Hazard Analysis Critical Control Point program, 65
    sanitation procedure, 241
    verification documentation, 83, 259, 294, 392, 400, 420

Tree nuts, tree nut derivatives, allergen program, 190
Tuna salad, *Listeria* in, 2

U

Urschel RA-A chopper, sanitation program, 226
U.S. Department of Agriculture, regulation by, 1
U.S. Food and Drug Administration, regulation by, 1
USDA. *See* U.S.Department of Agriculture
Utensils, 122–123

V

Vegan certificate, 44
Vendor relationship, quality control and, 16–17
Ventilation, sanitation standard, 137
Verification documentation
    employee good manufacturing practice, annual review, 151
    good manufacturing practice, training, 148
Vibrio, 403
Visitor good manufacturing practices, 155
Visitor sign in sheet, biosecurity program, 422
Visitors, good manufacturing rules, 152–153

W

Walls
    sanitation program, 227
    sanitation standard, 133
Warehousing, 126
Waste management, environmental responsibility program, 440–441
Wastewater, environmental responsibility program, 440
Weekly schedule, sanitation program, 237, 245
Weight control program, 37, 104, 205–212
    application, 206–207
    documentation, 209
    examples, 207, 211–212
    general information, 209
    overview, 209
    required program, 205
    responsibility, 209
    supplemental materials, 208–210
    theory, 205–206
    verification, 209
    weight control chart, 210, 212
    weights, 209
Wheat, wheat derivatives, allergen program, 190
Windows, sanitation standard, 133
Wood quality control program, 105, 387–394
    application, 388–389
    overview, 393
    prevention, 393
    supplemental materials, 390–394
    theory, 387–388
    verification documentation, training, 392

Y

Yearly quarterly semi-yearly schedule, sanitation program, 239
Yearly review, employee good manufacturing practice, 155
Yellow #5, allergen program, 190
Yellow #6, allergen program, 190
Yersinia enterocolitica, 50, 403